Lorenzo Magnani

Understanding Violence

Studies in Applied Philosophy, Epistemology and Rational Ethics, Volume 1

Lorenzo Magnani

Understanding Violence

The Intertwining of Morality, Religion and Violence: A Philosophical Stance

Author

Lorenzo Magnani
University of Pavia
Department of Philosophy
P.zza Botta 6
27100 Pavia
Italy
E-mail: lmagnani@unipv.it

ISBN 978-3-642-27020-8 ISBN 978-3-642-21972-6 (eBook)

DOI 10.1007/978-3-642-21972-6 ISSN 2192-6255

Studies in Applied Philosophy, Epistemology and Rational Ethics

Softcover reprint of the hardcover 1st edition 2011

Typeset & Cover Design: Scientific Publishing Services Pvt. Ltd., Chennai, India.

Printed on acid-free paper

9 8 7 6 5 4 3 2 1

springer.com

To my wife Anna

Revenge is a kind of wild justice, which the more man's nature runs to the more ought law to weed it out.

Francis Bacon, Of Empire

Preface

The explanation of the "genesis" of this book is simple. I have always directly seen, in the behavior of violent human beings and in my very own "mild" violent conducts, that they were basically performed on the basis of serious and firm *moral* convictions and deliberations, at different levels of consciousness and multimodally released (cognitively, rationally, emotionally, etc.): I think morality – and thus religion – and violence are strongly intertwined. At the same time I have always seen how human beings tend to avoid to analyze violence, for example liquidating it through a kind of easy "psychiatrization". To undertake a serious investigation of violence calls for a discrete amount of courage and sincerity: as a human being I can wishfully ignore my own violence, thanks to that "embubblement" I am illustrating in this book, but as a philosopher I cannot ignore it, even if – I hope – my "violent" deeds are extremely unlikely to make the headlines!

This kind of considerations triggered the personal exigency of knowing more about violence. Step by step this intellectual commitment has become more and more specific and theoretical, so that I soon thought violence had to be transformed in an autonomous subject of philosophical reflection. Why? Because we have to *respect* violence and – so to say, paradoxically – its "moral dignity" as a philosophical topic. Philosophers (myself included) are used to deal with clear and highly valued subjects like rationality, science, knowledge, ethics and so on, which are supposed to be endowed with an intellectual dignity *per se*. Philosophers seem to think that violence, *just because it is violence* appears as something trivial, bad, intolerable, confused, ineluctable, and marginal, not sufficiently interesting for them: as a matter of fact, history, sociology, psychology, criminology, anthropology, just to mention a few disciplines, seem more appropriate to study violence and to provide data, explanations, and causes. I am instead convinced that, at least in our time, philosophy is exactly what possesses the style of intelligence and intelligibility suitable for a fresh, impertinent, and deep *understanding* of such an intellectually disrespected topic. When dealing with violence, philosophy, still remaining an abstract discipline as we know it, paradoxically acquires the marks of a kind of irreplaceable and indispensable "applied science".

I think that it is time to attribute more philosophical dignity to violence, because violence is extremely important in the life of human beings, whether we are willing to accept it, or not. There is a philosophy of science and a philosophy of morality, a philosophy of biology but also a philosophy of arts, and so on, no reason to philosophically disrespect violence anymore: we must be proud because we are trying to establish a *philosophy of violence*, as an autonomous field of speculation, which can rescue violence from being held captive, like an embarrassing cultural Cinderella, in a poor, fragmentary, and often exoteric collection of philosophical thoughts or cold, just-so, scientific results.

This book might help up acknowledge that we are intrinsically "violent beings": such an awareness, even if not therapeutic (violence should be explained, and never *explained away*), could improve our possibility of being "*responsible* violent beings". By this I mean that it could suggest us ways of monitoring our rational and emotional reactions, in order to reduce the possibility (unavoidable) of errors and improve our aspiration to acquire a free and safer ownership of our destinies: on a more cognitive ground, this means individuating and, only after that, fortify those *cognitive firewalls* that could prevent violence from becoming overwhelming instead of being confined in just one, or few, dimensions.

I adopt the basic eco-cognitive perspective I already developed in my recent book *Abductive Cognition. Epistemological and Eco-Cognitive Dimensions of Hypothetical Reasoning* (Springer, Heidelberg/Berlin, 2009), which takes advantage of the results coming from various disciplines, from the area of distributed cognition to the biological research on cognitive niches, from the study on fallacies to the catastrophe theory, merging psychological, social, and evolutionary frameworks about the development of culture, morality, and religion, with the role of abductive cognition in illustrating hypothetical guessing and acknowledging insights from postmodernist philosophy to cognitive paleoanthropology and psychoanalysis.

This book inaugurates a new series that I am editing for Springer: SAPERE (*Studies in Applied Philosophy, Epistemology and Rational Ethics*). As it means to provide a fresh and provocative understanding of violence, I hope this book does justice to the aim of the series and spotlight a topic that has been mostly undermined and neglected in contemporary cultural settings.

For valuable comments and discussions I am particularly grateful to various academic colleagues who directly, but also indirectly, furnished various insights and ideas useful to improve my thoughts about violence, and to my collaborators Emanuele Bardone and Tommaso Bertolotti, who also helped in discussing, reshaping, and enriching various sections. Some sections of chapters two, three, five, and six, have been written in collaboration with them: 2.2.2, 3.3.2, 3.5, 4.6 (first part), 5.4, 6.5.2 with Emanuele Bardone, and sections 5.3.1, 6.3, 6.4, and 6.6.2 with Tommaso

Bertolotti. I am also indebted to my friends Enrichetta Buchli, psychoanalyst, and Giovanni Corsico, psychiatrist, both exceptionally skilled at detecting the violent aspects in the least suspicious of human behaviors. I discussed many parts of this book with my wife Anna to such an extent that I can say many sections were written together. This book is dedicated to her, in honor of her sentiment and persuasion that violence has to be considered seriously, also from the cultural point of view. The research related to this volume was supported by grants from the Italian Ministry of University and the University of Pavia. I also wish to thank Louise Sweet, a freelance editor who enhanced my written English. The preparation of the volume would not have been possible without the contribution of resources and facilities of the Computational Philosophy Laboratory (Department of Philosophy, University of Pavia, Italy). I am grateful to Springer for permission to include portions of previously published articles.

Along the book I will use *he*, *she*, *her* or *his* depending on the context. I do not think appropriate to use the now-conventional expression "she" or "he or she" when dealing with violence, which is certainly similar in males and females, but very often has to be more appropriately attributed to males.

April 2011

Lorenzo Magnani
Pavia, Italy

Contents

Chapter 1
"Military Intelligence"
Coalition Enforcement and the Evolutionary Origin of Morality and Violence

1.1 Philosophy: The Luxurious Supplement of Violence

In the paper "Philosophy – The luxurious supplement of violence", Bevan Catley (2003) says

> In many of the growing number of accounts of workplace violence there exists a particular sense of certainty; a certain confidence in what violence "really" is. With these accounts, philosophy appears unnecessary – and even luxurious – in the face of the obvious and bloody reality of workplace violence. [...] one outcome is an absence of a sense of curiosity about the concept of violence in many typical commentaries on workplace violence. Through a turn to philosophy it is suggested, we might possibly enquire into other senses of violence that may otherwise be erased. However, weary of simply "adding" philosophy, this paper also begins to sketch out some possible consequences a philosophy-violence connection might have for doing things with philosophy and organisation.

Workplace violence is only a marginal aspect of the widespread violence all over the world but I agree with Catley that philosophers usually do not consider this theme; he says they lack a "curiosity" about violence: this book aims at further filling this gap.

Similarly, in a recent book about the relationships between phenomenology and violence, which mainly addresses the problem of war, Dodd (2009, p. 2) says:

> It might strike one as a strange point of departure for a reflection on the obscurity of violence to raise the question of its properly philosophical character. Does not virtually any obscurity, not to mention profound questions of human existence, by definition invite philosophical reflection? This already begs the question. For perhaps it is instead the case that the problems of violence are not, in the end, all that obscure, even if they may be difficult to understand.

Furthermore, the existence of a conviction about a kind of stupidity of violence is illustrated:

> In its barest form, the stupidity of violence principle states that violence is and can be only a mere means. As a mere means, pure violence remains trapped, according to the

L. Magnani: Understanding Violence, SAPERE 1, pp. 1–33, 2011.
springerlink.com

> principle, within the confines of a very narrow dimension of reality defined by the application of means. Violence as such is thus blind; when taken for itself it is ultimately without direction. The practices of violence, however traumatic and extreme, fade into indefinite superficialities unless supported by a meaningful cause or end. To be sure, the stupidity of violence does not detract from the seriousness of the consequences of violence, the damage it inflicts – the shredded flesh, the famine and disease, the pain both physical and spiritual, and the shocking number of corpses that it leaves in its wake. [...] violence is stupid, in that it involves nothing more significant than what can be captured and organized in a technical fashion. [...] Likewise, the morality of war does not find, according to this principle, any chance of being expressed, so long as we only follow along the passage of the event of violence that has been so tightly reduced to this line of cause to effect (Dodd, 2009, pp. 11–12).[1]

The lack of curiosity is accompanied by a sense of certainty about "what violence is", Catley says, and I agree with him. There is a sense of certainty that suppresses its historical variability[2] and that only leaves a hypocritical sense of it as a merely deplorable outbreak on an otherwise peaceful scene. "Philosophy, it would seem, appears unnecessary – and even luxurious – in the face of the obvious and bloody reality of workplace violence" (ibid.)

In sum, we do not only need philosophers of science or moral philosophers but also "philosophers of violence". I also strongly believe that the problem of violence is no less important than the problem of morality and I plan to show how the two are strictly intertwined.[3]

1.1.1 Individual and Structural Violence

A kind of common prejudice is the one that tends to assign the attribute "violent" only to physical and possibly bloody acts – homicides, for example – or physical injuries; but linguistic, structural, and other various aspects of violence also have to be taken into account. However, even homicide itself is more complicated than expected, in fact recent research on the legal framework of homicide and on its biological roots shows many puzzling aspects. Brookman (2005) contends that it is not appropriate to think of homicide just as a kind of criminal or violent behavior, because the phenomenon is complex and socially constructed for the most part.

[1] Various classical philosophers concerned with the meaning of war and violence are usefully analyzed in the book: Arendt (multidimensional and subjective character of violence); von Clausewitz and Schmitt (political instrumentality of violence and "constitutive" violence); Heidegger, Sartre, Nietzsche, and Jünger (non-instrumentality and nihilism of violence); Patoĉka (violence and responsibility).

[2] Cf. also (Keane, 1996).

[3] However, a new focus of attention on violence in philosophy seems to be "in the air". Very recently McCumber (2011), in "Philosophy after 9/11", has stressed the need to reshape research in philosophy (and its related community) from various perspectives, for example defending Enlightenment, beyond the mere use of the "[...] standard arsenal of argument forms, of which our Analytical colleagues make excellent though wrongly exclusive use" (p. 28). The paper is part of a collection with the eloquent title *Philosophy and the Return of Violence* (Eckstrand and Yates, 2011).

Those who kill do so for very different reasons and under different circumstances, and some structurally originated killing is often hidden and disregarded, even from the legal point of view, such as in the case of multiple homicides perpetrated by corporations. He illustrates various explanations of homicide: psychoanalytic and clinical approaches, accounts from evolutionary psychology or social and cognitive psychology frameworks, sociological and legal aspects and the biological roots of killing.[4]

The extension of the meaning of the word violence brings up the need to reconsider our concepts of safety, ethics, morality, law and justice. May be the philosophy-violence connection (with the help of other related disciplines, both scientific and cognitive) will generate novel ideas and suggestions. Of course people and intellectuals are clearly aware that drug, alcohol, revenge, frustration, and mobbing behaviors are related to violent events, and so many aspects of linguistic and structural violence are acknowledged, but this acknowledgment is almost always fugitive and superficial. In sum, we see the spectacle of violence everywhere, but, so to say, the violence always out there, involves *other* human beings and we can stay distant from the theme of violence by adopting a simple and familiar – but practically empty – view of it. Indeed, it is implicit that, if we know for certain that we are the possible target of violent behavior, we *a priori* think of ourselves, spontaneously, as exempt from any contamination, in our supposed purity and immunity. We just hope not to become victims, but this is usually considered just a question of good luck, and, above all, we are not interested (even if we are, or think of ourselves as, philosophers) in analyzing the possible existence and character of *our* own (more or less) violent behavior. It is better not "to problematise our confidence in, or familiarity with, 'what violence is' " (Catley, 2003).[5]

Familiarity with violence involves a trivial and simple *sense* of violence as interpersonal, physical, and illegitimate, which can be clearly seen in the case of workplace violence as a:

> [...] deviant set of behaviours to be eradicated through a series of familiar strategic interventions. Workplace violence becomes reduced to a technically rational set of "procedural issues about workforce selection, early detection of potential troublemakers, adequacy of liability insurance, risk management and effective exclusion of potential as well as actual offenders" [...]. And it is this familiarity that erases questions about the constitution of violence that might lead us to ask other critical questions about the organisation of work and the work organisation beyond individual pathography. Arguably, the familiarity of violence as interpersonal and illegitimate has encouraged explanations of workplace violence to focus on the individual and the eradication of such deplorable behaviour. In these explanations, the focal point has tended around the exposition of the personality and motivations of the "perpetrator", typically with a view to profiling the violent individual (Catley, 2003).

[4] On the psychological explanations of mothers killing their children see the recent (McKee, 2006).

[5] Further details about the concept of structural violence, the problem of its possible legitimization, and the relationship between violence and justice in the case of "just war" are illustrated in (Coady, 2008, chapters two and three).

The passage is clear and eloquent. Violence tends to become an easy matter for psychologists, sociologists, etc., who make – sometimes vane – efforts to depict the structural aspects of situations and institutions and how they change (families, markets, nations, prisons, workplaces, races, genders, classes, and so on). However, what really matters is to describe how individuals respond in a dysfunctional way because of their personality, motivation, and (lack of) rationality. The full violent potential of structures, artifacts, institutions, cultures, and ideologies is marginalized and disregarded. For example we rarely acknowledge the fact that contemporary humans can – with scarce awareness – damage the future as well as the present causing future populations to be our "unborn victims", thanks to the consumption of energy, the pollution of air and water, or the demolition of cultural traditions. It is only the "abnormal" individual that is seen as prone to suddenly perpetrating violence, and only (right wing?) repressive policies that are considered extinguishers of the violent anomaly.

To summarize: it is the individual that is violent, but the inherent violence of structures (where for example injustice and relations of dominations are common, and violent attitudes are constantly distributed and promoted) is dissimulated or seen as a "good" violence, in the sense that it is ineluctable, obvious, and thus, acceptable. Every day in Italy at least three workers die in their workplace, and this seems natural and ineluctable. The potential harmfulness of organizational practices and workplace management leads to violent events and are rarely seen as causing them. When they are seen as violence-promoters, this consideration is a kind of "external" description which exhausts itself thanks to a huge quantity of hypocritical rhetoric and celebration. Again, we tend to focus on the familiar idea of violence as basically physical and we ignore or belittle other aspects of violence which neither leave their mark on the body nor relate back to it.

In his recent book Žižek (2009) reveals "the hypocrisy of those who, while combating *subjective violence*, commit *systemic violence* that generates the very phenomena they abhor" (p. 174). Language and media in modern life continually show a spectacle of brutality accompanied by a kind of urgent (but "empty") "SOS violence", which de facto systematically serves to mask and to obliterate the symbolic (embedded in language, as I will amply illustrate in the second part of this chapter and in the following one) and structural-systemic violence like that of capitalism, racism, patriarchy, etc. Žižek provides the following, amazing example of the intertwining between charity and social reproduction of globalized capitalism: "Just this kind of pseudo urgency was exploited by Starbucks a couple of year ago when, at store entrances, posters greeting customers pointed out that almost half of the chain's profits went into health-care for the children of Guatemala, the source of their coffee, the inference being that with every cup you drink, you safe a child's life" (p. 5). The instantiation of this hypocrisy is clear when analyzing the philanthropic attitude of what Žižek calls the *democratic liberal communist capitalists*, who absolve themselves as systemic violent monopolizers and ruthless pursuers of profit thanks to an individual policy of charity and of social and humanitarian responsibility and transcendental gratitude. Žižek also warns us that a cold analysis of violence typical of the media and human sciences somehow reproduces and participates in its horror; also empathising with the victims, the "overpowering horror of violent acts [...] inexorably functions as a lure which prevents us from thinking" (p. 3).

Another aspect of structural violence can be traced back from the analysis of the commodification of sexuality in our rich capitalist societies. The urgency of sex tends to violate the possibility of love and the encounter with the Other:

> Houellebecq (2006) depicts the morning-after of the Sexual Revolution, the sterility of a universe dominated by a superego injunction to enjoy [...] sex is an absolute necessity [...] so love cannot flourish without sex; simultaneously, however, love is impossible precisely because of sex: sex, which "proliferates as the epitome of late capitalism's dominance, has permanently stained human relationships as inevitable reproductions of the dehumanizing nature of liberal society; it has essentially ruined love". Sex is thus, to put it in Derridean terms, simultaneously the condition of the possibility and of the impossibility of love (p. 9).

To make another example, harassing, bullying, and mobbing are typical violent behaviors that usually involve only language instead of direct physical injury: that is, blood is not usually shed. Bullying can also be ascribed to such a category, but in a somewhat minor dimension since bullies often resort to both physical and linguistic threats. Of course, the sophisticated intellectual and philosopher – but not the average more or less cultured person – says "certainly, mobbing involves violence!", but this violence is *de facto* considered secondary, at most it is only worth considering as "real" violence when it is in support of actual events of physical violence. It is well-known that people tend to tell themselves that the mobbing acts they perform are just the same as cases of innocuous gossiping, implementing a process of dissimulation and self-deception.

1.1.2 Psychiatrization and Dehistoricization of Violence

A strategy that aims at weakening serious attention to violence is often exploited by the media (and is common in everyday people). The mass media very often label episodes of violence, for example terrorism, as something pathological: "the terrorists kill for the sheer pleasure of killing and dying, the joy of sadism and suicide", "they constitute a pathological mass movement". Following Wieviorka (2009, p. 77) we can say that these considerations only serve to obscure any hidden causes for the "horrifying attacks" and represent a form of medicalization or *psychiatrization* of a behavior that comes to be dismissed as simply insane.[6]

Another strategy that aims at playing down the problem of violence consists in labeling some violent episodes as "new", in a kind of "ahistorical individual pathology" where organizations and collectives are considered simply "affected"

[6] On the relationships between psychiatry and violence cf. below, chapter five, section 5.2.

by violence, and certainly not part of it. Here is an eloquent example: "To represent workplace violence as a 'new' event serves to deny and evacuate any historical relationship between violence and work. This is not to suggest that the 'object' of workplace violence has remained constant throughout time, but to inoculate against any imagined sense that violence has not been present in the organisation of work"(Catley, 2003). We know that this last statement is patently false just like in the case of bullying in schools – often marked as "new". Catley nicely concludes: "Consequently, the result of accounts of workplace violence that distance themselves from viewing workplace violence as historical and sociological in favor of an ahistorical individual pathology is a continuance of the representation of the organization as being *affected* by violence rather than possibly *contributing* to, constructing or reproducing violence" (ibid.)[7]

A meaningful recent analysis of the so-called "deadly consensus", which is at the basis of UK workplace safety is provided by Tombs and White (2009). The authors document the regulatory degradation as a result of its compatibility with neoliberal economic strategy. The degradation is justified by political and economical reasons, which rapidly become the *moral* grounds at the level of the individuals' legitimacy in disregarding safety. A subsequent analysis of empirical trends within safety enforcement reveals a virtual collapse of formal enforcement, as political and resource pressures have taken their toll on the regulatory authority. Furthermore, an increasing impunity with which employers can actually kill and injure (even if they do not *mean* to, as the end of their action) is also observed.

1.1.3 Victims and Media

In sum, we can say that in a common sense (as it emerges, for instance, in the mass media) the representation of violence is really partial and rudimental. In our current technological globalized world media and violence live in a kind of symbiosis (Wieviorka, 2009, p. 68) – just think about terrorism – this presents two aspects: the first is that victims are continuously exhibited and described, and this is positive, because they are not simply disregarded as collateral, like in the past, but are increasingly represented in the violence that they have suffered and which has damaged their moral and/or physical integrity. One can at least think "that the thesis that violence is functional loses its validity" (p. 160) since an effective depiction of violence halts the belief that violence has a (more or less) just reason and a reasonable scope. The second aspect is concerned with what the victims are like, so to speak, the "words" the journalists depict them with, and the fact that the subjective side of the victims is rarely presented, favoring empathy and the sense of the

[7] Amazingly, empirical research showed that in Australia around 26% of workplace bullying can be attributed to the 1% of the employee population, representing "corporate psychopaths". Corporate psychopaths, like toxic leaders, cause major harm to the welfare of others, trigger organizational chaos, and are a major obstacle to efficiency and productivity (Boddy et al., 2010; Boddy, 2010).

audience's psychological incorporation of the actuality of violence.[8] Thus a culture of fear is favored but, at the same time, violence may be unintentionally promoted and reinforced.

In intellectual settings, with the exception of literature and arts, and of other – so to say – more or less "technical" disciplines (theory of law, religious studies, some branches of psychology, sociology, psychoanalysis, paleoanthropology, evolutionary psychology, evolutionary biology, and criminology), the lack of curiosity for violence is widespread and violence is basically *undertheorized*. In the case of "technical" disciplines an interest in violence is very recent. As interestingly noted by Gusterson (2007), when – starting from the 1980s – "new wars" and local processes of militarization with high civilian casualty rates came about in Africa, Central America, the former Eastern block, and in South Asia, anthropologists showed a greater interest in various kinds of violence: a kind of "cultural turn", Gusterson says. They started to analyze terror, torture, death squads, ethnic cleansing, guerilla movements, fear as a way of life, permanent war economy, the problem of military people as "supercitizens", war tourism, the memory work[9] inherent in making war and peace, and also the problem of nuclear weapons and American militarism. They also stressed the role of current military apologetics in the USA, which favored and still favors degraded popular culture saturated with racial stereotypes, aestheticized destruction, and images of violent hyper-masculinity. In this perspective of a new widespread militarism (as a source of so much suffering in the world) it can be said that the "war on terror" has provided the occasion for a new intellectual commitment to the theme of violence debating the merits of military anthropology versus critical ethnography of the military. However, we have to say that all too often, especially in psychology and sociology, the analysis of violent behavior is exhausted by the description of the more or less vague characters of various psychopathological individuals or groups, where empirical data and practical needs are the central aim, but deeper structural and theoretical aspects are disregarded, while they could be crucial to a fuller understanding and subsequently to hypothesize long-term solutions.

1.1.4 Philosophy and Violence

I strongly agree with Catley that philosophy can become a "catalyst" for "energizing" our thinking about the multifaceted aspects of violence. However, we have to remember that of course "philosophy in the forms of its philosophical institutions and practices may too exert its own forms of violence. Rather than standing in

[8] On the need to construct, in "existentialist" critical criminology, a new concept of "event", as a guiding image able to offer an alternative approach to victimization, trauma, and identity cf. the interesting (Spencer, 2011, p. 49): "Trauma resulting from events of victimization can transmogrify singularities both in terms of their bodies and possibilities. Each singularity experiences selfhood in discrete ways and their histories are not reducible to selfsameness. This refusal of selfsameness moves victimology beyond the assertion of 'the victim' of a type of crime, to the plurality of experiences connected to manifold events of victimization and their attendant effects".

[9] A collective mobilization of memory about past injuries.

opposition to violence, philosophy may well reproduce that which it seeks to condemn. The faithful love of ideas harbors an affair with violence" (Catley, 2003). It is obvious that philosophy can afford to explore the problem of violence only if choosing a side to stand on; philosophy has to be aware that, given the inherent violent nature of knowledge and language I will soon illustrate in this book, it is not immune to it, and that every option chosen has the chance of producing violent effects. For example "Here one might think of the institutionalization of philosophy, where what is included and excluded in the canon is decided. Or the silencing effects of truth, where 'truth' is mobilized to close down dissenting positions. Philosophy becomes complicit in the sorts of domination enacted when the arbitrary takes on the status of the natural to close down alternative ways of knowing, courses of action, and subject positions in order to preserve dominant relationships of power" (ibid.): to make an example, how can we forget that often the presentation of a sophisticated philosophical argument about "creation" is experienced and "perceived" as a tremendous violence by some religious and/or uncultured people?

As I have already said in the preface to this book, philosophy has always displayed a tendency to "dishonor violence" by disregarding it. Step by step the philosophical intellectual commitment to violence has become more and more objective and abstract and is being transformed into an autonomous subject of philosophical reflection. Now we can say that philosophy has learnt to *respect* violence and – so to speak – its "moral dignity" as a philosophical topic.

An initial theoretical step in this book is to acknowledge that, often, structural violence is seen as *morally* legitimate: we have to immediately note that when parents, policemen, teachers, and other agents inflict physical or invisible violence on the basis of *legal* and/or *moral reasons*, these reasons do not cancel the violence perpetrated and violence does not have to be condoned. In chapter three[10] of this book I will describe how a kind of common psychological mechanism called "embubblement" is at the basis of rendering violence invisible and condoning it. Human beings are prisoners of what I call *moral bubbles*, which systematically disguise their violence to themselves: this concept is also of help in analyzing and explaining why so many kinds of violence in our world today are treated as if they were something else. Such a structural violence, which is in various ways legitimated, leads to the central core of my commitment to the analysis of various relationships between morality and violence.

It seems to me there is an increasing interest in the intertwining between morality and violence in concrete research concerning human behavior. For example Smith Holt and colleagues (2009, p. 4) contend that there is a common thread that emerges in various recent studies about killing: "[...] something about moral code, a religious or ethical belief enmeshed within a cultural context, determines one's stance on various types of killing and, indeed, on inhibitors to killing".

A philosopher I recently found to explicitly (seemingly, at least) acknowledge the strict link between morality and violence is Allan Bäck. He first of all contends that aggression is not necessarily a destructive or morally bad act:

---

[10] Cf. subsection 3.2.2.

> In contrast to aggression, I shall thus take "violence" in the basic sense to signify a certain sort of aggression, namely an aggressive activity to which judgments, of being good or bad, apply. In many moral and legal theories, such judgments require, among other things, considering the conscious of the moral agent(s). According to such views, if you are aware of what you are doing, will to do so, and could do otherwise, you are morally responsible for those acts. You might, then, be able to do something without being morally responsible, if you did not will the act, or were not aware of what you are doing, or were in a mental state of extreme duress of emotion. At any rate, I shall suppose that normal cases of human intentional action are subject to moral judgment and are chosen (Bäck, 2009, p. 369).

From the perspective of the perpetrator an action is violent when it is aggressive, it is "chosen" and when the victim would not want to suffer that harm (i.e., "the action unjustly violates the rights of the victim, where 'rights' signifies the morally ideal set of entitlements that the recipient of the action – typically a person – ought to have"). In such a way violence "contains" a moral component that "is associated with choosing to engage in actions that harm another person and attempting to force that person to act as you want". In this basic sense

1. violence – from the perspective of the perpetrator – is the fruit of moral deliberation (always endowed with its "idealistic" and "pure" halo),[11] but
2. usually, when we judge a human act to be violent we mean something *pejorative*, we need to add the condition that the selected choice is morally wrong, the agent ought not have made that choice. It is clear that this judgment of violence is of second order and derives from a moral judgment that usually does not belong to the perpetrator; indeed I strongly think the case of a perpetrator that consciously performs violence, by disobeying to *his own* moral conviction, is a rare case of a real mephistophelian behavior. It is clear that the second meaning of violence tends to *obliterate* the first one: we always forget that violent actions are usually performed on the basis of moral deliberations.

Taking advantage of some philosophical and interdisciplinary considerations, my aim is to attribute more sense to violence: I will link violence to a number of issues including ethical, juridical, cognitive, and anthropological ones. The following are the three first steps: 1) a definition of violence is required; 2) we need to emphasize how we can shed light on morality and violence through the so-called *coalition enforcement* hypothesis; 3) we have to deepen our knowledge about the moral and – at the same time – violent nature of language (for example, but not only, in the case of speech acts). I will consider these issues in the following sections illustrating the useful concept of *military intelligence*. To anticipate the content taking advantage of a kind of motto, I can say: "when words distribute moral norms and habits, often they also wound and inflict harm".

---

[11] (Baumeister, 1997).

1.2 Defining Violence: Violence Is Distributed and Always "Perceived" as Such

People, and especially analytic moral philosophers, usually ask for definitions. We can try to offer a definition of violence. Disregarding here the well-known definition of evil as a supernatural phenomenon, which is interesting from other cultural perspectives, we have to face its nature in terms of human and interpersonal relationships. First of all I can say that violence is always perceived as such directly by an individual (who eventually derives that way of perceiving from the shared culture of the group(s) to which he belongs): "The same act can count as violent or non-violent, depending on its context; sometimes, a polite smile can be more violent than a brutal outburst" (Žižek, 2009, p. 180). Furthermore, Taylor (2009) has recently emphasized the need – from an evolutionary perspective – to involve morality, at least as far as the definition of violence is concerned. She suggests that cruelty is a concept that has meaning only in the context of morality, and morality is an evolved property possessed only by humans (and rudimentarily by our closest evolutionary neighbors, as contended by many evolutionists and animal-cognition researchers). She contends that what is perceived as cruel is dependent on context and culture. Why, for instance, is targeted assassination by dropping a one-ton bomb from a plane on to the house of an alleged enemy not cruel, whereas a suicide bomber who immolates himself together with his victims is cruel?[12]

I have contended in the previous section that aggression in itself is not necessarily violent: a surgeon that causes pain to a patient by ordering chemotherapy is not violent in the "pejorative" sense of the term (and indeed that aggression is not perceived by the recipient himself in a morally pejorative sense). The mechanism of attribution of violence to an act, a situation, an event varies between cultures and approaches, and the judgments can be more emotional and less conceptual or vice versa.[13] Of course there are aspects of violence (like in the case of death from a bloody aggression, or in the case of incest) that are almost universally shared across cultures; even if we have to remember that, for example, those who commit suicide do not perceive killing themselves as an actually violent act, to them their tragic death is considered

[12] A full analysis of the concept of cruelty is given by Randolph Mayes (2009). The author defies the idea that cruelty is widely regarded to be a uniquely human trait, which follows from the received definition of cruelty as involving the intentional infliction of suffering together with the empirical claim that humans are unique in their ability to attribute suffering (or any mental state) to other creatures. It is argued that the intuitive appeal of this definition "[...] stems from our abhorrence of cruelty, and our corresponding desire to define cruelty in such a way that it is almost always morally wrong. Scientifically speaking this is an arbitrary condition that inhibits our attempt to study cruelty as a natural phenomenon. I propose a fully naturalized definition of cruelty, one that considerably expands the range of creatures and behaviors that may be conceived as cruel" (p. 21).

[13] Cognitive researchers have provided empirical evidence pointing to the fact that to metaphorically define or frame the issue of crime using the image of a predator yields systematically different suggestions with respect to those produced when crime is described as a virus, cf. (Thibodeau et al., 2009).

a relief. Moreover, killing oneself or killing others is often morally justified thanks to political or religious reasons both by the suicide/killer and by his group.[14]

1.2.1 Decent Violent People

Another important issue is related to the fact that human beings experience a pervasive difficulty in understanding perpetrators of evil. It is well-known that Hannah Arendt famously noted that the fact that normal, banal, decent people could commit violent and more or less atrocious acts, seems

> [...] to go against one's basic understanding of the world. Some exceptional explanation must be required, because it seems that evil deeds should be done by evil people, and yet many such deeds are committed by people who do not conform to the stereotypes of evil. Yet these stereotypes are one of the major obstacles to understanding evil. This is ironic, because the myth of pure evil was constructed to help us understand evil – but it ends up hampering that understanding. The myth is a victim's myth, and there is often a wide, almost impassable gap between the viewpoints of victims and perpetrators (Baumeister, 1997, p. 379).

Furthermore, thanks to the scientific research of the last few decades it is now clear that aggression, far from being an obvious exclusive instinctual endowment of human and non human animals, is mostly "learned" and specific to particular situations. Culture plays a great role, like the carnage of the last century clearly demonstrates, paradoxically showing that an increase of civilization has also increased some forms of violence.

Evil requires deliberate actions of one or more persons (or the unintentional intrinsic action mediated by an artifactual social structure), the suffering of another person, and the perception or judgment of either the second person or an observer.[15]

---

[14] An analysis of various aspects of suicide is given by Joiner (2007): he finds three factors that mark those most at risk of death: thwarted belongingness, morally perceived burdensomeness, and the ability for self-harm. More data and analyses about violent aggression are given in (Gannon et al., 2007). The chapter by Palmer (2007) clearly teaches how moral reasoning often contributes to aggressive violent behavior, through its interaction with a range of other social cognitive processes: the author acknowledges that moral reasoning and other social cognitive variables act as a mediating factor between socialisation experiences (particularly early parenting styles) and aggressive behavior. The biological counterpart (genes, neurotrasmitters, hormones, neurological impairment) of aggression – verbal, physical, sexual; with or without a weapon; impulsive or premeditated – is dealt with in (Nelson, 2006).

[15] Not every evildoer is an evil person but there is empirical evidence that a very large proportion of us is inclined to perform evil actions: following the detailed study illustrated in (Russell, 2010), which critiques the dispositional account of evil personhood, *S* is an evil person if *S* is strongly disposed to perform evil actions when in conditions that favor *S*'s autonomy.

Basically, it is the moral "perception"[16] human beings possess and activate which delineates "what is violent" and the effect of "victimization", but perceptions are variegated and likely to change. If I morally think incest is "bad", I can perceive incest as a violent sexual act; if I think patriarchal life is good, I can label women that rebel as violent/bad; but if I am a woman that rebels against patriarchal mentality, and so who adopts and activates a different morality (for example "feminist", or simply informed by the legal norms of Western societies), I am the one to perceive the patriarchal male behavior as violent/bad. To give another example, if I think that retaliation is morally good, I morally approve of the act of *killing the killer* by capital punishment. The patent violent aspects "we" – as people who do not agree with the death penalty – perceive in this action disappear insofar as we are individuals that morally endorse this kind of retributive revenge. What I will call the *moral bubble* is still operating:[17] obviously the retaliators see the performed violence "phenomenologically", as just a justified aggression, which is not morally considered a violence and so it can be disregarded as such. On the contrary, we have to repeat, the same action of killing the killer is usually seen as profoundly violent by people that do not "morally" favor the death penalty and retaliating by killing. In general the two groups conflict, and the conflict can in turn also become violent, as I will more clearly illustrate in the following section.

1.2.2 Honor and Institutional Cultures

Another example of conflicting moralities can be traced back from the distinction between "honor cultures" and "institutional cultures".[18] These cultures present various conflicting ideas for example about moral responsibility and desert and about punishment. Honor cultures favor revenge and thus pay more normative attention to the offended party, so that they stop – via some kind of retaliation – feeling outraged or resentful: such practices are aimed at preventing further abuse, disgrace, ostracism, decrease in status for both oneself and one's family. Conversely, institutional cultures favor retributive behaviors and discourage retaliation against

[16] I am taking advantage of the word perception in the general sense of the term, as an equivalent of judgment, feeling, reaction and so on, for example the reaction can derive from more or less "educated" emotions, or from more or less rational judgments, still related to the individual's hard-wired cognitive endowments and to his present plastic learned cognitive capacities, and of course from a mixture of both.

[17] Cf. chapter three, subsection 3.2.2.

[18] (Sommers, 2009). Lately, even psychologists and cognitive scientists interested in moral behavior (always reluctant to abandon the ethnocentric prerogative of "morality" according to the one imbedded in monotheistic religions or in modern civil mentality) acknowledge the fact that also honor and other cultures are characterized by *ethical* concerns. For example (Gigerenzer, 2010, p. 545): "Terrorists, the Mafia, and crack-dealing gangs run on moral principles (e.g. (Gambetta, 1996)). [...] I suggest that the heuristics underlying moral behavior are not the mirror image of the Ten Commandments and their modern equivalents, but embody more general principles that coordinate human group". Further insight on cultures of honor in southern United States, from and evolutionary psychological perspective, is provided by (Shackelford, 2005).

defectors, because of its significant costs and risks, and so normatively focus on the offender and his persecution (the offender has to be punished to deter him and others from cheating again).[19]

Both cultures incorporate moral behaviors that aim at enhancing cooperation, researchers contend that the honor cultures are typical of herding and frontier societies, but their characteristics "[...] may be present in inner city gang life, some tribal societies, outlaw and mafia cultures, and other types of environments as well" (p. 39), while the institutional ones are typical of agricultural communities and, in general, they are incorporated in strategies intended to promote the fitness of large groups. In this last case – often characterized by centralized forms of law-enforcement – massively discouraging free-riding is crucial for the interests of the whole group, not just for particular individuals, as it is in the case of honor cultures. It is from this perspective that Boyd, Gintis, Bowles and Richerson (2003) call the punishment exerted in these cultures "altruistic", which involves little sacrifice on the part of the punishers. In conclusion, honor cultures privilege the morality of retaliation while institutional cultures stress moral responsibility and desert.

Looking at honor cultures through the morality we are accustomed to (that is, as twenty-first century, Western human beings, citizens of well-established democracies), or through the abstract concepts of moral philosophy compels us to miss their intrinsically *moral* aspects. Of course I agree with Sommers, some honor killings are not perpetrated by moral monsters, like in the following case:

> This is when a family member murders a woman or girl because she has lost her virginity, which is seen as a stain on the family's honor. What makes these cases even more unfathomable is that often the murdered woman is a victim of rape. Most groups in Western cultures find this perspective not just appalling but utterly bewildering. Even if one believes that premarital sex is a mortal crime, how can anyone think it is appropriate to punish victims of rape? It is clearly not the rape victim's fault that she committed this "crime" and lost her virginity. The phenomenon becomes a little easier to understand if we imagine that the cultures that support this behavior do not have a strong notion of desert to begin with. If that is the case, it is almost irrelevant that the victims do not in any way deserve their fate. Note that in many cases, the family members are not moral monsters; they are often consumed with grief at the death of a beloved daughter or sister. But they will seldom repent the act. We might imagine that family members regard victims of rape as a fatal illness. That is why tremendous grief is consistent with a complete lack of repentance for the killings. The injustice of these deaths does not resonate to members of honor cultures nearly as much as it does to those of us in institutionalized cultures (Sommers, 2009, pp. 42–43).

[19] It is worth noting that honor cultures – the mafia can be considered a fitting example – contrarily to law-enforcement cultures, do not pay attention to the "[...] whole concept of responsibility and, in particular, the 'deservingness' of the offender. [...] The important thing is that the injured party retaliates against someone, someone who bears a connection to the offender. Otherwise, honor is lost. Of course, [if, for instance, the offense is a murder] the most suitable target is the murderer (as long as he is of equal status to the victim). But there is no prohibition against punishing relatives or associates of the offender, since the primary function of retaliation is to restore the reputation of the offended party" (Sommers, 2009, p. 42).

Moreover, it is important to stress the fact that people certainly share common objective moralities and systems of values (collective "axiologies", as I will describe in chapter four), that are widespread and dominant in their group/groups, but also possess individual moralities that they use on occasion, often related to subcultures that endorse violence. Baumeister (1997) usefully notes that various subcultures – more or less widespread and implicitly carried by mass media and moralistic gossip – of the supposed morality of the *irresistible impulse* constitute a subcultural morality that favors and justifies much violence.[20] An example: I can adopt in certain circumstances a modern morality, inspired by legal/civil norms, that does not contemplate mobbing, but at the same time I can adopt – more often without any awareness – in some other circumstances, ancestral different moralities,[21] like the scapegoat mechanism, and I can become a mobber, implicitly believing (contrarily to my legal/civil convictions – still present, but disengaged) in the basic morality and rightness of mobbing. In such cases of momentary mobbing behaviors, performed by otherwise "decent" people, the moral bubble is still operating, and so no awareness of the violence perpetrated against the mobbed individual is present, even if everyone knows that those same decent people who mob others do not like themselves being mobbed and would feel violated and harmed in the same case. Actually, moral bubbles seal an agent (the mobber, in this case) against self-contradiction and keep her calm and satisfied with having applied a presupposed – momentarily felt as such – morality.

The variety of moralities (and so the legitimacy of ancestral protomoralities, which interrupt and substitute a "civil" morality) has been recently acknowledged by a few authors interested in going beyond the abstractness of moral philosophy. For example, Sommers, whom I have already quoted above, considering the case of the concept of "moral responsibility" in the "honor" and "institutional" cultures, clearly notes:

> Some beliefs might be more or less rational within a particular environment, but no theory can describe the truth about responsibility in all environments. [...] At the very least, a defender of a universal condition-based concept of moral responsibility will have to answer the following question: why should the notions of moral responsibility and desert have any more objective or metaphysical status than the notions of honor and dishonor? If we are to say, with Kant, that it is an objective truth that a criminal deserves to suffer for his crime, then why should we not say that a family is truly dishonored, again in an objective sense, when they do not avenge a wrong committed by another tribe or individual? (2009, pp. 48–49).

From this perspective moral philosophers such as Mackie (1990) appear gracefully candid and abstract when they contend that while we may find disagreement about a wide range of moral issues, we also find cross-cultural agreement about certain central ethical principles:

> These principles [...] may then be applied in different ways depending on the culture; but the principles themselves – the "don't harm unnecessarily" principle, for example

[20] Cf. chapter five, section 5.1, this book.

[21] We can call them *protomoralities*, if we frame them in the light of modern, civil or post-biblical moral tradition.

> – are universally accepted. This objection might seem plausible regarding some moral values, but if my observations are correct, there is simply no widespread cross-cultural agreement about core principles of moral responsibility. Honor cultures have a diminished notion of deservingness, and what appears to be a vastly stripped down control condition, which institutionalized cultures consider to be the very essence of moral responsibility. And what principle could be more fundamental to Western notions of moral responsibility than the claim that only the culpable, only those who had control over their actions, deserve blame and punishment? (Sommers, 2009, pp. 47–48).

Furthermore, Sommers correctly notes that the two cultures are often intertwined in a particular community, so that the motive of honor and the motive of moral responsibility and desert are in reality the opposing ends of a continuum: "Naturally, there will be overlap – a notion of honor in institutionalized cultures, and a notion of moral responsibility in honor cultures" (p. 44). This further supports what I have illustrated above, concerning the fact that various subcultures (for example the honor subculture which still lives inside a dominant institutional culture) can constitute a subcultural morality that individuals can occasionally adopt, thus performing acts that appear "moral" in the light of that subculture, and "violent" in the light of the dominant culture. The following is an example: often honor morality is comfortable with the killing of a person that is not directly responsible for the offense that must be vindicated/punished, but is just somehow related to that offense (for instance because she is a relative, a friend or a partner of the offender): such a retaliation is immediately seen as an immoral crime in the light of our dominant morality of responsibility and desert.

A useful cognitive analysis of punishment is illustrated in a recent cognitive research.[22] The authors contend that there are three different systems of norm enforcement: revenge, punishment and sanction. Revenge is more related to social rituals, as a reaction to aggression, to make the aggressor's suffering repay the damage, and to deter further aggressions by changing the target's belief: it is scarcely institutionalized; punishment is still related to the deterrence of the offender, now considered as responsible, but also to *righting* the wrong in a possible retributive regime – more or less proportionate to the offense – to repay the damage; sanction, still a deterrent and retributive tool, also aims at *pedagogically* modifying the belief of the target by referring to an objectified established *norm* – social (justice), legal, or moral – and it is thus detached from the punisher's will.

I would note that some of the described aspects are often intertwined and so the useful distinction above would have to be rendered more dynamic, for example, often punishment in terms of revenge is inserted in a law, still in effect in a modern constitutional democracy. In Italy, a law, abolished in 1981 (that is, about thirty five years after the birth of the constitutional republican democracy) embodied the persistence of an archaic, patriarchal honor culture of revengeful punishment against the "offense" of unlawful sexual relations or other more violent crimes. For example, Article 587 of the law, part of the Rocco Code introduced during the Fascist regime, granted the reduction by one third of the penalty for retaliators who killed their wife, daughter, or sister to "save" the honor of their family. In detail, Article

[22] (Giardini et al., 2010).

587 of the Italian Rocco Code enabled consideration of the "cause of honor" in homicide or physical injuries, providing: "whoever discovers unlawful sexual relations (i.e. sexual relations outside marriage) on the part of their spouse, daughter, or sister and in the fit of fury occasioned by the offense to their or their family's honor causes their death, shall by punished with a prison term from three to seven years".[23]

1.2.3 Defining Violence

My friend and young colleague Jeffrey Benjamin White contends that moral perception typically has to do with how a person feels/thinks/judges that the world as a whole *should* be. He is right. He also adds that, what a person perceives demands a consistent account of all related things. Of course moral bubbles obey a kind of human exigency of ethical consistency – even if contingent: where, morally speaking, "inconsistent" a person ends up in a moral bubble, so that the inconsistency itself is not really "perceived".

The reader could suspect I am endorsing here a kind of moral relativism, to avoid dogmatism and ethical absolutism. This is not the case. I would say that I take advantage of a simple naturalistic approach, since I trust the results of scientific knowledge: exact and human sciences (and especially cognitive science) teach us results about perception, language, judgment, consciousness, etc., which also provide empirical and theoretical knowledge about the actual moral, violent, and criminal behavior of human beings. Of course science too adopts ontological constraints, but we have to remember that scientific knowledge changes and new ideas arise, so that the ontological presuppositions change too. Therefore, here naturalism is preferable to relativism.

In this naturalistic perspective human beings are embedded in *cognitive niches*[24] that – among other things – shape them both culturally and morally. They just behave as they can, depending on various constraints and limitations, biological (intelligence, sensitivity) and cultural ones (learning processes, information and available knowledge, etc.): how can we see deviance in this perspective? Deviance is a strong word and in general, reciprocal areas of deviance are individuated simply because different moral perspectives tend to consider each other as contrasting.

For example, let us pay attention to the moral taboo consisting in the prohibition of incest: following my relatively straightforward naturalism, a human organism seems to be wired to regard incest as "violent" because, if allowed, practiced and socially encouraged the resulting world would be the opposite of the one evolution has made us feel, think and judge as *right*. So it seems that evolution favors non-incestuous behaviors. All of this can be grounded in the evolution of the organism, to maximize survivability conditions, through individual actions, for every human group in the world as a whole. The world resulting from the practice of incest, in effect, would exist in a sort of alternative "morality", which lives in a kind of

[23] On the political and social role of legal punishment in modern states cf. (Hoskins, 2011).
[24] Cf. chapter four, this book.

potentiality being closed off to the agent as a morally permitted end to his action. But this alternative is very often activated and made real, like psychoanalysts teach us, and perceived by their endorsers as fair. Sexual incest is narrowly practiced but, more important and extremely widespread, narcissists can be typically "incestuous" (or "incestual" like Paul-Claude Racamier (1995) says) persons (even if not from the direct sexual point of view): they are persons who engage in a kind of relationships with parents, sons, and relatives, where the *incestual* character is clearly seen in the spontaneous perversion of the main command/prohibition. Those who are the object of incest are in turn *de facto* violently exploited and "treated only like a means", so that it is appropriate to see them as victims – but the narcissist, actually, does not perceive them as victims at all and perceives his behavior as moral and legitimate. Psychoanalysis sees these behaviors as ranging from pathological ones, where it is difficult to attribute a fully normal deliberative responsibility to the involved actors, to perfectly healthy ones, where the perpetrators responsibly choose the incestuous (or *incestual*) behavior, trusting its rightness and moral appropriateness.

In sum, even if incest can be classified as a possible maladaptive behavior from a strict Darwinian point of view, incest and incestual relationships can happen as they *actually* happen, trusted and justified by their participants. Moreover, since it is unlikely that cognitive/moral niches do evolve according to the strict Darwinian scheme, the maladaptive character of incest can be questioned too. Incest (and its psychic sub-products: incestual and incestuous) can substantiate individual moral attitudes, which do not necessarily result from a psychopathological/clinical background, which further guarantees the, so to say, "contingent" prosperity of individuals or small groups, at the expense of others who are more or less violated and exploited, for a limited lapse of time.

Let us see other evidence of the interplay of multiple moralities, and its interest from an evolutionary perspective. Jeffrey Benjamin White, the friend of mine I quoted above, continues

> Persons feel/think/judge that certain acts are morally permissible when in fact they seem contrary to survivability of both individual and group. For instance, consider the Chinese one-child-per-family practice. This practice has been proposed, recently by a Canadian writer as a possible solution to global energy/pollution/resource problems, as it has proven effective in China. However, her proposal has been met with anger, even ridiculed, by Canadians. In other words, perceived as "violent". Of course, there has been resistance to the practice in China, as well, but generally the attitudes there are different. The basis for these differences are thought to be a combination of both evolved neurology and enculturation. Western people typically focus on individuals as the most salient features of situations, while Eastern people tend to focus on the situations, themselves. Thus, weighing the situation above the self, the one-child-per-"my"-family policy is better received amongst most Easterners, while the same proposal is immediately dismissed by most Westerners. Of course, one can tell an evolutionary story about the source of these perceptual differences, this is easily done and quite solid. Even the individual deviations from enculturated norms can be derived from evolutionary accounts, as cultural norms and artifacts tend to reinforce lessons from natural evolutionary pressures, and individuals deviate from such when the instigating pressures are no longer perceived. So, in any event, this is not the issue. The issue is

that global pressures are not immediately perceived by any individual organism, and not immediately represented in any single culture, in any event! In fact, to respond to these pressures requires an idealization. In fact there seems to be two families of responses to these pressures. One focuses on the group, and the situation, and so maximizes global survivability. The other focuses on the individual, at the expense of the group, and so maximizes local survivability. For either, there are consequences. First, to pursue one is to endorse it, and not to endorse/pursue the other. Thus, one is defined as "mine" and "ours", and – consistently – the other as alien, if not sacrilegious. Second, from this it follows that there is something very active about morality, and very troubling for purely evolutionary accounts thereof, in the sense that we create the world, not merely respond to it.

The picture provided by Benjamin White is very interesting because it focuses on conflicting moral views. Before these conflicting moral views and deliberations my comment to the observation above further rejoins a critique to excessively reductionist evolutionary accounts of the development of human culture. It is difficult to frame human cultural behavior in a strictly Darwinian scheme,[25] I am basically inclined to disagree with evolutionary psychology, group selection theory, and with the old sociobiology as regards their ability to provide a full and stand-alone account of the development of culture, morals etc. Even if some insights of sociobiology and evolutionary psychology can be extremely fecund, I think that cultures do not evolve in a Darwinian way and that their *direct* Darwinian maladaptive character is difficult to see in the detailed and short-term case of culture.[26] Groups and individuals struggle against other groups and individuals, so to say egoistically. Both the cooperative and less cooperative ones are violent against the deviant inside the group and against other groups. Simply, cooperative individuals emphasize the overlapping between the moral visions (they for example can tend to reach a kind of "overlapping consensus", in Rawls' words (1987))[27] of how people feel/think/judge the world should be, while the others emphasize divisions.

Indeed, conflicts represent the main circumstance in which violence is "perceived as such". I have to stress that moral and other conflicts are not necessarily negative and they do not necessarily lead to violent outcomes. Unfortunately, like (Wieviorka, 2009, p. 10) observes, not every conflict[28] is negotiable, and there is always the possibility of violence as a way of solving it. Of course the idea of

[25] On the contrary, Mesoudi, Whiten and Laland (2004) try to demonstrate that cultural evolution has key Darwinian properties: culture is shown to exhibit variation, competition, inheritance, and the accumulation of successive cultural modification over time. Adaptation, convergence, and a loss or change of function could also be identified in culture.

[26] Clearly, rejecting the hard-core of evolutionary psychology & co. does not force one to refuse an evolutionary account of morality *as a whole*: we can recuperate an evolutionary account of the role of culture if we consider the co-evolution of cognitive niches and ecological niches taken together, as I will describe in chapter four.

[27] On the applicability and limitations of Rawlsian approaches to resolve ethical issues cf. (Doom, 2009).

[28] A rich analysis of the problem of controlling violence in modern societies (especially in the case of school shooting, terrorism, state crisis) is contained in the recent collection (Heitmeyer et al., 2011).

violence as ineluctable and positive is very widespread among human beings, also in the case of intellectuals: Georges Sorel (1999) – I would say, following Dodd (2009, p. 4) "with his crypto-fascist celebration of violence as a foundation for the moral character of the working class – has already noted, and defended, the disruptive and regenerative role of violence, as Italian Futurism did at the beginning of twentieth century". In this case violence "directly" and "intellectually" receives an attribution of moral worth.

It can be said that after the end of the total conflicts of the Cold War and workers' movements/class struggle, that social conflicts have become "limited", fragmented in a multitude of oppositions, and especially concentrated in individuals or small groups (*banlieue* young rioters, criminal gangs, terrorists, etc.). Furthermore, different sectors of an individual's everyday life relate to (often) discrepant worlds of meaning and experience, and this fact is also, following Barnett and Littlejohn

> [...] exacerbated by urbanization, or close contact among different people, and mass communication, or dissemination of confusingly diffuse realities. It is also exacerbated by political and territorial interests. [...] But we cannot have a community without a boundary to separate us from "outsiders". Often, like in Yugoslavia, the line between the community and the outsiders is political and ethnic. Sometimes it is territorial or regional. Sometimes the boundary is religious sometimes it is racial and sometimes it is ideological. Sometimes it is economic, gender based, or generational. And much of the time, it is a combination of these (1997, p. 108).

As I will illustrate further on in section 4.5 of chapter four these variegated aspects give rise to the multiplicity of the so-called *collective axiologies*, as generators of limited conflicts typical of the post-Cold War and of the post-class struggle era. The concept provides a suitable tool to describe the problem of conflicts in general and of their violent or peaceful resolution. In summary, in the aforementioned contemporary conflicts negotiation is difficult – because of the lack of an appropriate institutional channeling of the disagreement – and violence considerably probable: "from completely institutionalized conflicts to completely unbridled violence" (p. 25).

1.2.4 Various Kinds of Violence

A recent very useful collection edited by Bufacchi (2009) provides the opportunity to draw up a list of issues trying to illustrate how violence is variegated and how it can be described in the light of multiple perspectives. The list is not systematic but I think it facilitates the reader in avoiding the acceptance of limited and narrow perspectives on violence (for example "violence is physical assault" or "violence is depriving a person of her just freedom").

The modern state, as we know it, is founded on force (law), like Max Weber already clearly explained: furthermore, the state claims the monopoly of violence, and even if the exercise of violence is certainly an economic activity, violence is often wasteful.[29] Furthermore, violence is morally and/or legally and politically justified,

[29] (Dewey, 2009).

not only at the level of more or less legal institutions but also at the level of individuals (for example through law, revolutionary instances, terroristic "reasons",[30] shared religious norms or more personal religious convictions, etc.). In the case of individual convictions which lead to violent acts, these are variable and often contradictory but always momentarily supposed to be (by the agent himself) "moral". Moreover, the *progress* of civilization is not necessarily correlated with a diminution of all kinds of violence, like the twentieth century patently demonstrates (concentration camps, secret police, gas chambers, labor camps, liquidation of classes, races and whole populations, nuclear weapons, etc.) Violence in constitutional democracies tends to be highly regulated, and it appears justified or legitimated or simply impossible to avoid (Wolin, 2009); violence is physical and psychological and can take the form of positive and negative influences (for example, violence derives from social disorder, but also from a social order that causes a degree of poverty which in turn triggers further violence (Lee, 2009)).[31]

From another perspective, violence can be intended or unintended, manifest or latent (Audi, 2009; Garver, 2009) (or fruit of omissions – "negative actions" (Harris, 2009; Salmi, 2009)): especially, as I have already illustrated above, we have to distinguish between direct violence (where there is an actor that perpetrates violence) and structural violence (without any actor, but caused by institutions or economic systems (Harris, 2009), families (Galtung, 2009), which are more or less invisible examples of what I will call, later in this book, *violent mediators*). Finally, we have to reiterate that from the mere fact that something or someone is injured, destroyed or damaged (MacCallum, 2009), it does not follow that some violence was effectively committed: we always need to remember that violence is always labeled, contextual, and *perceived* as such.

1.3 Coalition Enforcement: Morality and Violence

1.3.1 Abduction, Respecting People as Things, and Animals as Moral Mediators

In this subsection I aim at introducing three key concepts that will be in the background of various argumentations contained in this book: 1) the importance of abduction, as a way of guessing hypotheses, in moral reasoning and in describing the role of fallacies in moral and violent settings; 2) the modernity – and urgency – of my moral motto "respecting people as things", also for understanding subtle cases of violence; 3) the philosophical usefulness of framing the problem of violence in a biological and evolutionary[32] perspective, where – so to say – the "animality of human animals" is a tool for avoiding traditional and robust philosophical illusions, prejudices, and ingenuities about violence.

[30] On the justification of violence in terrorism cf. (Nielsen, 2009).

[31] Various kinds of violent conflicts, intrapersonal, sexual, at the origins of wars, in labor relations, in natural "cosmic violence", are illustrated in (Jones and Fabian, 2005).

[32] Not in the sense of sociobiology and evolutionary psychology, though.

My study on hypothetical reasoning in terms of abductive cognition[33] has demonstrated that the human activity of guessing hypotheses – that is, abductive cognition – touches the important subject of morality and moral reasoning, and therefore violence. In the framework of "distributed morality", a term coined in my recent book *Morality in a Technological World. Knowledge as Duty*,[34] I have illustrated the role of abduction in moral decision, both in deliberate and unconscious cases, and its relationship with hard-wired and trained emotions.[35] Moreover, the result that the abductive guessing of hypotheses is partly explicable as a more or less "logical" operation related to the "plastic" cognitive endowments of all organisms and partly as a biological phenomenon, eventually led me to the rediscovery of animals as "cognitive agents" but also as being endowed with moral intrinsic value.

I also repeatedly argued that people have to learn to be "respected as things". Various kinds of "things"(for instance works of art and other cultural artifacts, institutions, symbols and of course animals) are now endowed with increasing intrinsic moral worth. When I say that people have to be respected as things I am referring to the fact that various "things" often have more intrinsic value than a human being[36] and so in these cases we can learn to re-attribute to humans the new moral value we have envisaged in those recently "dignified things".

Let us consider the case of animals that can become what I call "moral mediators". Darwin (1981) noted that studying cognitive capacities in non-human animals possesses an "independent interest, as an attempt to see how far the study of the lower animals throws light on one of the highest psychical faculties of man" – that is, the moral sense. Indeed, from this perspective, animals could play the role of moral mediators, because they mediate new aspects of human beings' moral lives.[37]

This line of thought traces back to Jeremy Bentham's question "Can they [animals] suffer?" and to their violent scapegoating in sacrificial rituals – like in Abel's sacrifice of an animal and Abraham's sacrifice of a ram in place of his son, but we can also see some recently advanced philosophical speculations that link morality to the problem of the violent relationships between human and non-human animals. In *The Animal that Therefore I Am*, Derrida (2008) expresses his intense opposition to our current concept of the "Animal":

> Confined within this catch-all concept, within this vast encampment of the animal, in this general singular, within the strict enclosure of this definite article ("the Animal" and not "animals"), as in a virgin forest, a zoo, a hunting or fishing ground, a paddock

---

[33] (Magnani, 2009).

[34] (Magnani, 2007b).

[35] The problem of the role of emotions (and their relationships with reason) in moral decision from the point of view of neuroscience is treated in (Moll et al., 2007; Koenigs and Tranel, 2007; Moll and de Oliveira-Souza, 2008).

[36] This means, as an example, that one might be more willing to hurt and/or destroy a human being than one of those other entities: the killing of a whale or the destruction of an ancient statue are perceived as much more violent (and hence they give rise to greater scandal) than the homicide of a man.

[37] I have treated the problem of moral mediators in (Magnani, 2007b), see also chapter four, section 4.4, this book.

> or an abattoir, a space of domestication, are all the living things that man does not recognize as his fellows, his neighbors, or his brothers. And that is so in spite of the infinite space that separates the lizard from the dog, the protozoon from the dolphin, the shark from the lamb, the parrot from the chimpanzee, the camel from the eagle, the squirrel from the tiger, the elephant from the cat, the ant from the silkworm, or the hedgehog from the echidna (p. 34).

Human language tends to seize the animal *in general*: any animal, an arbitrary animal that, as it cannot speak, has its identity denied. Following his own tenet about the violent role played by language Derrida continues unveiling the crime silently perpetrated against animals:

> No, no, my cat, the cat that looks at me in my bedroom or bathroom, this cat that is perhaps not "my cat" or "my pussycat", does not appear here to represent, like an ambassador, the immense symbolic responsibility with which our culture has always charged the feline race, from La Fontaine to Tieck [...] from Baudelaire to Rilke, and many others. If I say "it is a real cat" that sees me naked, this is in order to mark its insubstitutable singularity. [...] The confusion of all non human living things within the general and common category of the animal is not simply a sin against rigorous thinking, vigilance, lucidity, or empirical authority, it is also a crime (pp. 9 and 48).

Those reflections should bring us back the main problem of this book, the intertwining between morality and violence which is object of the following subsections. To frame the problem in an evolutionary and diachronic perspective is an initial useful step to insert it correctly in our philosophical scene. Studying the naturalistic basis of morality is clearly of great importance in increasing knowledge on morality in the direction already masterfully envisaged by Darwin.[38]

Greene (2003) usefully reminds us that – even if in Western culture many moral philosophers have regarded scientific research as irrelevant to their work because science deals with what is the case, whereas ethics deals with what ought to be – it is important to take advantage of those approaches that question the is/ought dichotomy, arguing that science and normative ethics are actually continuous. I have illustrated myself[39] how Searle is correct in considering "bizarre" that feature of our intellectual tradition, according to which true statements – so to say, fruit of "sound" inferences – that describe how things are in the world can never imply a statement about how they ought to be:[40] Searle contends that to say something is true is already to say you ought to believe it, that is *other things being equal*, you ought not to deny it. This means that normativity is more widespread than expected.

Besides, the problem of the abductive construction of extended cognitive niches offers a chance to see the role of human cognition in the management and correction of maladaptive artifactual niches, which immediately relates to the relationship

---

[38] The possible biological roots of morality, and so of punishment and *moral-based* violence, is discussed in depth in (Broom, 2009) (which also considers the moral role of religion), in (Waal et al., 2006) and, more recently in (Bekoff and Pierce, 2009): these books address the problem of individuating and describing the moral behavior of many animals.

[39] (Magnani, 2007b, chapter seven).

[40] I will return on this problem in chapter three, section 3.2.

between morality and knowledge in our technological world and to the role of the creative hypothetical reasoning employed in such a task. Furthermore, the analysis of the interplay between fallacies and abductive guessing I will present in chapter three will also illustrate how: 1) abduction and other kinds of hypothetical reasoning are involved in dialectic processes, which are at play in both everyday agent-based settings and rather scientific ones; 2) they are strictly linked to so-called smart heuristics and to the fact that very often less information gives rise to better performance; 3) heuristics linked to hypothetical reasoning like "following the crowd", or social imitation, are differently linked to fallacious aspects – which involve abductive steps – and they are often very effective. I will especially stress that these and other fallacies, are linked to what René Thom calls "military intelligence", which relates to the problem of the role of language in the so-called *coalition enforcement*, that is in the affirmation of morality and the related perpetration of violent punishment. It is in this sense that I will point out the importance of fallacies as "distributed military intelligence".

Indeed the aim of the following subsections is to clarify the idea of "coalition enforcement". This idea illustrates a whole theoretical background for interpreting the topics above concerning morality, violence, and hypothetical reasoning as well as my own position, which resorts to the hypothesis about the existence of a strict multifaceted link between morality and violence. The theme is further linked to some of the issues I will illustrate in the following chapter, where I make profitable use of Thom's attention to the moral/violent role of what he calls "proto-moral conflicts": I will fully stress that, for example, the fundamental function of language can only be completely understood in the light of its intrinsic *moral* (and at the same time *violent*) purpose, which is basically rooted in an activity of *military intelligence*.

1.3.2 Coalition Enforcement Hypothesis

The coalition enforcement hypothesis, put forward by Bingham (1999; 2000), aims at providing an explanation of the "human uniqueness" that is at the origin of human communication and language, in a strict relationship with the spectacular ecological dominance achieved by *H. Sapiens*, and of the role of cultural heritage. From this perspective, and due to the related constant moral and policing dimension of *Homo*'s coalition enforcement history (which has an approximately two-million-year evolutionary history), human beings can be fundamentally seen as *self-domesticated animals*. I think the main speculative value of this hypothesis consists in stressing the role of the more or less stable stages of cooperation *through morality* (and through the related unavoidable violence). Taking advantage of a probabilistic modeling of how individual learning – coupled with cultural transmission and a tendency to conform to the behavior of others – can lead to adaptive norms, Boyd and Richerson (2001) contend that it is possible for norms to guide people toward sensible behaviors that they would not choose if left to their own devices.

The hypothesis implicitly shows how both *morality* and *violence* are strictly intertwined in their social and institutional aspects,[41] which are implicit in the activity of cognitive niche construction. In an evolutionary dimension, coalition enforcement is enacted through the construction of social *cognitive niches* as a new way of diverse human adaptation. Basically the hypothesis refers to cooperation between related and unrelated animals to produce significant mutual benefits that exceed costs and are potentially adaptive for the cooperators.[42] Wilson, Timmel, and Miller (2004) aim at demonstrating the possibility that groups engage in coordinated and cooperative cognitive processes, thus exceeding by far the possibilities of individual thinking, by recurring to a hypothesized "group mind" whose role would be fundamental in social cognition and group adaptation. The formation of appropriate groups which behave according to explicit and implicit more or less flexible rules of various types (also moral rules, of course) can be reinterpreted – beyond Wilson's strict and puzzling "direct" Darwinian version – as the "social" constitution of a "cognitive niche",[43] that is a cognitive modification of the environment which confronts the coevolutionary problem of varying selective pressures in an adaptive or maladaptive way.[44]

In hominids, cooperation in groups (which, contrary to the case of non-human animals, is largely independent from kinship) fundamentally derived from the need to detect, control, and punish social parasites, who for example did not share the meat they hunted or partook of the food without joining the hunting party[45] (also variously referred to as free riders, defectors, and cheaters). These social parasites were variously dealt with by killing or injuring them (and also by killing cooperators who refused to punish them) *from a distance* using projectile and clubbing weapons. In this case injuring and killing are cooperative and remote (and at the same time they are "cognitive" activities). According to the coalition enforcement hypothesis, the avoidance of proximal conflict reduces risks for the individuals (hence the importance of *remote killing*). Of course, cooperative morality that generates "violence" against unusually "violent" and aggressive free riders and parasites can be performed in other weaker ways, such as denying a future access to the resource, injuring a juvenile

[41] North, Wallis and Weingast (2009) integrate the problem of violence into a larger framework that includes social sciences and history, showing how economic and political behavior in modern states are closely linked. "Natural states" introduce a political manipulation of the economy that allows the creation of privileged interests limiting the use of violence by a tit-for-tat mechanism, insofar as powerful, privileged individuals think it is safer not to threaten each other's privileges. Such a practice establishes a compromising trade-off between safety and moderate economic and political development. Modern societies generate and rely on open access to economic and political organizations, favoring political and economic growth through competition.

[42] On human and animal various types of altruism and on human cooperation through social institutions see (Tommasello, 2009).

[43] Cf. chapter four, this book.

[44] On the coevolution of intelligence, sociality, and language in the persspective of cognitive niches see (Pinker, 2010).

[45] (Boehm, 1999).

relative, gossiping to persecute dishonest communication and manipulative in-group behaviors or waging war against less cooperative groups, etc.[46]

In such a way group cooperation (for example for efficient collective hunting and meat sharing through control of free riders) has been able to evolve adaptively and to render parasitic strategies no longer systematically adaptive. Through cooperation and remote killing, individual costs of punishing are greatly reduced and so are individual aggressiveness and violence, perhaps because violence is morally "distributed" in a more sustainable way: "Consistent with this view, contemporary humans are unique among top predators in being relatively placid in dealing with unrelated conspecific nonmates under a wide variety of circumstances" (Bingham, 1999, p. 140). [I have to note, "contrarily to the common sense conviction", due to the huge amount of violence human beings are still faced with everyday!] Hence, it has to be said that humans, contrarily to non-human animals, exchange a fundamental and considerable amount of relatively reliable information with unrelated conspecifics (p. 144).

Here the role of docility is worth citing (which relates to the already recognized distressing human tendency to conform, displace responsibility, comply, and submit to the authority of dominant individuals, emphasized by social psychology (Dellarosa Cummins, 2000, p. 11)). According to Herbert Simon (1993, p. 156), humans are docile in the sense that their fitness is enhanced by "[...] the tendency to depend on suggestions, recommendations, persuasion, and information obtained through social channels as a major basis for choice". In other words, humans support their limited decision-making capabilities, relying on external data obtained through their senses from their social environment. The social context gives them the main data filter, available to increase individual fitness. Therefore, docility is a kind of attitude or disposition underlying those activities of cognitive niche construction, which are related to the delegation and exploitation of ecological resources. That is, docility is an adaptive response to (or a consequence of) the increasing cognitive demand (or selective pressure) on those information-gaining ontogenetic processes, resulting from an intensive activity of niche construction. In other words, docility permits the inheritance of a large amount of useful knowledge while lessening the costs related to (individual) learning.

In Simon's work, docility is related to the idea of *socializability*, and to altruism in the sense that one cannot be altruistic if he or she is not docile. In this perspective, however, the most important concept is docility and not altruism, because docility is the cognitive condition of the possibility of the emergence of altruism. I believe, at least in the light of the coalition enforcement hypothesis, that moral altruism can be correctly seen as a sub-product of – or at least intertwined with – the violent behavior needed to "morally" defend and enforce coalitions. I have said that groups need to detect, control and punish social parasites, that for example do not share meat (food as nutrient as hard to acquire), by killing or injuring them (and any cooperators who refuse to carry out punishment) and to this aim they have to gain the cooperation of other potential punishers.

---

[46] On the moral/violent nature of gossip and fallacies cf. (Bardone and Magnani, 2010).

Research on chimps' behavior shows that punishment can be seen as altruistic for the benefit of the other members of the group (and to the aim of changing the future actions of the individual being punished), often together with "the function of keeping the top ranking male, or coalition of males, at top, or preserving the troop-level macrocoalition that disproportionately serves the interests of those on top" (Rohwer, 2007, p. 805). The last observation also explains how altruistic punishment can serve individual purposes (and so it can be captured by individual selection models): in chimps it reflects the desire to maintain status, that is a new high position in the hierarchy. Rohwer's conclusion is that "certain behaviors that can be covered by the definition of altruistic punishment need not have arisen from group selection pressures [...] as the Sober and Wilson model assumes. Seen through the lens of the linear dominance hierarchy, it is reasonable to suspect that altruistic punishment may have originated primarily through individual selection pressures" (p. 803 and p. 810).[47]

I have said above that groups need to detect and punish social parasites by killing or injuring them (and any cooperators who refuse to carry out punishment) and to this aim they have to gain the cooperation of other potential punishers. This explains altruistic behavior (and the related cognitive endowments which make it possible, such as affectivity, empathy and other non violent aspects of *moral* inclinations) which can be used in order to reach cooperation. To control free-riders inside the group and guard against threat from other alien groups, human coalitions – as the most gregarious animal groups – have to take care of the individuals who cooperate. It is from this perspective that we can explain, as I have said above, quoting Bingham, why contemporary humans are not only violent but *also* very docile and "[...] unique among top predators in being relatively placid". Moreover, Lahti and Weinstein (2005) further emphasize the evolutionary adaptive role of morality (and so of cooperation) as "group stability insurance": the exigence of morality as group stability would explain the "viscosity" of basic aspects of the morality of a group and why morality is perceived as having an air of absolutism. I also add that this viscosity favors the "embubblement" I will illustrate in chapter three, subsection 3.2.2, which explains the obliteration – at the level of the agent awareness – of the actual violent outcomes of his moral actions.[48]

The problem of docility is twofold. First, people delegate tasks of data acquisition to their experience, to external cultural resources and to other individuals. Second, people generally put their trust in one another in order to learn. A big cerebral cortex, speech, rudimentary social settings and primitive material culture furnished the conditions for the birth of the mind as a "universal machine" – to use Turing's

[47] On the puzzling problem of the distinction between individual and group selection for altruism in the framework of multilevel selection cf. (Rosas, 2008): multilevel selection theory claims that selection operates simultaneously on genes, organisms, and groups of organisms. A history of the debate about altruism is given by (Dugatkin, 2008).

[48] On the intrinsic moral character of human communities – with behavioral prescriptions, social monitoring, and punishment of deviance – for much of their evolutionary history cf. (Wilson, 2002a, p. 62) and (Boehm, 1999).

expression. I contend that a big cortex can provide an evolutionary advantage only in the presence of a massive storage of meaningful information and knowledge on external supports that only a developed (even if small) community of human beings can possess. If we consider high-level consciousness as related to a high-level organization of the human cortex, its origins can be related to the active role of environmental, social, linguistic, and cultural aspects. It is in this sense that docile interaction lies at the root of our social (and neurological) development. It is obvious that docility is related to the development of cultures, morality, actual cultural availability and to the quality of cross-cultural relationships. Of course, the type of cultural dissemination and possible cultural enhancements affect the chances that human collectives have to take advantage of docility and thus to potentially increase or decrease their fitness.

As I have already noted, the direct consequence of coalition enforcement is the development and the central role of cultural heritage (morality and sense of guilt included): in other words, I am hinting at the importance of cultural *cognitive niches* as new ways of arriving at diverse human adaptations. From this perspective the long-lived and yet abstract human sense of guilt represents a psychological adaptation, *abductively* anticipating an appraisal of a moral situation to avoid becoming a target of violent coalitional enforcement. Again, we have to recall that Darwinian processes are involved not only in the genetic domain but also (with a looser and lesser precision) in the additional cultural domain, through the selective pressure activated by modifications in the environment brought about by cognitive niche construction. According to the theory of cognitive niches, coercive human coalition as a fundamental cognitive niche constructed by humans becomes itself a major element of the selective environment and thus imposes new constraints (designed by *extragenetic information* on its members).[49]

Some empirical evidence (presented by Bingham (2000)) seems to support the coalition enforcement hypothesis: fossils of *Homo* (but not of Australopithecines) show, on observation of skeletal adaptations, how selection developed an astonishing competence in humans relating to the controlled and violent use of projectile and clubbing weapons (bipedal locomotion, the development of *gluteus maximus* muscle and its capacity to produce rotational acceleration, etc.). The observed parallel increase in cranial volume can be related to the increased social cooperation based on the reception, use and transmission of "extragenetic information". Moreover, physiological, evolutionary, and obstetric constraints on brain size and structure indicate that humans can individually acquire a limited amount of extragenetical information, that consequently has to be massively stored and kept available in the external environment.

Usually it is said that Darwinian processes operating on genetic information produce human minds whose properties somehow include generation of the novel, complex adaptive design reflected in human material artifacts *sui generis*. However, these explanations

---

[49] On the concept of extragenetic information cf. chapter six of (Magnani, 2009), and chapter five, section 5.2.3, this book.

> [...] fail to explain human uniqueness. If building such minds by the action of Darwinian selection on genetic information were somehow possible, this adaptation would presumably be recurrent. Instead, it is unique to humans. Before turning to a possible resolution of this confusion, two additional properties of human technological innovation must be recalled. First, its scale has recently become massive with the emergence of behaviorally modern humans about 40,000 years ago. Second, the speed of modern human innovation is unprecedented and sometimes appears to exceed rates achievable by the action of Darwinian selection on genetic information (Bingham, 1999, p. 145).

Hence, a fundamental role in the evolution of "non" genetic information has to be hypothesized. Appropriately, coalition enforcement implies the emergence of novel extra-genetic information, such as large scale mutual information exchange – including both linguistic and model-based[50] communication between unrelated kin. As mentioned above, it is noteworthy that extra-genetic information plays a fundamental role in terms of transmitted ideas (cultural/moral aspects), behavior, and other resources embedded in artifacts (ecological inheritance). It is easy to acknowledge that this information can be stored in human memory – in various ways, both at the level of long-lived neural structures that influence behavior, and at the level of external devices (cognitive niches), which are transmitted indefinitely and are thus potentially immortal – but also independently of small kinship groups. Moreover, transmission and selection of extragenetic information is at least partially independent of an organism's biological reproduction.

1.3.3 The Origin of Moral/Social Norms, Social Dominance Hierarchies and Violence

It is important to show how moral norms, cooperation, and social dominance hierarchies can be accounted for from an evolutionary point of view. A great deal of important research has been done recently in this field.

The evolutionary account of social norms in terms of cognitive niches allow us to try to overcome the still puzzling question of their genetic or extra-genetic origin: some of them may not be learned behaviors at all (e.g, incest avoidance), but a vast area of social patterns of behavior seem self-evidently learned. Bandura (1977) basically sees them as the result of social learning. O'Gorman, Wilson and Miller (2008) report that some researchers seem to contend that social norms are not due solely to incidental group boundaries or selfish herds, but to an ancient phylogenetic history, given the fact that some animal species demonstrate a kind of conforming behavior (following the herd), to avoid exposure to predators; others suggest that they may result from an evolved strategy to avoid the costs of individual learning

[50] Examples of model-based cognition are constructing and manipulating visual representations, thought experiment, analogical reasoning, etc. but it also refers to the cognition animals can get through emotions and other feelings. Charles Sanders Peirce already acknowledged the fact that all inference is a form of sign activity, where the word sign includes various model-based ways of cognition: "feeling, image, conception, and other representation" (CP, 5.283).

and/or to react to a fluctuating environment; others that norms are related to the fact that certain behaviors may achieve some form of symbolic status for a group,[51] in such a way that the very *achievement* of learning the behavior is important to be considered part of the group.

The perspective that is closer to the coalition enforcement hypothesis is the one that sees norms as the product of cooperative group behavior, where some norms can be elevated to moral norms or rules,[52] or to between-group selection regulators.[53] It is worth noting that punishment of norm violators[54] is seen as a near-universal trait of humans,[55] through gossip, exclusion and expulsion from groups, or murder. O'Gorman, Wilson, and Miller note that a great part of this research leads to the prediction that there is a conformity bias in human cognition "facilitating enhanced recall of normative information".[56] These authors have also provided empirical evidence that human beings have superior recall for normative/social information than for non-normative information, also providing a limited cross-cultural endorsement of the hypothesis, and favoring the conviction about their likely hardwiring in the human brain and a consequent possible explanation in evolutionary terms.

Recent research acknowledges the role of social dominance hierarchies in cooperative activities of human and non-human animals (like social mammals), also stressing the deontic importance of moral behavior and intertwined violent components. Evidence from comparative, developmental, and cognitive psychological research show how they shaped the evolution of the human mind and so of human social institutions. Pressures that derive from living in hierarchical groups shaped basic concept and cognitive devices (which are not interpreted as innate modules but instead as favored by a kind of "biological preparedness", a propensity to develop them in a combination of genetical and environmental circumstances). These concepts and cognitive devices are in turn related to the various stages of development of cognitive niches, crucial factors to survive in those hierarchic settings. They are: 1) recognizing and reasoning transitively about dominance relations (making rank and kinship discriminations), 2) fast-track learning of social norms (permissions, prohibitions, and obligations), 3) detecting violations of social norms and codes – also called *deontic effect* – (cheating and deceptions to flout social norms and to have better access to reproductive opportunities through monitoring reciprocal obligations based on resource sharing) 4) reading intentions of others and so predicting their behavior, especially – again – to favor detection of violations (but also to allow compassion and social codes based on something other than social norms).[57]

---

[51] (Dessalles, 2000).

[52] (Boehm, 1999).

[53] (Wilson and Kniffin, 1999).

[54] Zaibert (2006) illustrates that current state legal punishment is tied to the moral practice of blaming, and that the very practice of placing the blame is what presumptively provides many of the intuitions that we have as regards punishment.

[55] (Brown, 1991).

[56] (O'Gorman et al., 2008, p. 72).

[57] (Dellarosa Cummins, 2000).

We have to add that perceived violations were already studied as the most common cause of aggression in primate groups:[58] without a capacity for violation detection high-ranking individuals cannot monopolize resources and maintain fruitful alliances, and preserve the peace. Especially in human collectives, violation detection is the most precious tool for ensuring that social norms (implicit and explicit) are honored thus facilitating social regulation, and to promote the emergence of altruism as a stable strategy.

It is easy to derive from this picture that transgressions are often seen as *moral* transgressions by the detectors but not necessarily by the transgressors. In modern humans violence activated by the detectors to control transgression is seen by them as *morally* justified but it is often seen as a mere violence by the transgressors. Is this not what happens in our current society when a murderer thinks he "did the right thing", for example to retaliate, contrarily to "other" human beings that instead think it is moral to execute him in the legal form of capital punishment? The former indeed thinks he has performed a "moral" murder: revenge is part and parcel of various kinds of human moral behaviors seen in the light of social, historical, and anthropological perspectives, but the wrong he wanted to revenge was not thought so by the rest of the coalition.

Dellarosa Cummins is implicitly aware that social dominance hierarchies are benign but also potentially negative, and the following consideration in some sense mirrors, from the point of view of a cognitive evolutionary psychology, my philosophical contention about the strict relationship between morality and violence:

> In their most benign form, hierarchies reflect nothing more than leadership qualities among a few individuals. In their more malevolent form, such hierarchies can constitute rigid and profound socioeconomic stratification. From a socio-political perspective, perhaps the most important contribution that cognitive psychologists can make is to investigate which environmental contingencies trigger or hinder the expression of these potentially destructive biological predispositions (2000, p. 22).

Other aspects of the importance of cooperation (and so of punishment and violence) are also analyzed in the field of evolutionary and cognitive studies. The cognitive importance of the internalization of phonemes for cooperation has been recently stressed by cognitive paleoanthropological research.[59] The so called "enhanced working memory" (EWM) (and its executive functions) can be plausibly traced back about 30.000 years in hominids: the authors show that it seemed to coevolve with the birth of a *phonological storage capacity* along with consequent language and other modern reasoning abilities such as planning, problem solving/algorithm manipulation, analogy, modeling, holding inner representations, tool-use, and tool-making. In particular, an increase in phonological storage could also have aided cross-modal thinking (and so abductive hypothetical cognition) and the social tasks required by the need for coalition enforcement: "[...] enhanced phonological storage may have freed language from the laconic and its confinement to present tense and simple imperatives to rapidly-spoken speech and the use of

[58] (Hall, 1964).

[59] (Coolidge and Wynn, 2005).

future tense – the linking of past, present, and future, and the use of the subjunctive [...]. Although real enemy's actions might be anticipated, imaginary enemies could be envisioned and other intangible terrors could be given life. Great anxieties could arise with novel vistas (e.g. the meaning of life, thoughts of death, life after death, etc.)" (p. 22). Consequently more room for morality, punishment, and subsequent more sophisticated chances and methods for perpetrating violence would be opened up!

Finally, Castro and Toro (2004) contend that the birth of moral judgements also coevolves with the cultural transmission itself, as a whole cumulative inheritance system. Following the model of the dual inheritance theory and gene-culture coevolution, they suggest that there seems to be an increase, either qualitative or quantitative, in the efficiency of imitation as the fundamental factor able to explain the transformation of primate social learning in a cumulative cultural system of inheritance as happens during *hominization*. The authors contend that more efficient imitation is necessary but not enough for this transformation to occur and that the key factor enabling such a transformation is that some hominids developed the deontic capacity to approve or disapprove of their offspring's learned behavior.[60] Such a capacity to approve or disapprove of one's offspring's behavior makes learning both less costly and more accurate, and it transformed the hominid culture into a system of cumulative cultural inheritance similar to that of humans, although the system would still partake of a prelinguistic nature. In this perspective, axiological and moral aspects result completely intertwined with the evolution of culture as a whole.

1.3.4 Cooperation, Altruism, and Punishment: An Epistemological Note

West, Grifin and Gardner (2008, p. 415) correctly say that, from an evolutionary perspective, social behaviors – and hence *moral* ones – are those affecting the fitness of the individual that enacts that particular behavior and of other individuals as well, as a consequence. They also usefully note that, over at least four decades, an enormous amount of theoretical and empirical literature has developed on this topic, as demonstrated by the illustration I have given in the previous two subsections, but "the progress is often hindered by poor communication between scientists, with different people using the same term to mean different things, or different terms to mean the same thing. This can obscure what is biologically important, and what is not". The clear discussion is related to the semantic confusion about concepts such as altruism and weak altruism, cooperation, mutualism, punishment, reciprocity and strong reciprocity, and group selection. In the end they emphasize – and I strongly

[60] In (Castro et al., 2010) the authors add that the social approval/disapproval of behavior is also adaptive because it tends to homogenize behaviors, beliefs and values of the groups whose members interact in a cooperative and docile way – for mutual benefit. It is assumed that an essential aspect of man is his condition as *Homo suadens*: if a behavior is approved of then it is a good behavior.

agree with them – it is at least necessary to distinguish between *proximate* (mechanism) and *ultimate* (survival value) explanations of behavior. Even if I think the described confusion is truly spectacular, I am sure it does not affect my philosophical concern, which addresses general issues related to the intertwining between morality and violence. I think my analysis of this intertwining is just fecundated by this interdisciplinary and often embarrassing epistemological situation, because philosophy, which does have neither direct scientific aims nor too many epistemological worries, can focus on some general aspects that can be extracted and shared – and thus further critically enlightened – by almost any of the aforementioned approaches.[61]

Finally, recent research aims at increasing cognitive knowledge on the relationship between cooperation and punishment taking advantage of an abstract Bayesian modeling of a population of learning agents, studying the effect of innate behavioral dispositions (such as *prosociality*) combined with the effect of learning (as a response to contingent punishment).[62] A relationship which in a sense partially reverberates my theme concerning the link between morality and punishment. The research stresses the fact that the coevolution of punishment and prosociality depends critically upon the "manner" in which perpetrators learn from punishment: for example the capacity to learn from punishment can allow both punishment and prosocial behavior to evolve by natural selection. The modeling indicates interesting levels of evolutionary selective optimality in the balance between innate prosociality and learning. Costly punishment of antisocial behavior would be adaptively favored when it causes others to act prosocially in future interactions, that is when agents can learn to avoid punishment. The authors themselves point out the limit of this kind of modeling, which relates to its abstractness and to the fact that it is unlikely to represent prosociality as just linked to a mere expectation of contingent reward or punishment, while it is clear that prosocial behavior also has an intrinsic utility.

Rosas (2009, p. 555), still in an evolutionary perspective, further studies the relationship between reciprocity (cooperation) and punishment and concludes that hierarchy among them cannot be established. Indeed, we find out that reciprocity is the core of behaviors driven by *norms* prescribing either *goodness* or *badness*: "The mechanism for reciprocity in humans emerges as a meta-norm that governs both retaliation and punishment", whereas punishment appears basically aimed at inculcating norms. Reciprocity frames cooperators as those who pursue a higher-level form of "moral" punishment, unselfish and emotionally driven, which does not merely look primarily at the balance between costs and benefits for the perpetrators.

Finally, the possible coevolution between the so-called "moral signals" and indirect reciprocity is analyzed by Smead (2010). The author finds that it is possible for moral signals – as an unlikely accident of the evolutionary process' starting point – to evolve alongside indirect reciprocity but, in the absence of some external *exogenous* pressure aiding the evolution of a signaling system, such a coevolution is

[61] For an illustration of the puzzling theoretical problems related to evolution, cooperation, altruism, and punishment, in the case of a study on scientific, philosophical, and theological reflections on the origin of religion, cf. (Schloss and Murray, 2009).

[62] (Cushman and Macindoe, 2010).

unlikely: "Thus, for the use of 'moral signals' to evolve in this setting, it is important that signals be subject to selective pressure apart from indirect reciprocity. Furthermore, when such exogenous pressures are included, interesting possibilities arise. Different populations can evolve different 'moral systems' that vary in the way they use moral signals'. These simple models show that even if moral language plays a crucial role in cooperation through indirect reciprocity, it does not follow that this setting has provided the selective pressure driving the evolution of a moral language" (p. 50).

Chapter 2
The Violent Nature of Language
Language Is a Tool Exactly Like a Knife

2.1 The Violent Nature of Language: Abduction, Pregnances, Affordances

The concept of violence can be usefully enriched taking advantage of Thom's theory of morphogenesis, based on the catastrophe theory. The reader has to be patient, the concepts of catastrophe theory are difficult but I think they are rewarding, because they permit us to grasp the naturalness of violence as an ordinary semiophysical process. Furthermore, in the framework of catastrophe theory it is easy to understand the constitutive violent nature of language, as we already stressed in the previous chapter taking advantage of a merely philosophical perspective. *Nevertheless*, should the reader find this theory excessively complicated or fatiguing, she could pass over its detailed explanation and get acquainted with it from the subsequent references and, maybe, eventually come back to the more complete description I am about to undertake.[1]

To understand the basic tenets of catastrophe theory it is useful to exploit the concept of abduction, which refers to the role in human and non human animal cognition of "guessing hypotheses". First of all we need to clarify the notions of pregnance and salience, which play an important role in this theory. An example of instinctual (and putatively "unconscious") abduction (i.e. hypothetical cognition) is given by the case of certain cognitive abilities embodied in animals. These abilities are in turn capable of leading to some appropriate behavior: as Peirce said, abduction even takes place when a new born chick picks up the right sort of corn. I have contended that this is an example of spontaneous abduction – analogous to the case of other hardwired unconscious/embodied abductive processes in human beings:

[1] That is, the reader could directly proceed to subsection 2.1.4 (Violence: "Vocal and Written Language is a Tool Exactly Like a Knife"), where I directly deal with language as an essentially violent tool and skip the parts in which I introduce Thom's theory, about how it relies upon the concepts of *saliences and pregnances* (2.1.1), its relevance in an eco-cognitive – and proto-moral – dimension (2.1.2), the relevance of Thom's theory for the formation of language (2.1.3), and the the link between language and biological factors (2.1.3.1).

L. Magnani: Understanding Violence, SAPERE 1, pp. 35–64, 2011.
springerlink.com

> When a chicken first emerges from the shell, it does not try fifty random ways of appeasing its hunger, but within five minutes is picking up food, choosing as it picks, and picking what it aims to pick. That is not reasoning, because it is not done deliberately; but in every respect but that, it is just like abductive inference.[2,3]

I think the concept of pregnance, introduced by Thom (1972; 1980) on the basis of Wertheimer's Gestaltic concept of *Prägnanz*, can shed further light on a kind of morphodynamical "physics" of abduction, first of all in the case of its instinctual hardwired aspects.

Pregnance and salience, as I will show, are key concepts which can be seen in the light of both the instinctual and plastic nature of abductive hypothetical cognition. As they acquire their meaning in a very naturalistic intellectual framework, they are also useful to propose a unified perspective on moral and violent abductive processes and other various cognitive/psychic ones, seen as basic physico-biological events, endowed with a profound eco-cognitive significance. The pregnance affects an organism, and the related abductive/hypothetical response is promptly triggered; in gregarious animals this response is often a proto-moral/proto-violent one.[4] Pregnances are genetically transmitted but can also be actively created for example through learning and high cognitive capacities, through the formation of multiple forms of hypothetical intelligence.

2.1.1 Saliences and Pregnances

The complicated – and at first sight obscure – concept of *pregnance* is based on the concept of *salience*, which emerges in the dynamical framework of the "semiophysical" perspective. First of all we can say that in general, phenomenal discontinuities are perceived by organisms as *salient forms* (for example, in the auditive case, the eruption of a sound in the midst of silence), that is, as contextual effects between forms: "The simplest feature is the punctual discontinuity geometrically represented

---

[2] Cf. the article "The proper treatment of hypotheses: a preliminary chapter, toward and examination of Hume's argument against miracles, in its logic and in its history" [1901] (in (Peirce, 1966, p. 692)).

[3] Cf. also the following related passage: "How was it that man was ever led to entertain that true theory? You cannot say that it happened by chance, because the possible theories, if not strictly innumerable, at any rate exceed a trillion – or the third power of a million; and therefore the chances are too overwhelmingly against the single true theory in the twenty or thirty thousand years during which man has been a thinking animal, ever having come into any man's head. Besides, you cannot seriously think that every little chicken, that is hatched, has to rummage through all possible theories until it lights upon the good idea of picking up something and eating it. On the contrary, you think the chicken has an innate idea of doing this; that is to say, that it can think of this, but has no faculty of thinking anything else. The chicken you say pecks by instinct. But if you are going to think every poor chicken endowed with an innate tendency toward a positive truth, why should you think that to man alone this gift is denied?" (Peirce, CP, 5.591).

[4] Some non-human animal behaviors can reasonably be called proto-moral, to lessen the anthropomorphic aura of the adjective "moral".

by a point dividing the real straight line **R** into two half lines" (Thom, 1988, p. 3). Discontinuities *out there* in the environment are basically *translated* into other more or less amplified discontinuities in the subjective sensorial state, as a kind of "echo" or "shock" of the physical environment *within* an organism. In the case of sensory systems, salience of course is at the basis of the first possibility of perceiving individuated forms. In this case perception can be appropriately influenced by a certain form of "concept, that is to say a class of equivalence between forms referent to the same concept": the lack of the concept can annihilate the grasping of the individuated form, especially when analysis proceeds from the whole to the parts.

The term pregnance can be applied both to physical and biological phenomena. It can further clarify the distinction between the instinctual chicken abduction above and other plastically acquired abductive ways of cognition:

> So we will get this general pattern of a world made up of salient forms and pregnances – salient forms being objects, very often individuated, that are impenetrable to one another, and pregnances being occult qualities, efficient virtues that emanate from source-forms and invest other salient forms in which they produce visible effects (that is the so-called "figurative" effects for the organisms invested) (Thom, 1988, p. 2).

We have to note that when Thom calls pregnances "occult qualities", that is just a metaphor: actually Thom thinks that pregnances are not occult and mysterious qualities at all, because they could be accounted for as fully explainable psychological phenomena in neurological and biological terms, and they can also be made intelligible through mathematical models. The description of the processes affected by pregnance activity aims at providing what Thom calls a "protophysics, source and reservoir of all permanent intuitions, of all those archetypal metaphors that have nourished man's imagination over the ages" (p. 3).

Thom further says: "*Pregnances* are non-localized entities emitted and received by salient forms. When a salient form 'seizes' a pregnance, it is invaded by this pregnance and consequently undergoes transformations in its inner state which can in turn produce outward manifestations in its form: we call these *figurative effects*" (p. 16). To clarify the two concepts of salience and pregnance the following two examples can be of some utility [the wide range of events covered by the two concepts is testified by the fact that the first example does not have any cognitive/psychological significance]: 1) an infection (pregnance) contaminates healthy subjects (representing the "invested" form: salience). These subjects in turn re-emit the same infection (pregnance) into the environment. In this case pregnance has in itself a material/biological support (for example a virus) – as a mediator, which in turn is transmitted thanks to a suitable medium (for example air or blood); 2) worker honeybees communicate with each other by means of signs (through the iconic movements of a dance) – pregnance – that express the site where they have found food in order to inform the other conspecific individuals – the invested salience – about the location. In this second case the pregnance is transmitted – mediated – through undulatory sounds and light signals and produces a neurobiological effect at the destination, that is, in other words, a "psychic" effect [of course we can use in this case the expression "psychic" only if we admit, in a mentalistic and unorthodox way, that honeybees are endowed with *a kind of* animal psyche: an example regarding a cat or a boy would have been more convincing for the reader...].[5] However, to better grasp the concept of pregnance further analysis is needed.

2.1.2 Pregnances as Eco-Cognitive Forms: The Naturalness of Proto-Morality and Violence

In general, in the case of salient forms, their impact on the organism's sensory apparatus "remains transient and short lived" (Thom, 1988, p. 2), so they do not have relevant long-term effects on the behavior of the organisms. In the case of cognitive events, if we adopt the perspective of the affordances,[6] we can say that salient forms – contrary to pregnant forms, "afford" organisms without triggering *relevant* modifications either at the level of possible inner rumination or in terms of motor actions. Thom says that when salient forms carry "biological significance", like in the *form of prey* for the hungry predator, or the predator for its prey, or in the case of sex and fear, or when a salient form is invested by an infection, the reaction is much bigger and involves the freeing of hormones, emotive excitement and behavior (or an immune response in the case of the infection) devoted to attracting or repulsing the form: *salient forms of this type are called pregnant.*

However, in the *cognitive* case, we still have to deal with pregnances. In this case, *pregnances*, no matter whether due to innate releasing processes or to complicated, more or less stable internal learnt processes and representations (or pseudorepresentations,)[7] are triggered by a very small sensory stimulus (a stimulus "with a little figuration, an olfactory stimulus for instance" (p. 6). Hence, they represent a relationship with certain *special* phenomenological aspects, that are of course stable to different extents and so can appear and disappear. At some times and in some cases the special sensitivity to pregnances is disregarded. Like in the case of affordances, this variability and transience can be seen at the level of the differences of pregnance sensitivity among organisms and also at the level of the same organism at subsequent stages of its cognitive and biological development. We can say that a pregnant stimulus is – so to say – *highly diagnostic* and a trigger to initiate abductive cognition, like in the case of the hardwired pregnance occurring to our Peircean chicken and its food: the chicken promptly reacts when perceiving it. When a pregnance affects an organism, the abductive reaction can be promptly triggered. Finally, we have to

[5] Fields in physics are the true paradigm of *objective pregnances* in modern science, because in that case we are theoretically able to calculate their variation in space-time thanks to a mathematical description (based on an explicit geometrical definition of space-time) (Thom, 1988, p. 32).

[6] If we acknowledge that environments and organisms evolve and change, and also both their instinctual and cognitive plastic endowments, we may argue that affordances can be related to the variable (degree of) "abducibility" of a configuration of signs: a chair affords sitting in the sense that the action of sitting is a result of a sign activity in which we perceive some physical properties (flatness, rigidity, etc.), and therefore we can ordinarily infer that a possible way to cope with a chair is sitting on it. Cf. (Magnani, 2009, chapter six).

[7] On pseudorepresentations, see below in this section.

recall that the pregnant character of a form is always relative to a receiving subject (or group of subjects), just as in the eco-psychological case of affordances.

Pregnances can be abductively activated or created. When a bell ringing is repeated often enough together with the exhibition of a piece of meat to a dog, thanks to Pavlovian conditioning the alimentary pregnance of meat spreads by contiguity to the salient auditive form, so that the salient form, in this case the sound of the bell, is invested by the alimentary pregnance of the meat; here the metaphor of the invasive fluid – even if exoteric – can be useful: "So we can look on a pregnance as an invasive fluid spreading through the field of perceived salient forms, the salient form acting as a 'fissure' in reality through which seeps the infiltrating fluid of pregnance" (p. 7).[8] The propagation can also occur through similarity, taking advantage of the mirroring force of some features. Once the reinforcement is established, the bell – Thom says – refers *symbolically* in a more or less stable way, to the meat. Of course extinction of pregnances through absence of reinforcement is possible, when an organism moves away for a long time from the source form or when the invested salient form is associated with another pregnant form still in absence of reinforcement. From this point of view the "symbolic activity" is seen as fundamentally linked to biological control systems in two ways: 1) it is an extension of their efficacy (new favorable cognitive abductive chances – new pregnances – are added); 2) an internal simulation concerning the relationship between the food and its index, the bell, is implemented, so that the door is opened to the formation of multiple forms of abductive semiotic cognition (and /or intelligence):

> The fact that initially, as in the Pavlovian schema, this stimulation is no more than a simple association, does not stop us from considering that we have the first tremors in the plastic and competent dynamic of the psychism of [the actant] of an external spatiotemporal liaison interpreted not without reason as causal. [...] Hence, from the beginning, the situation is not fundamentally different from that of language [...]. Only these fundamental "catastrophes" of biological finality have the power of generating the symbols in animals (pp. 268–269).

As I have already indicated Thom sees pregnances not only as innate endowments (like in the case of the basic ones seen in birds and mammals: hunger, fear, sexual desire), but also as related to higher-level cognitive capacities, which also involve the role of *proto-morality* (in non human animals) and *morality* in (human beings). "When animal pregnance is generalized in the direction of human conceptualization 'conceptual' or individuating pregnances will be revealed, the nature of which is close to 'salience' " (p. 6). At this point it should be clear that I maintain we can synthetically account for both these processes in terms of different kinds of abductive hypothetical cognition. For example, Thom observes, reverberating the view of visual perception as semi-encapsulated,[9] that "[...] it is doubtful whether

[8] To explain the formation of pregnances Thom exploits the classical Pavlovian perspective. More recent approaches take advantage of Hebbian (Hebb, 1949) and other more adequate learning principles and models, cf. for example (Loula et al., 2010).

[9] Perception is informationally "semi-encapsulated", and also pre-wired, i. e., despite its bottom-up character, it is not insulated from plastic cognitive processes and contents acquired through learning and experience, cf. (Raftopoulos, 2009).

genetics alone would be able to code a *visual* form [...]. Whence the necessity of invoking cultural transmission, linked with the social or family organization of the community" (p. 10). In gregarious animals the signals (which also have to be seen as referring to the explanation of the origin of the "pregnance-mirroring" functions of human language) are a vector of pregnances insofar as they transfer a pregnance from one individual to another, or to several others. In such a way they favor teaching and learning, working to constitute the collectively shared behavior needed for example to capture food and to ward off predators. In this perspective of gregarious animals pregnances are de facto immediately related to the emergence of kinds of proto-moralities relying on shared proto-axiological features.

When an organism – through abductive cognition – traces back a symbolic reference to a "source" form [in Thom's sense as indicated above], often a motor reaction becomes necessary to bring satisfaction:

> In a social group, one individual's encounter with a source form S may give rise to a dilemma: whether to pursue the "individual interest" which consists in using the regulatory reflex that will result in selfish satisfaction, or to follow the altruistic community strategy by uttering the cry that will carry the pregnance S to the other members of the community; such a cry is then the signal by which the signal P of S experienced by individual 1 can be transferred to another individual 2 (p. 12).

Thom himself nicely adds that this kind of proto-moral conflict resonates with the more clearly "moral" conflict of civilized societies "This dilemma exists well and truly in our society. Witness the scruples most honest citizens have in making true declarations of their taxable revenues" (Thom, 1988, p. 12).

An example is provided by the case of a signal (or a proximal "clue"), which transfers the pregnance of fear in birds, which further prompts the motion of taking flight but that also incurs the risk of attracting the predator's attention. Animals perceive the pregnant sign/clue (for example tracks or excreta of the predator), and then emit a further sign (cry) that mirrors that sign/clue and its pregnance. The establishment of a proto-morality immediately depicts behaviors and reactions that are exposed to punishment and violence.

From the point of view of the functions of human language Thom sees the birth of the "genitive" as the syntactical form that denotes the proximity of a being whilst denying its immediate presence. This syntactical form permits us to emit and receive alarm calls[10] which provide individuals (and the group) with an adequate defense. From this perspective the presence of a pregnant sign associated with a form S can be considered as a fundamental kind of *concept* or class of equivalence between salient forms, which incorporates a primary, rudimentary and prompt abductive power.

As I have already stressed, the *cultural* acquisition of a sensitivity to source forms has to be hypothesized in both humans and various animals. In these cases pregnance transmission occurs, beyond the hardwired cases, thanks to the presence of suitable artifactual cognitive niches[11] (such as human natural languages),

[10] Also, in many animals alarm calls/cries are the analogue of the second-person singular imperatives typical of human natural languages (Thom, 1980, p. 172).

[11] Cf. chapter four, section 4.1.1, this book.

functioning as *pregnance mediators*, where plastic teaching and learning is possible. These cognitive niches make plenty of cognitive tools available, that in turn make the organisms who acquire them able to *pregnantly* manage signs (which consequently gain a special "meaning"). This process is clearly illustrated by the description of various aspects of "plastic" – and not merely hardwired – cognitive skills in animal abduction and by the relevance of the "mediated" character of several affordances.[12] In these last two cases both cognitive skills and sensitivity to suitable affordances require cultural learning/training imbued in appropriate cognitive niches.

In chapter five of my book on abductive cognition[13] I have emphasized that fleeting and evanescent internal pseudorepresentations (beyond reflex-based innate releasing processes, trial and error or mere reinforcement learning) are needed to account for many animal "communication" performances even at the rudimentary level of chicken calls: Evans says that "[...] chicken calls produce effects by evoking representations of a class of eliciting events [food, predators, and presence of the appropriate receiver] [...]. The humble and much maligned chicken thus has a remarkably sophisticated system. Its calls denote at least three classes of external objects. They are not involuntary exclamations, but are produced under particular social circumstances" (Evans, 2002, p. 321): in Thom's words, these calls are of course pregnant signals which can be learnt, which in turn play a proto-moral and a kind of "deontological" role by triggering reactions that are implicitly considered good. Of course in the case of animal cheating, analogous calls trigger reactions that are basically negative for the receiver's welfare.[14]

Chickens form separate representations when faced with different events and they are affected by prior experience (of food, for example). These representations are mainly due to internally developed plastic capacities to react to the environment, and can be thought of as the fruit of learning. Many animals (especially gregarious ones) go beyond the use of sound signals in their cognitive performances, they for example *reify* and delegate cognitive/semiotic roles to true pregnant external artificial "pseudorepresentations" (for example landmarks, urine-marks, etc.) which artificially modify the environment to consequently become an affordance for themselves and other individuals of the group or of other species.

2.1.3 Mental and Mindless Semiosis and the Formation of Language, Habits, and Violent Conflicts

I have said that various animals possess a subsystem consisting of neuron activities that mainly aims at furnishing models of the exterior space reflecting the interesting objects found there (for example prey, food, predators) and the position of the body – spatial representations – in relation to these objects, thus allowing appropriate and fruitful spatial competition. In humans and many animals what I call *semiotic*

[12] (Magnani, 2009, chapters five and six).
[13] (Magnani, 2009).
[14] Deception in animals is illustrated in chapter three, subsection 3.3.1, this book.

brains are brains that are engaged in making or manifesting or reacting to a series of signs. Through this semiotic activity they can be at the same time engaged in "being minds" and in thinking intelligently. Thom's example of the Pavlovian dog is striking, when the dog is conditioned to recognize the sound sign "the ringing of a bell" as a pregnant alimentary form "[...] this transformation only occurs in the psyche of the conditioned dog. Nothing in the phonic structure of the signal has changed intrinsically; the only objective aspect concerns the dog itself, the subject, whose psyche is going to run through a whole associative chain of the category Γ_p in order to reach the source form. (When this chain includes actions on the part of the subject, then we have the principle of training or dressage)" (Thom, 1988, p. 14). Humans have conditioned the dog to acknowledge *material*, physical events (like the the ringing of a bell) which, as Thom says, become a pregnant sign in the dog's "psyche". This is a psyche that consequently manifests the existence of a kind of "mind" absorbed in reacting to signs (and further able to make up and manifest other related signs),[15] that is engaged in – so to say – "being a mind".

It is important to stress again that Thom provides a mathematical model of many of these semiophysical processes. In the case of Pavlovian conditioning the ground state of the receiving subject can be seen as a non-absolute minimum, surrounded by a ring of basins the bottoms of which are lower than the ground state. The perception of a form as pregnant is such because it creates a "tunnel effect", well-known in quantum mechanics, which precipitates the representative point into a peripheral basin therefore liberating energy and constituting an excited state (cf. Figure at p. 15 of (Thom, 1988)). The so-called attractors[16] are at play, as the indetermination of the pregnant form is due to the attraction held by it in a space of forms, where the excited states have to play a biological regulation and a subsequent regulative action looking for satisfaction.

Pregnance (which in the present Pavlovian case is just the memory of an earlier satisfaction) is the *abductive anticipation* of the satisfaction at play, an anticipation that implies a "virtual" affectivity (investment of the subject) that has an effect. The hungry animal is excited: once the food/prey appears, is caught and ingested, the "actual" affective satisfaction lifts the level of the excited state basin and possibly annihilates it so that the animal returns to the ground state. In the general case of the interplay predator/prey a distinction has to be made between potential investment by a pregnance (when the animal is looking for the prey) and actual investment "which

[15] For example the dog can emit the cry for food, a salient form which is also a pregnant form for other dogs or for humans.

[16] Some dynamical systems are very complex and behave non-linearly and erratically, jumping from a point in the state space to another very different point in a brief time (like in the case of the states of the atmosphere). However, notwithstanding these sudden changes, a dynamical system has a series of states into which a system repeatedly falls, the so-called attractors, which are stationary. A system can have a lot of attractors, contemplating more than a single stable state. The transition from one attractor to another can be viewed as analogous to a sort of *phase transition* (as in the case of water that then becomes ice: in dynamical systems small local changes could lead the system to a qualitatively different state).

comes in after the perception catastrophe" (Thom, 1988, pp. 15–16). Following my semiotic interpretation it is clear that pregnant forms/signs are "dialogic" both at the intrasubjective (in some cases – human beings for example – organisms are aware of them) and at the intersubjective level, while at the same time their conceptual significance can only be grasped in an eco-cognitive framework.

These perspective pregnances consist in intertwined internal and external semiotic processes. If a pregnant alimentary form "only occurs in the psyche of the conditioned dog", and in turn salient signals can be emitted and then seen as pregnant by other individuals, following Thom we can say that every element of the dog's psyche has something corresponding to it in the world. Peirce analogously said that "[...] man is an external sign. That is to say, the man and the *external sign* are identical, in the same *sense* in which the words *homo* and *man* are identical":[17] according to Peirce, and also from my perspective, signs include feelings, images, concepts, and other representations, which means that model-based aspects of both animal and human cognition are central and constitute a significant – pregnant – phenomenal manifestation of all organisms. In such a way, animals' *material* salient signals become available in a social processing to other animals (and, in part, to humans) for pregnant interpretations and consequently *guide our actions in a positive or negative way*.

In Thom's terms, salient forms become pregnant for the receiver because they mirror the object through resemblance/isomorphism [icons], contiguity [indexes], or conventionality [symbols], just as Peirce contended. The social interplay between saliences and pregnances established by Thom is a process of "sign action", or *semiosis*, and does not necessarily refer to an actual person or mind, but can also be attributed to an animal interpreter and, furthermore, to the obviously "mindless" cases of triadic semiosis (sign, object, interpretant), or even to plant behavior. Indeed we can recognize that a sign has been interpreted not because we have observed a mental action but by observing another related material sign. Finally, unconscious or emotional interpretants are also typically widespread in human beings.

As anything can be seen as a pregnant sign, the collection of potential signs may encompass virtually anything available within the agent, including all data gathered by its sensors. In the context of the science of complexity, *pregnant semiosis* can be depicted as an emergent property of a system. Emergent properties constitute a certain class of higher-level properties, related in a certain way to the microstructure of a class of systems, that thus becomes able to produce, transmit, receive, compute and interpret signs of different kinds.

In summary, abduced pregnances mediate salient signs and work in a triple hierarchy: feelings, actions, and concepts. They are partially analogous to Peirce's "habits"– and both proto-morality and morality consist in habits – that is, various generalities as pregnant responses to a sign. They are the effect of a process involving signs. In Peircean terms, the interpretant produced by the sign can lead to a feeling (*emotional* interpretant), or to a muscular or mental effort, that is an *energetic* interpretant of some kind of action (not only outward, bodily action, but also

---

[17] (Peirce, CP, 5.314).

purely inward exertions like those “mental soliloquies strutting and fretting on the stage of imagination”.[18] Of course sign(al)s are subject to modifications, because they are always, so to say, incomplete, with the possibility of taking any particular feature previously unknown to their interpreters, so that their possible pregnant character can be further revised or withdrawn. In human beings some signs (for example linguistic and/or symbolic in a Peircean sense) are pregnant for the receiver in so far as they are immediately endowed with abstract meaning and live in semiosis with a longer or shorter stability in the hierarchy of iconic, enactive, and symbolic communication. They give rise to “thoughts” related not only to intellectual activity but also to initiating ethical action, as a “modification of a person’s tendencies toward action”[19]

In ontogenetic affect attunement between human infants and their caregivers the interplay is mainly model-based and mostly iconic (taking advantage of the force of gestures and voice). Hence, meaningful words are present, but the semiotic propositional flow is fully “understood” only on the part of the adult, not on the part of the infant, where words and their meanings are simply being learnt. Let us focus on this problem related to language acquisition in the light of catastrophe theory. The child is, in a way, “invaded” by a great number of words, that can be characterized as *potentially pregnant* and their presence is commonly thought to be at the basis of the distinction between human and other animal communication. Prior to the model-based interplay related to face and gestures (that reaches the baby when it is just a few days old) there is the main (alimentary) pregnance, when the child is breast-fed. Following Thom, it can be hypothesized that in this process the pregnance of a baby’s own body, step by step, becomes autonomous “if the organic coenaesthesis is linked up with the external image of the body (mirror stage)” (Thom, 1988, p. 23). A similar process is occurring in the case of the formation of the assessment of the autonomy of external objects in so far as they are the object of action, so that we witness “the mother’s original pregnance exfoliating into a great number of beings and objects” (ibid.)

The child receives some words uttered by the mother, who, at the same time, through *deixis*, emits a quantum of her pregnance to invest the object indicated (through contact or the pointing finger), thus coupling it with the sound. The heard word acts as a tunnel effect (cf. above) and a stabilizer, to individuate the pregnance of the object: “[...] the possibility the child has of causing the object to appear by saying the word reinforces the autonomy of this pregnance which eventually dissociates itself completely from that of the mother” (ibid.) Deixis, which externalizes a gesture related to a word, creates a transitory pregnance which invests the objects shown and, by conveying the designating word, invades the receiver’s psychism through internalization and neural fixation (towards eight or nine months the child is able to reproduce the same behavior as the mother asking her the name of the

[18] (Colapietro, 2000, p. 142).

[19] (Peirce, CP, 5.476). I have devoted part of chapter three of (Magnani, 2009) to the clarification of the relationship between the semiotic approach and abduction, taking advantage of Peirce’s original results.

object and pointing to it). This is how the "meaning" of the word is firmly grasped and internalized.

The syntactical aspects of natural languages,[20] such as for example the genitive – see subsection 2.1.2 above – or the management of the verb,[21] can be further explained in terms of archetypal morphological space-time processes susceptible of being described in mathematical topological terms. For instance "Upon hearing an order the cerebral dynamics suffers a specific stimulus s, which sends it into an unstable state of excitation. This state then evolves towards stability through its capture by the attractor A, whose excitation generates the motor execution of the order by coupling the motor neurones" (Thom, 1980, p. 172). For example, the verbs of feeling *to fear* and *to hope* express that an actant subject admits in internal co-ordinates "a morphology of the future, which is accepted with repulsion or attraction" (p. 211).

2.1.3.1 Language, Biological Factors, and Sacrifices

An example of the morphological isomorphism between language and biological functions, which reflects a world of conflicts and possible violence, is the following: the verb as it appears in sentences like "The cat eats the mouse" reproduces at the linguistic level the biological transition between the virtual investment of a subject by a pregnance and the satisfaction resulting from the act analogous to the one I have described above in the Pavlovian case. Of course when the "verb" is used to describe the process of inanimate nature, part of this process – intentionality and satisfaction – tends to disappear.

Thom's analysis also sheds further light on the cognitive role of discreteness seeing it as intrinsically intertwined with the emergence of saliences and pregnances. The discrete decomposition of naturally continuous phenomena is "not just an illusion of the mind" (Thom, 1988, p. 25), but can be grasped in terms of the distinction between salience and pregnance. This distinction is fundamental even if it tends to be blurred sometimes:

> [...] spreading and ramifying objects such as smoke from fire, or chemical diffusions (odors), are pregnances which sometimes have salient aspects; the phase transitions of matter, from salient solid to purely pregnant gas, provide an example of the modification of bodies which can be interpreted as due to conflicts between local pregnances. [...] (The form of a solid is a "figurative effect" of its "individuating pregnance" and at the same time its definition) (Thom, 1988, p. 25).

These "modifications of bodies" are at the roots of many cognitive frameworks that are usually called "magic", when seen in the perspective of scientific mentality: bodies easily become part and parcel of religious or proto-religious cognitive

[20] Basic syntactical mechanisms are intended by Thom as simulated copies (defined on an abstract space) of the fundamental biological functions such as predation and sexuality.

[21] For example a verb transfers a pregnance from subject to object and so constitutes an attractor of the cerebral dynamics.

systems, which in turn participate of the proto-moral or moral commitment of a human community:

> So it is easy to understand why exceptional natural forms (*lusus naturae*), such as rocks resembling human faces, were endowed with a pregnance that is half religious, half magic, an undifferentiated pregnance in which Durkheim saw the origin of the modern notion of energy (ibid.)[22]

In this case (magic), the imaginary world of pregnances is controlled by the will of man (or rather of certain men, *magicians*, expert in efficient practices), who are the central mediators of religious/moral rituals, duties, and punishments; in the case of science, control is determined by the internal generativity of formal language describing external situations, a generativity over which man has no hold, once the initial conditions are laid down (p. 33). Furthermore, Thom points out that long before the appearance of Greek geometry, men had become aware of the relatively inflexible constraints imposed by geometry and mechanics in our ambient space:

> If space-time had been a fluctuating and plastic entity deformable *ad libitum*, there would have been no need for those special gifts, the talents of the magician, to reveal distortions. The fact that such a distortion required a specialist intervention did show that even the "the average man" was fully conscious of the exception character indeed properly miraculous of these actions at a distance. In modern terms, it could be said that there was established the concept of a relatively stable and regular space-time in its ground state, but which takes deviant "excited" forms (p. 135).

To "excite" the forms of the regular space-time a supplementary energy has to be impressed, and this always was the task of sacred rituals, more or less endowed with moral but also brutal outcomes, such as sacrifices:

> In order to realize these excited forms, it is necessary to breathe into the space a supplementary "energy", or a "negentropy" which will channel a multitude of local fluctuations in a prescribed manner. Such was the aim of rituals and magical procedures, which frequently involved the sacrifice of living animals. It is as if the brutal destruction of a living organism could free a certain quantity of "negentropy" which the officiant will be able to use in order to realize the desired distortions of space-time.
>
> We can see how little the conceptual body of magic differs basically, from that of our science. Do we not know in the theory of the hydrogen atom for example, that the energy level of a stationary state of the electron is measured by the topological complexity of the cloud which this electron makes round the nucleus? In the same way, certain quantum theorists such as Wheeler, tried to interpret quantum invariants in terms of the topology of space-time. And, in General Relativity, the energy density of the universe is interpreted as a geometric curvature (pp. 135–136).

---

[22] In the subsection 4.8.3 of chapter four I will describe, from a psychoanalytic perspective, how not only natural pregnant objects like the above rocks resembling a human face, but also intentionally and creatively built artifacts such as cognitive/affective mediators, can be seen in the framework of the "psychic energy" flow. Those artifacts, in so far as they are aspects of human cultures and parts of the related cognitive niches, are clearly seen as "transformers of (psychic) energy", and consequently potential external moral and violent carriers.

2.1.4 Violence: "Vocal and Written Language Is a Tool Exactly Like a Knife"

In *Excitable Speech. A Politics of the Performative* (Butler, 1997, p. 5 and p. 7) says that "if language can sustain the body, it can also threaten its existence" and "it enacts its own kind of violence". Language certainly constructs subjectivity, but also disassembles identity for example by positing subjects as inferior. It does not simply endorse, legitimate or justify physical violence, contrarily to the "received view" about its role: "A tradition whereby language is transparent, the route to truth and freedom where speech is the antithesis of violence (Hanssen, 2000). Such a representation troubles the distinction that the solution is simply 'more communication', where conflict can be resolved through clearer or more eloquent articulation. Violent speech acts also problematize any suggestion that incidents of bullying or harassment, which might be significantly verbal in their nature, are somehow 'less serious'" (Catley, 2003).

Let us return to the problem of language and violence, taking advantage of Thom's intellectual framework. The further ontogenetic syntactical development of language in humans is linked to the communicative target – as it is driven by *social necessity* – of extracting the "meaning" incorporated in pregnances received from humans and artifacts through various sensory systems based on undulatory sounds, light, and direct contact: basically auditory, visual [written texts], tactile [for example, Braille cells]. Obviously, the acquired linguistic syntactic competence allows humans to affect other people with subjective pregnances at will. A linguistic sentence like "Peter sends John a letter by post" where the target is of course to *inform* the hearer, has the precise biological realization of precipitating "his logos into a new stationary state, a new *form*" (Thom, 1980, p. 183). Indeed any message is in a state of morphological instability, which then stabilizes itself in the recipient in a locally stable network of behavior. It is interesting to stress that from Thom's semiophysical perspective the elementary mathematical structure (elliptic umbilic singularity) of this sentence also describes messages that are externalized without the help of uttered or written sentences. Semiotic information takes advantage of very different material realizations, for instance model-based and through artifacts. It is extremely interesting to note that a special instance of the same mathematical morphological structure is the extreme event of non-linguistic "communication" regarding a message that has the aim of brutally capturing or destroying the receiver: here the message is a projectile, a *bullet*, the messenger being a firearm used by an aggressive and violent human being.

From this perspective we can say that vocal and written language can be a tool exactly like a knife;[23] the material prolongation of an organ. It is a vector of pregnances of biological origin, the possible support of a subsequent moral/violent action and the outward extension of an organic activity.[24]

[23] Language being a tool is already a key point of the perspective established in the framework of distributed cognition, cf. (Wheeler, 2004; Clark, 2008), related more to its cognitive and epistemic virtues than to moral and violent ones.

[24] On the analysis of language in the systemic terms of catastrophe theory cf. (Thom, 1988, chapters two and eight) and (Petitot, 1985, 1992, 2003).

We can also stress that, in principle, both brain events, motor – like writing and speaking – and inner thought responses, are material/biological processes of the organism, the first originating action at the phenomenal level of the individual, the second originating thought, which we hypothesize in its physicochemical aspects. In both cases, so to speak, the organism "acts". Secondly, both "processes" are subsequently modified by calculation based on the experiential outcomes, so that the action or thought chosen can be revised. From this neural naturalistic perspective the *quasi*-ontological dichotomy between actions and thoughts, even if justified at a different, not neural, epistemological level – typical of the received philosophical and cognitive tradition – vanishes: obviously, both actions and thoughts are, so to say, "actions" of the organism, both the phenomenal and non-phenomenal levels are "performed" in the organism's body. This conclusion is also acknowledged by Edelman who, in the perspective of the so-called neural Darwinism,[25] quotes Peirce:

> Peirce pointed out that sensations are immediately present to us as long as they last. He noted that other elements of conscious life, for example, thoughts, are actions having a beginning, middle, and end covering some portion of past or future. This fits our proposal that thought has an essential motoric component reflected in brain action but not in actual movement. [...] This view of thought as being essentially motoric is consistent with the known interactions of the frontal and parietal cortex with basal ganglia, the subcortical regions involved in motor programs (Edelman, 2006, p. 123 and p. 168, footnote 3).

We have said above that, following Thom, vocal and written language can be a tool exactly like a knife: he further stresses that a word is at the stage of emission an "action" in the literal sense of a "muscular motor field (a chreod in Waddington's sense) affecting the muscles of the thorax, the glottis, the vocal chords and the mouth" (Thom, 1980, p. 236). In so far as a spoken word is an action, it is always an action which can be perceived by an agent as violent.

Let us come back to the mother/child interplay related to language acquisition. Once the reciprocal use of a "word" is abductively stabilized, it constitutes the expected linguistic "attunement" to the mother, (the "communication" is established), and at the same time the appropriate (maternal) cognitive link to the external environment. The process is slow, the fruit of a progressive abductive activity of subsequent "linguistic (abductive) hypotheses" [words] externalized by the two agents, until the plausible and acceptable one is reached. In this process, the external manifestation [sound/word] of the baby is established as a commitment to the external world, which in turn testifies a possible shared coherent flux of communication with the mother. A side effect is also the sharing of affection that can be mediated by words and which is at the basis of further social expression of emotions in a language-based way as well. Roberts (2004) states that children form knowledge and expectations about the symbolic functioning of a particular word through routine events where model-based perceptual and manipulative aspects of cognition are predominant and provide suitable constraints. Finally, in the general dialectic interplay of human adult communication, the formation of new words can be traced

[25] Neural Darwinism is dealt with in subsection 5.2.3 of chapter five, this book.

back to the general interplay between source(s) – both biological and artificial – and the receiver. Words are frequently used with meanings different to their original ones, so "tensions" are introduced, which can lead to potentially successful neologisms in further communication and to the extinction of old terms. In the interplay of communication, the meaning tension can be exploited with the aim of deceiving, of course, an issue that is looked at in a variety of ways in the present book.

As a counterpart to the ontogenetic description above, the phylogenetic aspects of attunement have to be taken into account. From the phylogenetic perspective deixis can be traced back to the effect of light and heat given the fact that "Biological pregnances have turned little by little into objective pregnances, physical in modern terms" (Thom, 1988, p. 25). It can be hypothesized that the abductive mechanisms of propagation and investments of subjective pregnances, acting on organisms like us, have been mirrored and represented externally onto things, in the interplay between internal and external representations, with a considerable epistemic and practical success. I agree with Thom: he says it would have been easy for our ancestors to detect that light behaves like a pregnance which has its own source forms and which, when emitted, invests and transforms the objects on which it falls: objects which in turn emit a similar pregnance. Adopting a different lexicon we can say, following Jung, that first of all rough external representations (Jung says "observations"), merely reached through senses, were rapidly "internalized" and became "mythologized". "Summer and winter, the phases of the moon, the rainy season, and so forth, are in no sense allegories" of those primary observations, because they become "symbolic expressions of the inner" [i.e, pregnances] and "become accessible to man's consciousness by way of projection, that is mirrored in the events of nature" (Jung, 1968a, p. 6).[26] Later on, a mechanism or subsequent projections of internal representations outside in the external environment and subsequent re-internationalizations are implemented.

If language can be in itself violent, this means that it is not only involved in interpersonal violence as it occurs, but that it is a violence carrier and mediator. If we see language as a structure that affects human behavior, it is easy to see its nature of violence mediator in itself: it is often easy to trace a violent speech act back to its actor, but this is only one half of the story. Language (both spoken and written) can also be seen as a systematic and patterned designer of a framework of violent relationships, in which agents not always perceive (on their own, from their first person perspective) the violence of their acts, both in nature and effects. So to say, using Catley's words, language carries a violence that is always at our disposal in our "healthy society" because it "inhabits a peaceful scene" (2003).[27]

2.1.4.1 Language as an "Invisible Violent Hand"

I have already stressed in chapter one that there is a structural aspect of violence that cannot be disregarded. We can add that there is a structural *invisible violent*

[26] Cf. also chapter four, section 4.8, this book.

[27] Cf. also (Anglin, 1998).

hand – often disguised as a gentle cluster of speech forms – that distributes harmful, abusive, destructive and damaging roles, commitments, inclinations and habits. Certainly language carries part of what Bourdieu calls "symbolic violence", where an everyday "gentle, hidden exploitation is the form taken by man's exploitation of man, whenever overt brutal exploitation is impossible" and where we face "the gentle, invisible form of violence, which is never recognized as such, and is not so much undergone as chosen, the violence of credit, confidence, obligation, personal loyalty, hospitality, gifts, gratitude, piety" (1977, p. 192). When you receive gentle words always remember the motto: "*timeo Danaos et dona ferentes*", which warns you of the possible hidden harmful intentions.[28] In sum, violence is not only embedded in individualized instances of physical violence, but as far as language is concerned – and in many aspects of our cultural "cognitive niches", such as for example institutions and technological artifacts, as we will see in the following chapters – violence displays the character of being just "built in", so that one can never totally prescind from violence.[29]

In sum, let us note that when I said that language is in itself violent and demonic, I was mainly concerned with its effects at the level of interpersonal violence, more or less systemic and hence not necessarily intentional; however we have to remember that psychoanalysts such as Jacques Lacan stressed how language is a violence carrier and mediator even if seen in the perspective of the individual psyche. To be clearer, language, which is the very medium of cooperation and non-violence, also involves unconditional violence: Hegel already observed that there is something violent in the symbolization of a thing, thanks to the "essencing" effect of language, which equals its mortification. Heidegger uses the notion of "ontological violence", which concerns the founding gestures of the new communal world of one people (and so their new *ethos*, realized by poets, thinkers, and statesmen).

Again, ontological violence is clearly seen as the space opened up for mere ontic but also physical violence.[30] "Language is the first and greatest divider", Žižek says (2008, p. 3). In this respect, the biblical episode of the *Tower of Babel* told in the Book of Genesis (chapter eleven, NIV) immediately comes to the mind. Mankind had "one language and a common speech" (v. 1) and started to build the highest tower as a symbol and guarantee of their might, but God's reaction to this act of pride was dreadful: "If as one people speaking the same language they have begun to do this, then nothing they plan to do will be impossible for them. Come, let us go down and confuse their language so they will not understand each other" (vv. 5-7). *Confusing* the language was the only needed step to create incommunicability and hence violent divisions among men. The Hebrew word *babel* meant precisely *confused*: the link between confusion, dazzlement and irritation leading to violence

[28] Cf. also (Bourdieu, 1991, 2001; Thompson, 1984).

[29] On violence which resorts to the so-called emotional abuse cf. the recent (Jantz and McMurray, 2009). The book stresses how emotional abuse can be performed through words, actions, neglect, and also by means of spiritual tools (taking advantage, so to say, of the control of God). Various dangers are illustrated: they can affect the sense of self, or result in (indirect) physical threats and damaged relationships.

[30] (Žižek, 2009, p. 52 and p. 58).

is common even in our own experience. As a matter of fact the Book of Genesis often refers to this violent role of language: even at the very beginning of history (as told in chapter two, NIV) God brings all the animals to Adam for him to give them a name: "Now the LORD God had formed out of the ground all the beasts of the field and all the birds of the air. He brought them to the man to see what he would name them; and whatever the man called each living creature, that was its name. So the man gave names to all the livestock, the birds of the air and all the beasts of the field" (vv. 19-20). Language is here the primary sign of domain: man takes possession over Creation because he puts names – as signs of division – upon all beings. In a way reminiscent of ancient victory parades in which captives and goods where shown to the king, the triumphal parade of animals is brought to Adam, appointing him sovereign and guardian of the Creation. Language is man's first instrument of control *as of* intelligibility, and therefore must be totally *his*: the unconditioned acceptance shown by God to the names chosed by Adam witnesses this original delegation of power.

I have said that "language is the first and greatest divider": this effect is also clear if we consider that "essences" are a basic result of the work of language, so that human beings who do not share similar languages can live in incommensurable worlds: "For Catholics the castrato voice was one the very voice of angels prior to the Fall. For us today, it is a monstrosity." Secondly, acknowledging Lacan's ideas, Žižek tells us that it is man that "dwells" in language in a sort of constitutive passivity, and not language that dwells in humans: "man is a subject caught and tortured in language",[31] language is the "big Other", and, again, human animal does not "fit" language:

> [...] [man] dwells in a torture-house of language: the entire psychopathology deployed by Freud, from conversion-symptoms inscribed into the body up to total psychotic breakdowns, are scars of this permanent torture, so many signs of an original and irremediable gap between subject and language, so many signs that man cannot ever be at home in his own home. This is what Heidegger ignores: this dark torturing other side of our dwelling in language – and this is why there is also no place for the Real of *jouissance* in Heidegger's edifice, since the torturing aspect of language concerns primarily the vicissitudes of libido (Žižek, 2008, p. 3).

An eloquent explanation of this kind of violent effect of language, which constitutively affects the individual psyche, is illustrated by the case of mutilations, where parts and/or aspects of the real body suffer – so to say – from the "signifier" effected by the language activity. The semiotic pressure established by language constrains the body and so its structures of desire:

> The rituals of initiation assume the form of the changing of form of these desires, of conferring on them in this way a function through which the subject's being identifies itself or announces itself as such, through which the subject, if one can put it this way, fully become a man, but also a woman. The *mutilation* serves here to orientate desire, enabling it to assume precisely this function of index, of something which is realized and which can only articulate itself, express itself, in a *symbolic beyond*, a beyond

[31] (Lacan, 1966, p. 276).

> which is the one we today call being, a realization of being in the subject (Lacan, 1959, quoted in (Žižek, 2008), p. 5).

Finally, another example of the violent effect of language is racism. The *being* of blacks (*as* blacks) is a socio-symbolic one: when blacks (or anyone else) are treated *as* inferior by white people, this "does indeed make them inferior at the level of their socio-symbolic identity". The white racists ideology, carried by their socio-linguistic acts and habits, exercises a performative – violent – efficiency. It is important to note it is not only a mere "interpretation" of what blacks are, like a benevolent and ingenuous intellectual could pathetically contend, but an interpretation that directly and violently generates the *very being* and *socio-semiotic existence* of the interpreted subjects (Žižek, 2009, p. 62).

2.1.5 The Role of Abduction in the Moral/Violent Nature of Language

In a study concerning language as an adaptation the cognitive scientist and evolutionary psychologist Pinker (2003, p. 28) says: "[...] a species that has evolved to rely on information should thus also evolve a means to exchange that information. Language multiplies the benefit of knowledge, because a bit of know-how is useful not only for its practical benefits to oneself but as a trade good with others". The expression "trade good" seems related to a moral/economical function of language: let us explore this issue in the light of the coalition enforcement hypothesis I have introduced in the first chapter.

Taking advantage of the conceptual framework brought up by Thom's catastrophe theory on how natural syntactical language is seen as the fruit of social necessity,[32] its fundamental function can only be clearly seen if linked to an intrinsic *moral* (and at the same time *violent*) aim which is basically rooted in a kind of *military intelligence*. Thom says language can simply and efficiently transmit *vital* pieces of information about the fundamental biological oppositions (life – death, good – bad): it is from this perspective that we can clearly see how human language – even at the level of more complicated syntactical expressions – always carries information (pregnances) about moral qualities of persons, things, and events. Such qualities are always directly or indirectly related to the survival needs of the individual and/or of the group/coalition.

Thom too is convinced of the important role played by language in maintaining the structure of societies, defending it thanks to its moral and violent role: "information has a useful role in the stability or 'regulation' of the social group, that is, in its defence" (Thom, 1988, p. 279). When illustrating "military" and "fluid" societies he concludes:

> In a military type of society, the social stability is assured, in principle, by the imitation of the movement of the hierarchical superior. Here it is a question of a slow mechanism where the constraints of vital competition can impose rapid manoeuvres on the group.

[32] Cf. subsections 2.1.2 and 2.1.3.

> Also the chief cannot see everything and has need of special informers stationed at the front of the group who convey to him useful information on the environment. The invention of a sonorous language able to communicate information and to issue direction to the members of the group, has enabled a much more rapid execution of the indispensable manoeuvres. By this means (it is not the only motivation of language), one can see in the acquisition of this function a considerable amelioration of the stability of a social group.
>
> If language has been substituted for imitation, we should note that the latter continues to play an important role in our societies at pre-verbal levels (cf. fashion). In addition, imitation certainly plays a primary part in the language learning of a child of 1 to 3 years (pp. 235–236).

I have illustrated in chapter one that, in human or pre-human groups, the appearance of coalitions dominated by a central leader quickly leads to the need for surveillance of surrounding territory to monitor prey and free-riders and watch for enemies who might jeopardize the survival of the coalition.

This is an idea shared by Thom who believes that language becomes a fundamental tool for granting stability and favoring the indispensable manipulation of the world "thus the localization of external facts appeared as an essential part of social communication" (p. 26), a performance that is already realized by naming[33] (the containing relationship) in divalent structures: "X is in Y is a basic form of investment (the localizing pregnance of Y invests X). When X is invested with a ubiquitous biological quality (favorable or hostile), then so is Y" (ibid.) A divalent syntactical structure of language becomes fundamental if a *conflict* between two outside agents has to be reported. The trivalent syntactical structure subject/verb/object forges a salient "messenger" form that conveys the pregnance between subject and recipient. In sum, the usual abstract functions of syntactic languages, such as conceptualization, appear strictly intertwined with the basic *military* nature of communication.

I contend that this military nature of linguistic communication is intrinsically "moral" (protecting the group by obeying shared norms), and at the same time "violent" (for example, killing or mobbing to protect the group). This basic moral/violent effect can be traced back to past ages, but also when we witness a somehow *pre-human* use of everyday natural language in current mobbers, who express strategic linguistic communications "against" the mobbed target. These strategic linguistic communications are often performed thanks to hypothetical reasoning, abductive or not. In this case the use of natural language can take advantage of efficient hypothetical cognition through gossip, fallacies and so on, but also of the moral/violent exploitation of apparently more respectable and sound truth-preserving and "rational" inferences. The narratives used in a dialectic and rhetorical setting qualify the mobbed individual and its behavior in a way that is usually thought of by the mobbers themselves (and by the individuals of their coalition/group) as moral, neutral, objective, and justified while at the same time hurting the mobbed individual in various ways. Violence is very often subjectively dissimulated and paradoxically considered as the act of performing just, objective moral judgments and of

[33] It is important to stress that pregnant forms, as they receive names, tend to loose their alienating character.

persecuting moral targets. In sum, *de facto* the mobbers' coordinated narratives harm the target (just as if she were being *stoned* in a ritual killing), very often without an appreciable awareness of the violence performed.

This human linguistic behavior is clearly made intelligible when we analogously see it as echoing the anti-predatory behavior which "weaker" groups of animals (birds, for example) perform, for example through the use of suitable alarm calls and aggressive threats. Of course such behavior is mediated in humans through socially available ideologies (differently endowed with moral ideas) and cultural systems. Ideologies can be seen as fuzzy and ill-defined cultural mediators spreading pregnances that invest all those who put their faith in them and stabilize and reinforce the coalitions/groups: "[...] the follower who invokes them at every turn (and even out of turn) is demonstrating his allegiance to an ideology. After successful uses the ideological concepts are extended, stretched, even abused", so that their meaning slowly changes in imprecise (and "ambiguous", Thom says)[34] ways, as we have seen happens in the application of the archetypical principles of mobbing behavior. That part of the individual unconscious we share with other human beings – i.e. *collective unconscious*[35] – shaped by evolution – contains archetypes like the "scapegoat" (mobbing) mechanism I have already mentioned.

In this cognitive mechanism, a paroxysm of violence focuses on an arbitrary sacrificial victim and a unanimous antipathy would, mimetically, grow against him. The process leading to the ultimate bloody violence (which was, for example, widespread in ancient and barbarian societies) is mainly carried out in current social groups through linguistic communication. Following Girard (1977; 1986) we can say that in the case of ancient social groups the extreme brutal elimination of the victim would reduce the appetite for violence that had possessed everyone just a moment before, leaving the group suddenly appeased and calm, thus achieving equilibrium in the related social organization (a sacrifice-oriented social organization may be repugnant to us but is no less "social" just because of its rudimentary violence).

This kind of archaic brutal behavior is still present in civilized human conduct in rich countries and is almost always implicit and unconscious, for example in that racist and mobbing behavior I have already quoted. Let me reiterate that, given the fact that this kind of behavior is widespread and partially unconsciously performed, it is easy to understand how it can be implicitly "learned" in infancy and still implicitly "pre-wired" in an individual's cultural unconscious (in the form of ideology as well) that we share with others as human beings. I strongly believe that the analysis of this archaic mechanism (and of other similar *moral/ideological/violent* mechanisms) might shed new light on what I call the basic equivalence between engagement in morality and engagement in violence since these engagements, amazingly

[34] From this perspective the massive moral/violent exploitation of equivocal fallacies in ideological discussions, oratories, and speeches is obvious and clearly explainable, as I will illustrate below (chapter three, section 3.2).

[35] I have illustrated the speculative hypothesis of collective unconscious in chapters four and five of my book on abductive cognition (Magnani, 2009). Cf. also chapter four, subsection 4.8.5, in this book.

enough, are almost always hidden from the awareness of the human agents that are actually involved.

Recent evolutionary perspectives on human behavior, taking advantage of neuroscience and genetics[36] have also illustrated the related process of *otherisation* – which decisively primes people for aggression – as a process grounded in basic human emotions, i.e. our bias towards pleasure and avoidance of pain. Perceiving others as the "others" causes fear, anger or disgust, universal "basic" responses to threats whose physiological mechanisms are relatively well understood. It is hypothesized that these emotions evolved to enable our ancestors to escape predators and fight enemies. Of course the otherisation process continues when structured in "moral" terms, like for example in the construction of that special other that becomes a potential or actual scapegoat.

It is worth mentioning, in conclusion, the way Thom accounts for the social/moral phenomenon of scapegoating in terms of pregnances. "Mimetic desire", in which Girard (1986) roots the violent and aggressive behavior (and the scapegoat mechanism) of human beings can be seen as the act of appropriating a desired object which imbues that object with a pregnance, "the same pregnance as that which is associated with the act by which 'satisfaction' is obtained" (Thom, 1988, p. 38). Of course this pregnance can be propagated by imitation through the mere sight of "superior" individuals[37] in which it is manifest: "In a sense, the pleasure derived from looking forward to a satisfaction can surpass that obtained from the satisfaction itself. This would have been able to seduce societies century after century (their pragmatic failure in real terms having allowed them to escape the indifference that goes with satiety as well as the ordeal of actual existence)" (ibid.)

Recent cognitive research stresses the influence that *intentional gaze* processing has on "object processing": objects falling under the gaze of others acquire properties that they would not display if not looked at. Observing another person gazing at an object enriches it of motor, affective, and status properties that go beyond its chemical or physical structure. We can conclude that gaze plays the role of transferring to the object the intentionality of the person who is looking at it. This result further explains why mimetic desire can spread so quickly among people belonging to specific groups.[38]

Grounded in appropriate wired bases, "mimetic desire" is indeed a sophisticated template of behavior that can be picked up from various appropriate cultural systems, available over there, as part of the external cognitive niches built by many human collectives and gradually externalized over the centuries (and always transmitted through activities, explicit or implicit, of teaching and learning), as fruitful ways of favoring social control over coalitions. Indeed mimetic desire triggers envy

[36] (Taylor, 2009). Taylor's book also provides neuroscientific explanations on how brains process emotions, evoke associations, and stimulate reactions, which offer interesting data – at least in terms of neurological correlates – on why it is reactively easy for people to harm other people.

[37] Or through the exposure to descriptions and narratives about them and their achievements.

[38] (Becchio et al., 2008). On gaze cueing of attention cf. also (Frischen et al., 2007), who also established that in humans prolonged eye contact can be perceived as aggressive.

and violence but at the same time the perpetrated violence causes a reduction in appetite for violence, leaving the group suddenly appeased and calm, thus achieving equilibrium in the related social organization through a *moral effect*, that is at the same time a *carrier of violence*, as I have illustrated.

Mimetic desire is related to envy (even if of course not all mimetic desire is envy, certainly all envy is mimetic desire): when we are attracted to something the others have but that we cannot acquire because others already possess it (for example because they are rival goods), we experience an offense which generates envy. In the perspective introduced by Girard envy is a mismanagement of desire and it is of capital importance for the moral life of both communities and individuals. As a reaction to offense, envy easily causes violent behavior. From this point of view we can psychoanalytically add that "[...] the opposite of egotist self-love is not altruism, a concern for common good, but envy and resentment, which makes me act against my own interests. Freud knew it well: the death drive is opposed to the pleasure principle as well as to the reality principle. The true evil, which is the death drive, involves self-sabotage" (Žižek, 2009, p. 76). I will come back to deal with the problem of envy, mimetic desire, and the scapegoating mechanism in chapter six, framing it in the perspective of religion, morality, and violence.

2.2 The Awesome and Perverse Academic Face of the Relationships between Morality and Violence

The reviewer of a book is in the typical position of possible use of language as a weapon. A personal example, which further stresses the relationship between morality and violence, is the clearest one to illustrate the issue of the ambiguity of language as a form of "military intelligence". Recently I have received from a colleague, J. Forge (a typical exemplar of what I call a *tourist of philosophy*), a review (Forge, 2009) of my previous book: *Morality in a Technological World. Knowledge as Duty*.[39] The reviewer writes: "I have to say that I found Magnani's book so infuriatingly difficult to follow that I was tempted several times simply to give up reading it". Then he tries – in a few lines – to "explain" his statement, where he clearly shows he was just not able to grasp the connection between casuistry, moral deliberation, John Searle's account of reasons, and my synthesis in terms of abduction I had illustrated in that book. That is, he just repeats his initial puzzling statement. He does not provide an opposite view or a correction of my treatment of these subjects with appropriate argumentations, or any other intellectual consideration. He just says he did not understand the book. Even if the book got a lot of good reviews (some of them were not devoid of criticism but were always informed and sensibly disputed), I immediately "perceived"[40] his words – even if not coming from an "authority" – as violent against me, my book and work (and may be against Cambridge University Press). A violence also "lived" as a kind of "bullying" against philosophy, a modest

[39] (Magnani, 2007b).

[40] On the problem of the "perception" of violence, which is related to the problem of its definition, cf. section 1.2, in the first chapter of this book.

threat to my reputation, addressed without reasons and informed argumentations, and also an implicit affirmation of some people's right – which of course I do not consider proper – to say whatever they want even if incompetent and ill-prepared in a subject simply because, for instance, they are retired, older and comfortable with their achievements, like in this case.

Of course I can easily guess the reviewer did not think he was being violent against me, and this is obvious in the framework of my philosophical perspective because, as I contend in this book, violence is always distributed and only *perceived* as such. I can also add that maybe he "thought" he was behaving "morally", freely expressing his own opinion, according to the typically (even if too often hypocritical) Anglo-American principle of intellectual freedom. In the meantime, as I was involved in writing this book, I immediately considered this event as a mild example of the clashing of moralities and of the likely emergence of violence from such conflict.

I wrote a reply and I sent it by email to the journal *Metascience* – to the person I supposed to be the editor. He was no longer in charge and also kindly told me, immediately, that the journal did not accept replies. Later on, the editors in charge – still kindly – confirmed that policy. Then, I sent an email to withdraw my reply because, not having the possibility to reply, I had decided to insert the whole issue in this book. On that occasion I also surprisingly discovered that actually the journal not only would not accept replies to the published reviews but it rejects unsolicited manuscripts at all, so my reply had not been considered *a priori*. No problems, I decided to insert the reply in a section of my book about morality and violence, but I felt myself wounded again, because it was impossible for me to celebrate what I considered my right to reply and to open a free discussion.

To summarize my concern and to make use of this event to further explain some of the main philosophical concerns of this book I can develop two lines of argumentation: the first one relating to the behavior of the reviewer, the second to the behavior of the journal. Of course the reader will find this case rather trivial and can ask herself: Lorenzo Magnani, who cares? Why react? Why make such a fuss about it? The reader might also think I personally over-reacted. He is right, the example is trivial and modest and the violence I felt very low and negligible, but it is just when analyzing these bloodless interplays, in which the feeling of being *violated* is experienced by an individual, that we can easily describe some basic philosophical problems concerning the interplay between morality and violence, and the practically "invisible" potential role played by language. The language used in the review, which negatively affected my "psyche", will be quickly shown to be a "tool exactly like a knife", as I have illustrated above in subsection 2.1.4, employing an expression used by Thom.

In my book, object of the review, I maintain that *knowledge is a duty*, and also that "people have to be respected as things", of course when the things in questions are endowed with a value that seems to overcome that of human beings, as happens increasingly in our technological and post-capitalistic era. Yes, I state here again my conclusion, following an Enlightenment tenet: at least in our technological world – full of "objectified" scientific knowledge – rational and ethical knowledge *is* a duty.

After having read the review I was even more convinced that my claim (which is also a suggestion) is appropriate when dealing with ethical problems in our epoch. The reviewer basically and explicitly said he had not understood anything of my book and so, performing what I have seen as an unintentional perversion (cf. the following subsection), he concluded that it was me who had not understood anything about ethics, technology, etc. The reviewer candidly testifies that the book is difficult, then he quickly, without further specification, contends that the quoted chapters need a coherent structure and a coherent discussion. I wonder: which ones? Where? How? No answer can be found in the review. The reviewer continues, "Abduction may be supposed to somehow elucidate casuistry or inform Searle's work, but that was not clear to me". As I will illustrate later on in the last subsection of this chapter, the reviewer performed the fallacy called *ad ignorantiam*, which can easily, even if not necessarily, generate violent effects in the receiver.

Let me consider the philosophical *hard-core* of this situation. The reviewer does not express good intellectual and academic argumentations, but just an ambiguous use of language, mediated by a technological artifact, that is, an international journal. In absence of sufficient knowledge and appropriate competence to understand my book, the argumentation quickly becomes – to me as the receiver – something related to an offensive kind of "military intelligence". Of course the reviewer does not think he is offending the author without reason, he thinks he is just expressing his own opinion, and the freedom of speech, and we all know how "sacred" the freedom of speech is in our academic establishments. Anyway, the result seems to be: Magnani is too complicated, and so rubbish!

On my side, I think the reviewer simply did not have the sufficient philosophical knowledge to understand the sophisticated epistemological, logical, cognitive, and ethical problems addressed in the book, which other colleagues perfectly understood. Consequently, I think that his disappointing (at least for me) review paradoxically confirms one of the contentions of my book *Morality in a Technological World*: to avoid producing unethical and violent effects in the receiver (from my perspective, I consider it unethical to attack the work of others without clear arguments) make sure you possess sufficient knowledge and preparation to be able to deal with certain subjects. Knowledge is a duty, especially when dealing with complicated artifactual situations involving reviews, books, international academic journals and publishers. Why do we have to treat philosophy and philosophers in such a superficial way? Why is it legitimate to judge a "philosophical" book without in depth knowledge of philosophy? As I have already said, I contended in my book that, in our technological world, the adoption of the correct knowledge helps solving ethical problems generated by technology itself. Indeed, an international journal *is* a technological tool (and of course a sophisticated intellectual artifact); as I contend in my book that, like any other artificial thing, it should not be used for fallacious communication, *de facto* expressing merely superficial emotional feelings, such as the following, here expressed taking advantage of an "unsound" syllogism: premises 1) I did not understand anything, 2) I feel frustrated; conclusion 3) I attack you to overcome my frustration and my feelings of inadequacy by diminishing you.

2.2.1 *Perversion, Violence, and Narcissism*

Many psychoanalysts, for example Paul-Claude Racamier (1995), would quickly recognize in such behavior, typical of many reviewers, something "perverse", perfectly healthy, but that we can technically call "narcissistic", a behavior that is usually perceived as aggressive by the human targets (even if sometimes the victim remains a passive victim, perhaps – and this is the worst case – devoid of any awareness of being a victim). In general, the targets of a narcissist individual feel themselves, to use a Kantian concept, "treated like means" to the benefit of others (in our case to a good end! At least from the perspective of the narcissist person, the emotional relief of the reviewer). What is philosophically central is the following consideration: the narcissistic behavior normally is not intentionally and consciously aggressive, because the subject is embedded in a kind of *moral bubble*;[41] the narcissist just considers that her behavior is right and moral, and even when she is cautioned by the targeted humans about the violence perpetrated, she is not able to comprehend it, she is not able, Carl Gustav Jung would say, to "incorporate" [or "acknowledge"] the effect of her behavior on other human beings, thus becoming aware of the possible related violent aspects.

As I have already contended, the narcissists usually perceive their own behavior as completely moral, using justifications of any kind. For example, in the case of the reviewers, the fact that expressing their own narcissistic (I repeat, not seen as such from one's own first-person perspective) opinion they can be *utterly* justified by the abstract idea of intellectual freedom. They do not see that their opinion is not informed and competent, but just emotional, simply legitimated by some marginal facts, for example that they are old academicians: they might also think "I have the right of expressing my opinion because I am old". Ok, it is moral to respect elderly people and to perform a bit of over-respect of ourselves if we are old, but it is just one kind of old-fashioned patriarchal morality in the case of academic matters (still working everywhere), which other people can oppose. For example, adopting another morality, I can say: ok, we need to respect old people, but we also have to respect the fact they have to take care of expressing informed opinions, and to act accordingly, so, if I am the editor of a journal, I cannot permit old "outdated" professors to publish what they want, also possibly – so to speak – bullying other colleagues without serious reasons. If we add to our old fashioned moral perspectives the importance of the value "knowledge as duty", we see that this extended moral approach can illustrate how the lack of knowledge leads to unintentional harm, like in the case of our reviewer.

At this point, and thanks to this trivial example, I would guess that the reader of this book can see more clearly the status of the conflicts of moralities and their violent potentiality. The reviewer acted morally and so thought his behavior was moral. The target, that is, me in the case above, embedded in another morality, perceives the review as a violence, and if the target shares his complaint with the reviewer, he acts morally and so considers his behavior to be moral, but possibly the communicated complaint is *in turn* perceived as molesting by the reviewer. Both agents think

[41] Cf. chapter three, subsection 3.2.2.

of themselves as moral and both perform actions that are reciprocally perceived as violent!

I repeat, in these endeavors the human being that harms others through language does not necessarily see her own narcissism and so does not even see that the others can feel hurt by her narcissistic behavior: with slightly better ethical knowledge, as I contend, the narcissist could recognize her feelings and avoid harming other human beings. When the quoted reviewer says: "I have to say that I found Magnani's book so infuriatingly difficult to follow that I was tempted several times simply to give up reading it", the reviewer would have been more ethically coherent (of course in my opinion) to give up, and not to write that superficial review, recognizing his personal intellectual limits. But in this case he would have adopted *my* moral point of view, not his own: the reviewer thought exactly the opposite, and wrote the review, thinking it was his right to do it, freely expressing his opinion. What was just a free opinion to him, to me is just a perverse use of language: I do not write a review when I am far from appropriately understanding the subject. I can anticipate here[42] that I favor (and suggest) moral deliberations that try to avoid moral conflicts. Often, to "stoically" abstain from acting (for example dominating our own emotions) can diminish what I call the "overmorality" of a human collectivity, which can otherwise cause an insurgence of conflicts and possible violent outbursts.

The second and final order of considerations concerns the behavior of the Journal *Metascience*, which refused the reply I wanted to publish. First of all the journal put my book in the hands of a person not appropriately qualified to that aim, and this is already an astonishing act performed by an academic journal that aims at depicting itself as in tune with the Anglo-American international standard of quality (peer review, quality and appropriateness of reviewers, etc.); 2) what is even worse is that *Metascience* not only does not accept replies (without any exceptions, they say), but also they do not accept unsolicited papers. I was very surprised to see that kind of policy, and it was even more paradoxical to see those policies emerging from that Anglo-American mentality that – in abstract – values so correctly not only the intellectual freedom, but *also* the freedom of discussing and favoring peaceful controversies. Of course I immediately felt they did not share my moral view of academic life, and I felt morally molested because punished without serious reasons. They were embedded in another morality, full of those technicalities, policies and that trivial academic bureaucracy I do not like. In conclusion: the journal does not accept replies of any kind, and also it does not accept unsolicited papers. For this reason I decided to write this section, which also plays the role of a kind of reply, but basically takes advantage of this unfortunate story to illustrate some aspects of the relationship between morality, violence, and language.

2.2.2 Ad Ignorantiam *Fallacy as a Semantic Attack*

The so called *argumentum ad ignorantiam* is that argumentative move in which the arguer makes use of what she or he *does not* know: what may sound like a

[42] See chapter six, subsection 6.5.1 for further considerations.

non-sense is actually a very widespread form of reasoning among human beings. The most common example goes as follows:

1. If there were a meeting in my department I would know it.
2. I do not know such a thing.
3. Thus there will not be a meeting today in my department.

This is commonly referred to be a fallacy, because my ignorance is not probative. It can only prove that I do not know some data. For example, in the just mentioned example, there might be a number of reasons why I do not know whether the meeting is going to happen. They are de facto hypotheses I can formulate thanks to that kind of reasoning called "abduction" we already encountered in this book, which is a typical reasoning in presence of incomplete information.[43] 1) May be my colleague sent me an email I never received, because she did not type my email address correctly, and she forgot to pay attention to the delivery failure message she should have received back. Carelessness in this case could be the reason explaining my ignorance. 2) Maybe my colleagues did not inform me about the meeting because they noticed I have not shown up during the week, and they thought I took a week off work to complete the last chapter of my book. In fact, I often complain about the fact that some meetings are completely pointless. Or, 3) they did not inform me on purpose, because they simply wanted to try to mob me.

We have to note that an *ad ignorantiam* argumentation first of all has some virtue. For instance, it allows us to make a decision even in the absence of knowledge about a certain event or situation. For instance, I can easily withdraw the abductive "hypothesis" that I have not been informed because my colleague wrongly typed my email address, because she usually phones me. Besides, she is very meticulous and responsible, and she always sets a return receipt option for such emails. She would have let me know in the case she did not receive any confirmation from me. I would withdraw the second hypothesis, because I usually work at home. And, as for the third hypothesis, I would discard it as well, because I do not have problems with my colleagues. As one can see, my ignorance does not leave me wholly in the dark. Some abductive guessing is still possible to fill the information gaps. As I have mentioned, an *argumentum ad ignorantiam* might turn out to be useful in all those situations in which I lack some crucial information for making a certain decision.

The *argumentum ad ignorantiam* has been studied in the area of decision–making under the name of recognition heuristic (Goldstein and Gigerenzer, 2002). Recognition heuristic is the strategy according to which a person is inclined to select the option she recognizes. This is indeed a trick performed by our cognitive system, which takes the lack of recognition as probative, even when it is not. It is worth noting here that the recognition of an *ad ignorantiam* as a fallacy is context-based. For instance, in the Western justice system the defendant is innocent as long as the prosecution does not prove her guilty of the crime she is alleged to have committed. In this sense, the prosecution bears the burden of proof, which is set *asymmetrical* by the principle of the so-called presumption of innocence (Walton, 2002, p. 13).

[43] On abductive cognition cf. my recent (Magnani, 2009).

In science we have something different and worth citing here. If conclusive evidence has not yet been provided, then the hypothesis at stake is neither true or false, but still to be determined. It is a very human abductive attitude to seek for belief attainment, since doubt is irritating, but natural modern science aims at settling not for any belief, but the ones that can be empirically tested and proved, and call into doubt any other. Therefore, the burden of proof is borne by those seeking confirmations as well as by those seeking falsifications. Now, let us turn back to the case of the reviewer of my book illustrated in the previous subsection. Indeed, his argumentation can be easily labeled as an *ad ignorantiam*. He basically wrote that my book does not have a point, just because he did not really understand it. Here again we have the basic structure of an *ad ignorantiam*:

1. If the book were worth reading, I would understand it.
2. But I did not understand the book.
3. So the book is not worth reading.

Indeed, the conclusion – the book is not interesting – is perfectly legitimate. I have no claims on that. The problem is the "structure" of the kind of reasoning he deployed to say so. Like in the case of the meeting, I would say that there might be a number of reasons he did not quite get what I wrote in the book. I quoted that he wrote he was going to give up reading it. Would it be a sufficient reason for considering my book like a collection of inconclusive and loosely connected thoughts?

I have just mentioned that the *context* is crucial for assessing the fallacious nature of an argumentative move. First of all, an international journal such as *Metascience* is an intellectual space, in which different opinions and arguments "should" be rationally discussed. This alone would make the reviewer's remarks quite inappropriate for publication. In fact, as one accepts to write a review for an international journal, she implicitly assumes a commitment to *relevancy* (which in turn, at least in this case, is a commitment to knowledge: again, knowledge is a duty): in an intellectual setting, the arguer should be committed to deploy information that is as relevant as possible with respect to the matter in discussion.

From a philosophical and logical perspective, a piece of information is "relevant when it has a direct bearing on the truth or acceptability of the conclusion" (Tindale, 2007, p. 23). For example, in order to support a certain position the introduction of information harming more or less explicitly one's reputation should be avoided in a rational discussion. Why? Because they would be simply irrelevant. My sexual habits, the university I work for, my age, my mother tongue, my Italian name, would not be information to deploy in a debate about, for instance, the relationship about casuistry and abduction and likewise, one's ignorance, just as in the case of my reviewer. Judging a particular argument as confusing just because one did not quite get it is simply irrelevant in the face of the remarks and comments he could provide about a certain intellectual topic like, again, the relationship between casuistry and abduction. It proves nothing at all. In turn, as already briefly noted, I cannot judge that colleague as generally "ignorant" just because he lacks the knowledge on those topics pertaining to abduction and casuistry, this lack of specific knowledge is irrelevant with respect to the general judgment about his intellectual capabilities or competence.

From a philosophical perspective the reviewer of my book turned his ignorance – in my moral perspective – into a more or less violent "semantic attack" (Thompson, 2007), aiming at cognitively hacking at the ideas expressed in the book.[44] Basically a semantic attack can be considered as an argumentative move meant to manipulate the meaning a person (i.e. the reader) assigns to a given content. An *ad ignorantiam* – like other fallacies – can be used as a powerful device for such a purpose. Michaels, in his book *The Doubt is Their Product* (2008), knowledgeably reported the use of doubt as a powerful means for fabricating sophisticated semantic attacks and cognitive hacking relying on ignorance. He listed a number of examples in which the absence of the so-called smoking gun has been employed in court, for instance, by the Tobacco industry and other various manufacturers to plea not guilty in the face of an allegation. Doubt may be successfully employed, for instance, for establishing a fake controversy, questioning the reliability of a determinate study, objecting to a certain method of gathering data, and so on and on.

More generally, some epistemic virtues and practices – generally used in the scientific arena for disputing theories and hypotheses at conferences – can serve this purpose. On some occasion, a single study – sometimes referred to in the form of *gossip* – may dispute a theory only on the basis of the fact that it would falsify all the results reported so far. Sometimes one may even resort to an overemphasis on clarity – which is more a rhetorical virtue than an epistemic one – in order to call into doubt a certain theory or idea, no matter what it is, whether scientific or philosophical. My reviewer himself seemed to rely on the alleged absence of clarity as an argumentative move for discrediting the ideas presented in my book. But, here again, would that be relevant? As a researcher and philosopher I strongly maintain that I have indeed a strong commitment to communicating my ideas in such a way that the reader is somehow facilitated in the process of reading. However, clarity should not be intended in a merely rhetorical sense but in a more theoretical one: in this perspective clarity does not necessarily imply the content to be "simply" easy to understand. This would be a kind of cognitive paternalism for which I would literally lead the reader hand in hand. It seems to me that some reviewers – including mine – are affected by a bias, which might be called the "easy-to-read" bias. According to this, the reader is inclined to positively review a book (or an article) on the basis of its "perceived" easiness, not on its content.

---

[44] Various aspects of ignorance – and the fundamental problem of the possible bad consequences of "moral ignorance" – are illustrated in (Proctor and Schiebinger, 2008). The book stressed various aspects that link ignorance to both positive and negative (for example violent) outcomes: examples from the realms of global climate change, military secrecy, female orgasm (sic!), environmental *denialism*, Native American paleontology, theoretical archaeology and racism are described, also with the aim of showing how ignorance can present a violent potential ("Doubt is our product" is the tobacco industry slogan). On the several ways in which "the product defense industry" ambiguously exploits scientific (and pseudoscientific) arguments to undermine public health protections, corrupt the scientific record, and mislead the public cf. (Michaels, 2008).

In conclusion, science and even some values coming from the tradition of the Enlightenment may be instrumentally faked bearing in mind the purpose of violently deceiving or – even worse – harming others just like the case of the Tobacco industry reported in Michaels' book. The possible "military" violent use of fallacies and of abduction (hypothetical cognition) will be further illustrated in the following chapter.

Chapter 3
Moral Bubbles: Legitimizing and Dissimulating Violence
Distributing Violence through Fallacies

3.1 Violence, Language, and "Writing"

Among the philosophers, Derrida is the only one to clearly analyze and explore the link between violence and writing, offering reflections of great value on the subject.[1] The result echoes, from a strict philosophical perspective, the link between morality, violence, and language I described in the previous sections. The structure of the trace of writing (or difference) reflects violence in the sense that the common and obvious violence is just the vestige of a more fundamental and constitutive "arche-violence". The structure of violence is itself – to use Derrida's intense philosophical expressions – "marked" by the very structure of the "trace" of "writing" as the "production" of "presence". This primal acknowledgement of violence in writing is what allows a distinction between philosophy and violence, so that the former may recognize the latter. However, a second violence is perpetrated: a compensatory violence is also at stake in writing, which plays the role of systematically erasing the traces of the primordial violence of the "trace", as a kind of – supposed – counter-violence whose violence consists in the *denial of violence* (Grosz, 1999). It is "the violence that describes and designates itself as the moral counter of violence. This is the violence that we sometimes name the law, right or reason" (1999, p. 10). The familiar and obvious violence, the horrifying violence – so to say, "empirical" violence – we are everyday quite used to, rests upon, participates in, and is considerably made possible by the first two prior senses of violence described above.

Usually obvious violent settings and situations, for example family or workplaces,[2] rest upon and repeat violence "as inscription and the containment of inscription" (Catley, 2003), that is they affirm the violence of the systems which constitute them and the "moral"[3] necessity for such violent systems:

[1] Cf. (Derrida, 1976, 1981, 1990).

[2] On violence in family and workplace cf. (Hirigoyen, 2000).

[3] Force, violence, writing not only give birth, but also distribute and transform even that violence which cannot be called such. "The arche-writing is the origin of morality as of immorality. The non-ethical opening of ethics. A violent opening" (cf. "The violence of the letter. From Lévi-Strauss to Rousseau" [1974], in (Derrida, 1976, p. 140)).

L. Magnani: Understanding Violence, SAPERE 1, pp. 65–113, 2011.
springerlink.com

> [...] war, participates in both these modes of violence (violence as inscription, violence as the containment of inscription, the containment of violence). Mundane or empirical violence reveals "by effraction" the originary violence, whose energy and form it iterates and repeats; yet it "denudes" the latent or submerged violence of the law, whose transgression it affirms, while thus affirming the very force and necessity of the law (Grosz, 1999, p. 10).

Hence, violence is constitutive of politics, of thought, of knowledge, because all these practices are based on and justified by moral options and orientations that institute more or less constrained or preferred choices, conflicts and possible dominant relationships of power, and in turn favor (against others) certain behavior followed by individuals and groups.

Philosophy is also inserted in the vicious circle above, it repeats what it opposes, and this explains what I have illustrated in the first section of the previous chapter: the lack of interest in violence is in reality the other face of the denial of violence through the establishment of various "moral" counters of violence. For example when philosophy celebrates logos, justice, and rationality, but also when it celebrates the opposite antidote, such as for example intuition, irrationality, or the pathetic myth of a lost past purity, it always selects and establishes a choice that is surrounded by a moral option, in itself always potentially or actually demarcating violent outcomes in possible recipients. Derrida (1981) proposes to transform philosophical reflection in a kind of *pharmakon* – rather than a simple antidote – to violence, a *pharmakon* aware of being both a remedy and a poison.

In some sense this book follows Derrida's suggestions: I aim at showing and displaying violence where it is present, even if hidden, and at the same time warning myself (and every reader) that violence is traceable back to my (our) own door. What I am saying is not – in turn – innocently peaceful. Indeed I will describe all through the book that violence is considerably linked to morality, but teaching people this philosophical reflection, together with its consequences, can still be perceived and experienced as violent. After all when we act morally we "want" to believe we are acting in a non violent way *a priori*, and we "want" to preserve the *moral bubble* we are in, which permits us to erase the possible violence we are dealing with. That kind of *embubblement* will be better analyzed further in this chapter, but let us consider how, as a matter of fact, the adoption of any belief (and especially moral beliefs), induces a state of inertia that makes discarding such belief a less than ideal option: the reason is simple and can be traced down to one of Peirce's fundamental insights: "[...] the action of thought is excited by the irritation of doubt, and ceases when belief is attained; so that the production of belief is the sole function of thought" (Peirce, 1987, p. 261). And belief "[...] is something that we are aware of [...] it appeases the irritation of doubt; [...] it involves the establishment in our nature of a rule of action, or, say for short, a habit" (p. 263). Doubt and belief are antonyms in a genetic interplay: doubt is one of the most undesirable states of the human mind, and the natural way to placate doubt is the generation of *belief*. Once the habit of believing something has been acquired, its suppression would helplessly lead to a new emergence of the doubt it did placate: to find a new (if any) belief to placate requires the employment of cognitive and psychic energies, with the subsequent emotional distress. Hence, any direct or implicit attempt to eradicate one's belief can only be perceived as threateningly violent.

Of course a last question arises: are we condemned to stay in the company of violence, as if in a kind of paralysis? Various philosophers maintain that we can avoid and diminish the contact with violence (both ours and the one coming from "out there") in our lives, taking advantage of various moral and intellectual tools, some of which I will describe later on in this book. Even though we are always implicated and complicit with violence, we can and must nevertheless *act*.[4] Of course the recipes adopted are always "moral", insofar as they make "a stand for the best". For example, a tempered action is possible: within the scope of *also* taking advantage – when available and practicable – of the aristocratic[5] intellectual resources of a "philosophy of life", centered on ourselves and our relationships with other human beings and our cognitive niches.[6] Such an aristocratic proposal is for example embedded in the idea that we have only to prepare a better "futurity": "Derrida, following Levinas, seems to suggest an alternative economy, which exceeds the very notion of economy. It too, like violence, inscription, or writing, goes by many names in Derrida's writings. Among the more resonant of these is the Other, which he also describes in terms of the gift, hospitality, donation, generosity, or ethics" (Grosz, 1999). The gift would be in some sense outside the law, and "gives time" (so that it would be beyond the field of violence: a possible future, "a temporality in excess of the present and never contained within its horizon, the temporality of endless iteration" (ibid.)

The still moral beyond-violence encounter with the Other could also be seen from the perspective of Levinas' understanding of the ethical relation as a structure other than the Greek conception of the relation between self and other. The encounter with the other is neither representation nor limitation, nor conceptual relation to the same: it is, to use Levinas' speculative but clear lexicon, the "dis-inter-ested relation" (1985, p. 52). The infinitely-other (endowed for Levinas with an obvious and exclusive religious meaning) cannot be bound by a concept, cannot be thought of on the basis of a horizon; for a horizon is always a horizon of the same. It is an "irreducible intentionality, even if one must end by seeing that it ruptures intentionality". When in the encounter with the other, the subject "consciously" adopts the "responsibility for her" as "being-for-the-other". In the perspective of a practice of the generous morality of the gift, the subject "gives time" and "saves time" opening up the future: "time is not the achievement of an isolated and lone subject, but [...] it is the subject's very relationship with the Other" and a "relationship to unattainable alterity", which, in a certain sense, is still the subject, still *me* (1985, pp. 32, 57, 61). Levinas eloquently contends "I say, in *Otherwise than Being* (1981) that responsibility is initially a *for the Other*. This means that I am responsible for his very responsibility. [...] But justice only has meaning if it retains the spirit of dis-inter-estedness which animates the idea of responsibility for the other man". Again:

[4] (Kondo, 1995, p. 97).

[5] By "intellectual" I explicitly mean that it is not put forward as a solution accessible to everyone.

[6] On the concept of cognitive niche cf. the following chapter.

> Responsibility is what is incumbent on me exclusively, and what, humanly, I cannot refuse. I am I in the sole measure that I am responsible, a non-interchangeable I. I can substitute my-self for everyone, but no one can substitute himself for me. Such is my inalienable identity of subject. It is in this precise sense that Dostoyevsky said: "*We are all responsible for all men before all, and I more than all the others*" (Levinas, 1985, pp. 96, 99, 101).

Similarly, Kaplan (2001) illustrates the tension in liberal institutions between intentional agency, responsibility, and justice, in the perspective of individual actions, which can be classified as moral insofar as they go beyond normal rules and standards: reflections from Levinas, Foucault (2006), and Bonhoeffer (1965) are developed to show how justice in liberal states is not guaranteed by the confluence of jurisprudence, individual rationality, and social rules of responsibility. Like Levinas, Bonhoeffer – who was involved in the conspiracy to assassinate Hitler – provocatively contended that the radical responsibility of the individual is always to another person, not to an ethical principle.

Finally, the moral encounter – beyond-violence – with the Other could also be seen from the sophisticated viewpoint of a Waldenfels' (2006; 2007b) phenomenological approach, which takes "radical" otherness as something which withdraws from one's own experience and exceeds the limits of our common orders: "Radical otherness is something extra-ordinary, arising in my own body, situated between us and striking us before we look for it" (Waldenfels, 2007a, p. 69).[7] A "responsive phenomenology" is proposed: "The alien occurs and affects us by way of a sort of *pathos*, which precedes, provokes, and requires our *response*. Both pathos and response cross each other without converging, they are separated from each other by a hiatus which I call *diastasis*" (Waldenfels, 2007b, p. ix). The other is not just another ego or subject, nor is he/she an object, but someone who is like me but who is at the same time incomparable: this fact comes from what Waldenfels calls "*the doubling of myself in and by the Other*" (2007b, p. 82), a process by which my sense of my own *alienness* is reinforced by the duplicative alien nature of the Other: "I feel myself seen before I see the other as someone who sees things, and feel addressed before hearing the other utter certain words, and therefore, strangely, I perceive myself from elsewhere. Perhaps this should not seem so strange. After all I receive my name from others, not to mention my habits and characteristics. Even my language comes from others" (Saban, 2011, p. 108).

It is interesting to note that the phenomenological approach offers new perspectives on the role of "listening and answering", of psychiatry and its renovation, and on the inter-religious dialogue. For example, psychiatry must confront us with a peculiar sort of pathological otherness – the sick person as an *alien* per se – which in transcultural ethno-psychiatry is further doubled to an otherness of a higher degree – the sick person as a *cultural* alien (beyond the "scientific" universality of psychiatric nosology). How can psychiatry take the intercultural Other into account without sacrificing its otherness to universal points of view? Is there an alternative

[7] On the phenomenological approach to the analysis of otherness cf. also (Liebsch and Mensink, 2003).

to the extremes of fundamentalism and globalism, which tend either to repress otherness or to level it?

In chapter six I will study to a deeper level the paradoxical character of all the philosophical, religious, political, etc. (all intrinsically moral) commitments to a foreseeable beyond-violence state. I will stress that on one side appeals to a beyond-violence morality are certainly a way of explaining the genesis of violence and of curing it but that, unfortunately, they are also inclined to a kind of constitutive perversion, since the new appeal becomes the very condition of possibility of further conflicts that can lead to violent outcomes.

3.2 Fallacies as Distributed "Military" Intelligence

In the first chapter I described the general framework of the so-called *coalition enforcement hypothesis*, a perspective stressing the intrinsic *moral* (and at the same time *violent*) nature of language (and of abductive and other hypothetical forms of reasoning and arguments intertwined with the propositional/linguistic level). In this perspective language is basically rooted in a kind of *military intelligence*, as maintained by Thom (1988). Indeed we have to note that many kinds of hypothesis generation (from abduction to hasty generalization,[8] from *ad hominem* to *ad verecundiam*) are performed through inferences that embed formal or informal fallacies.

Of course not only language carries morality and violence, but also motor actions and emotions: it is well-known that overt hostility in emotions is a possible trigger to initiate violent actions. The "moral" role of emotions is well known and I think their potentially "violent" nature goes without saying.[9] de Gelder and colleagues (2004) indicate that observing fearful body expressions produces increased activity in brain areas narrowly associated with emotional processes and that this emotion-related activity occurs together with activation of areas linked with representation of action and movement. The mechanism of emotional contagion (fear) hereby suggested may automatically *prepare* the brain for action.

In Thom's terms, essentially, language efficiently transmits *vital* pieces of information about the fundamental biological opponents (life – death, good – bad) and hence it is intrinsically involved in moral/violent activities. This conceptual framework can shed further light on some fundamental dialectical and rhetorical roles played by the so-called *fallacies*, that are of great importance to stress some basic aspects of human abductive cognition. In the following subsection I will consider some of the roles played by fallacies that can be ideally related to the intellectual perspective of the coalition enforcement hypothesis mentioned above.

---

[8] This fallacy occurs when a person (it is witnessed in animals as well, for example in mice, where the form of hypotheses making can be ideally modeled as a hasty induction) infers a conclusion about a group of cases based on a model that is not large enough, for example just one sample.

[9] I have addressed the moral character of actions, both spontaneous/immediate and planned, in (Magnani, 2007b, chapter six).

Of course the "military" nature is not evident in various aspects and uses of syntactilized human language.[10] It is hard to directly see the coalition enforcement effect in the many *epistemic* functions of natural language, for example when it is simply employed to transmit scientific results in an academic laboratory situation, or when we gather information from the Internet – expressed in linguistic terms and numbers – about the weather. However, we cannot forget that even the more abstract character of knowledge packages embedded in certain uses of language (and in hybrid languages, like in the case of mathematics, which involves considerable symbolic parts) still plays a significant role in changing the moral behavior of human collectives. For example, the production and the transmission of new scientific knowledge in human social groups not only operates on information but also implements and distributes roles, capacities, constraints and possibilities of actions. This process is intrinsically moral because in turn it generates precise distinctions, powers, duties, and chances which can create new between-groups and in-group violent (often) conflicts, or reshape older pre-existent ones.

New theoretical biomedical knowledge about pregnancy and fetuses usually has two contrasting moral/social effects, 1) a better social and medical management of childbirth and related diseases; 2) the potential extension or modification of conflicts surrounding the legitimacy of abortion. In sum, even very abstract bodies of knowledge and more innocent pieces of information enter the semio/social process which governs the identity of groups and their aggressive potential as coalitions: deductive reasoning and declarative knowledge are far from being exempt from being accompanied by argumentative, deontological, rhetorical, and dialectic aspects. For example, it is hard to distinguish, in an eco-cognitive setting, between a kind of "pure" (for example deductive) inferential function of language and an argumentative or deontological one. For example, the first one can obviously play an associated argumentative role. However, it is in the arguments traditionally recognized as fallacious, that we can more clearly grasp the military nature of human language and especially of some hypotheses reached through fallacies.

Let me reiterate what I have already stressed in the first chapter: Searle considers eccentric that aspect of our cultural tradition according to which true statements, fruit of deductive sound inferences, that also describe how things are in the world, can never imply a statement about how they *ought* to be. According to Searle, to say that something is true is already to say you ought to believe it: other things being equal, you ought not to deny it. This means that normativity is more widespread than expected. It is in a similar way that Thom clearly acknowledges the "general" (and intrinsic) "military" (and so moral, normative, argumentative, etc.) nature of language, by providing a justification in terms of the catastrophe theory, as I have described in the previous chapter.

Woods contends that a fallacy is by definition considered "a mistake in reasoning, a mistake which occurs with some frequency in real arguments and which is

[10] The military/violent nature is obviously manifest for instance in hateful, racist, homophobic speech.

characteristically deceptive" (Woods, 2012).[11] Traditionally recognized fallacies, like hasty generalization and *ad verecundiam*, are considered "inductively" weak inferences, while affirming the consequent is a deductively invalid inference. Nevertheless, when they are used by actual reasoners or "beings like us", that is in an eco-logical[12] and not in an aseptic – ideal and abstract – logical framework, they are *no longer* necessarily fallacies. Traditionally, fallacies are considered to be *errors*, *attractive* and *seductive*, but also *universal*, because humans are prone to commit them. Moreover, they are "usually" considered *incorrigible*, because the diagnosis of their incorrectness does not cancel their appearance of correctness: "For example, if, like everyone else, I am prone to hasty generalization prior to its detection in a particular case, I will remain prone to it afterwards" (ibid.)

Woods calls this perspective the traditional – even if not classical/Aristotelian – "EAUI-conception" of fallacies. Further, he more subtly observes

> [...] first, that what I take as the traditional concept of fallacy is not in fact the traditional concept; and, second, that regardless whether the traditional concept is or is not what I take it to be, the EAUI-notion is not the right target for any analytically robust theory of fallacyhood. [...] But for the present I want to attend to an objection of my own: whether the traditional conception or not, the EAUI-conception is not even so a sufficiently clear notion of fallacyhood. [...] If the EAUI-conception is right, it takes quite a lot for a piece of reasoning to be fallacious. It must be an error that is attractive, universal, incorrigible and bad. This gives a piece of reasoning five defences against the charge of fallaciousness. Indeed it gives a piece of erroneous reasoning four ways of "beating the rap". No doubt this will give some readers pause. How can a piece of bad reasoning not be fallacious? My answer to this is that not being a fallacy is not sufficient to vindicate an error of reasoning. Fallacies are errors of reasoning with a particular character. They are not, by any stretch, all there is to erroneous reasoning (2012, chapter three).

If we adopt a sharp distinction between strategic and cognitive rationality, many of the traditional fallacies – for instance, a hasty generalization – demand an equalizing treatment. They are sometimes cognitive mistakes and strategic successes, and in at least some of those cases it is *more rational* to proceed strategically, even at the cost of cognitive error. As a matter of fact, hasty generalization or the fallacy of *affirming the consequent* instantiate the traditional concept of fallacy (for one thing, it is a logical error), but there are contexts in which committing the *error* is smarter than avoiding it.

According to Woods' latest observations the traditional fallacies – hasty generalization included – do not really instantiate the traditional concept of fallacy (the EAUI-conception). In this perspective it is not that it is "sometimes" strategically justified to commit fallacies (a perfectly sound principle, by the way), but rather that in the case of the *Gang of Eighteen* traditional fallacies they are just *not* fallacies. The distinction is subtle, and I can add that I agree with it in the following sense: the

[11] Of course deception – insofar as it is related to *deliberate* fallaciousness – does not have to be considered as part of the definition of what a fallacy is (Tindale, 2007).

[12] That is when fallacies are seen in a social and real-time exchange of speech-acts between parties/agents.

traditional conception of fallacies adopts – so to speak – an *aristocratic* (idealized) perspective on human thinking that disregards its profound eco-cognitive character, where the "military intelligence" I have quoted above is fundamentally at play. Errors, from an eco-cognitive perspective, certainly are not the exclusive product of the so-called fallacies, and in this wider sense, a fallacy is an error – as Woods says – "that virtually everyone is disposed to commit with a frequency that, while comparatively low, is nontrivially greater than the frequency of their errors in general".

My implicit agreement with Woods' "negative thesis" will be clearer in the light of the account of language's military nature I will provide in the following sections and I will broaden in the following chapter. My account will focus on those ideas: 1) human language possesses a "pregnance-mirroring" function, 2) in this sense we can say that vocal and written language can be a tool exactly like a knife; 3) the so-called fallacies, are linked to that "military intelligence", which relates to the problem of the role of language in the so-called *coalition enforcement*; 4) this "military" nature is not evident in various aspects and uses of syntactilized human language. As already stressed above, it is hard to directly see the coalition enforcement effect in the many *epistemic* functions of natural language (too easily supposed to be universally "good"); furthermore, we cannot forget that even the more abstract character of the knowledge packages embedded in certain uses of language – for example the "logical" use – still plays a significant role in affecting the moral behavior and it is still error-prone from many perspectives.

I have also to reiterate that, in this sense, it is hard to distinguish, in an eco-cognitive dimension, between a kind of "pure" (for example deductive) inferential function of language and the argumentative or deontological one, with respect to their error-enabling effects. So to say, there are many more errors of reasoning than fallacies: I feel like suggesting that, as far as those arguments traditionally recognized as fallacious are concerned, it is simply much easier than in other cases, to grasp their potential erroneousness, for example when they constitute hypothetical cognition. In sum, from an eco-cognitive perspective, when language efficiently transmits positive *vital* pieces of information (through so-called fallacies) and acts in the cognitive niche in a way that benefits the agent(s), it is still hard to label the related reasoning performance as *fallacious*: therefore, from now on, unless stated otherwise I will *always* refer to "fallacy" and "fallacies" according to this particular agent-based and eco-cognitive significance, making use of those words as flexible labels that *regroup* a typical series of arguments instead of insisting on the pejorative connotation usually conveyed by the word "fallacy" when used in a strictly traditional logical analysis.

Some important features concerning fallacies immediately arise. In real *agent-based* and *task-oriented* reasoning the agents' access to cognitive resources is limited by relevant constraints, such as:

- bounded information
- lack of time
- limited computational capacity.

On the contrary, logic, which Gabbay and Woods consider a kind of theoretical/institutional ("ideal", I would say) agent, is occurring in situations characterized by more resources and higher cognitive targets.

3.2.1 Epistemic Bubbles: Distributing Aggressive Fallacies

It is now necessary to stress that fallacies (and the hypothetical reasoning in which they are embedded) are usually exploited in a *distributed* cognitive framework, where moral conflicts and negotiations are normally at play. This is particularly clear in the case of *ad hominem*, *ad verecundiam*, and *ad populum* arguments. We can see linguistic reasoning embedded in arguments which adopt fallacies, as distributed across

- human agents, in a basically visual and vocal dialectical interplay,
- human agents, in an interplay with other similar agents mediated by artifacts, external tools and devices (for example books, articles in newspapers, media, etc.)

From this perspective the mediation performed by artifacts causes additional effects: 1) other sensorial endowments of the listener – properly excited by the features of the artifact – are mobilized; these artifacts in turn 2) affect the efficacy of the argument, in as much as artifacts (i.e. mass-media) have their own highly variable cognitive, social, moral, economical, and political character. For example, the same *ad hominem* argumentation can affect the hearer in a different way depending on whether it is witnessed in a television program (maybe already considered to be trashy and merely entertaining) or instead heard in a real-time interplay with a friend. It is obvious that in the second case different cognitive talents and endowments of the listener are activated:

- positive emotional attitudes toward the friend can be more powerful and other areas of brain storing information and knowledge – stored in conscious but also in unconscious memory – are activated and made available to the agent. Both these aspects can affect the negotiatory interaction, which of course can also acquire
- the character of a dialectical process involving frequent feedback, absent for instance when one reads an article in a newspaper, in a more passive way (in this case the one-way "rhetorical" effect prevails).

Some of the following aspects are typical of agent-based reasoning and they are all features characterizing fallacies in various forms. They can consequently be seen as good scant-resource adjustment strategies. They are:

1. risk aversion (beings like us feel fear!),
2. guessing ability and generic inference,
3. propensity for default reasoning,
4. capacity to evade irrelevance, and
5. unconscious and implicit reasoning.

Gabbay and Woods also contend that in this broader agent-based perspective any of the five conditions above may remain unsatisfied, for example: i) fallacies are not necessarily errors of reasoning, ii) they are not universal (even if they are frequent), iii) they are not incorrigible, etc. Paradoxically, fallacies often *successfully* facilitate cognition (and hypothetical thinking) (*Abundance thesis*), even if we obviously acknowledge that, actually, beings like us make mistakes (and know they do) (*Actually Happens Rule*). In sum, if we take into account the role of "fallacies" in actual human behavior, the cognitive acts they imply show a basic, irreducible, and multifarious argumentative, rhetorical, and dialectic character.

In conditions of scarce available knowledge, "belief" (independently of the strength of its foundation) seems sufficient for human groups to survive and flourish and indeed belief is *cognitively cheaper* than *Knowledge* (with a capital *K*, as we might say) for the very reason that belief *simulates* knowledge and conceals error.

The anti-intellectual approach to logic advanced by Woods' agent-based view is nicely captured by the *Proposition 8* (*Epistemic Bubbles*): "A cognitive agent *X* occupies an epistemic bubble precisely when he is unable to command the distinction between his thinking that he knows *P* and his knowing *P*" and its *Corollary 8a*: "When in an epistemic bubble, cognitive agents always resolve the tension between their thinking that they know *P* and their knowing *P* in favour of knowing that *P*" (Woods, 2012). Hence, we know a lot less than we *think* we do. Moreover, it is fundamental to stress that, when epistemic bubbles change, the distinction between merely apparent correction and genuinely successful correction exceeds the agent's command and consequently the cognitive agent, from his own first-person perspective, is encouraged in thinking that what happened was a genuinely sound correction. In sum, detection of errors does not erase the appearance of goodness of fallacies.

This Humean skeptical conclusion is highly interesting because it shows the specific and often disregarded "fragile" nature of the "cognitive" *Dasein*[13] – at least of contemporary beings-like-us. I also consider it fundamental in the analysis of fallacies from the point of view of military intelligence.

Various studies address the problem of the intertwining between the feeling of knowing and knowing, which is of course related to our concept of epistemic bubbles and to the cognitive capacity to conceal error. Burton (2008, p. 12) first of all notes that the feeling of knowing is commonly recognized by its *absence* (that is when we do not know something we feel we should know) and in any case it is strictly related to the widespread phenomenon know as cognitive dissonance: "In 1957, Stanford professor of social psychology Leon Festinger introduced the term cognitive dissonance to describe the distressing mental state in which people find themselves doing things that don't fit with what they know, or having opinions that do not fit with other opinions they hold".

[13] I am using here the word *Dasein* to refer to the actual and concrete existence of the cognitive endowments of a human agent. It is a German philosophical word famously used by Martin Heidegger in his magnum opus *Being and Time*. The word *Dasein* was used by several philosophers before Heidegger, with the meaning of "existence" or "presence". It is derived from *da-sein*, which literally means being-there/here, though Heidegger was adamant that this was an inappropriate translation of *Dasein*.

Finally, a quotation by Woods is noteworthy and self-evident:

> *Proposition 12* (*Inapparent falsity*) The putative distinction between a merely apparent truth and a genuine truth collapses from the perspective of first-person awareness, i.e., it collapses within epistemic bubbles. *Corollary 12a.* As before, when in an epistemic bubble, cognitive agents always resolve the tension between only the apparently true and the genuinely true in favour of the genuinely true. *Corollary 12a* reminds us of the remarkable perceptiveness of Peirce [...] that the production of belief is the sole function of thought. What Peirce underlines is that enquiry stops when belief is attained and is wholly natural that this should be so. However, as Peirce was also aware, the propensity of belief to suspend thinking constitutes a significant economic advantage. It discourages the over-use of (often) scarce resources (2012).

Hence, truth is "fugitive" because one can never attain it without thinking that one has done so; but thinking that one has attained it is not attaining it and so cognitive agents lack the possibility to distinguish between merely apparent and genuine truth-apprehension: "One cannot have a reason for believing *P* without thinking one does. But thinking that one has a reason to believe *P* is not having a reason to believe *P*". It can be said that fallibilism in some sense acknowledges the perspective above and, because of its attention to the propensity to guess (and thus also to abduce) and to error-correction, it does not share the error-elimination paroxysm of the syllogistic tradition.

The fugitive character of truth and, more generally, the process of embubblement provides a strong epistemological claim regarding corrigibility, namely, and the chance to correct our beliefs or – at least – those that might have negative effects on us as problem-solvers and decision-makers. According to Woods the process of embubblement is not corrigible, unless one dismisses a bubble to enter a new one. The issue about corrigibility seems to have something in common with another issue, namely, *de-biasing*. Both bias and epistemic bubble seem to point to some flaws or blind spots affecting our cognition and the way it operates. Accordingly, it seems that there are no measures able to correct such errors resulting from biases and embubblement. Even though, during the last decades, research on decision-making has brought to light several phenomena concerning biases and fallacies, the issues related to de-biasing are still quite unexplored. However, we find in literature a number of strategies that might be the basis for future research on decision making and epistemic bubbles.

Larrick (2009) listed three main strategies of de-biasing: motivational strategies, cognitive strategies, and technological strategies. Motivational strategies are prescriptive corrections basically relying on the creation of incentives to push people to adopt a sort of preemptive self-criticism. Incentives might be created on the basis of making people accountable for their decision – even economically. Cognitive strategies are based on the adoption of habits that are epistemologically more robust like considering the opposite, revising one's belief in face of falsifications, and so on. Some other cognitive strategies try to take advantage of training on normative rules that meet higher standards of behavior. Finally, technological strategies aim at providing the agent with additional cognitive capabilities. For instance, the use of certain technologies (for instance, Decision Support Systems) may improve the quality of the decision-making process by altering the task one should face.

It is worth mentioning here the case of *tunnel vision*.[14] This case is interesting, as it also shows how the collective dimension of a fallacious or biased decision-making process is related to military intelligence. According to its proponents, tunnel vision refers to a compendium of fallacies and biases potentially affecting the criminal justice system in all its parts and components: investigators, prosecutors, judges, and defense lawyers. Usually, tunnel vision is called to explain (and thus – as far as it is possible – to prevent) wrongful convictions. Even though it is hard to generalize, there are however recurrent patterns involving various cognitive distortions like, for instance, filtering out relevant information, treating evidence as probative – even if it is not – just because they support our conclusion, bandwagon effect and so on. Such cognitive distortions affect the whole process and they reduce the accuracy of the whole process leading to conviction. I contend that tunnel vision is not the product of fraudulent attempts to hack the criminal justice system. Conversely, it should be considered as a form of distributed military intelligence, in which the violent elements related to the criminal justice system emerge more clearly than in any other cases.

3.2.2 Morality and Moral Bubbles: Legitimizing and Dissimulating Violence

A basic aspect of the human fallacious use of language, as far as its military effects are concerned, is the *softness* and *gentleness* granted by the constitutive capacity of fallacies to *conceal errors* – especially when they involve abductive hypothesis guessing. Being constitutively and easily unaware of our errors is very often intertwined with the self-conviction that we *are not at all violent and aggressive* in the argumentation we perform (and in our eventual related actions). As I just contended, human beings use the so-called fallacies because they often work out in a positively vital way – even if, of course, they involve violent aspects: if they are eco-cognitively fruitful, we cannot call them fallacies anymore, at least not in the canonical pejorative connotation. I believe that the issue of the violence embedded in fallacies provides us with a significant clue about the existence of something akin to a *moral bubble*, that is very homomorphic with the epistemic bubble, in which an agent is "trapped". One should never forget how:

- unawareness of our error is often accompanied by lack of awareness regarding the deceptive/aggressive character of our speech (and behavior).

Woods says: "*Proposition 11* (*Immunization*) Although a cognitive agent may well be aware of the Bubble Thesis and may accept it as true, the phenomenological structure of cognitive states precludes such awareness as a concomitant feature of our general cognitive awareness", and, consequently "Even when an agent *X* is in a

[14] (Findley and Scott, 2006; Findley, 2010).

cognitive state S in which he takes himself as knowing the Bubble Thesis to be true, S is immune to it as long as it is operative for X". In short, a skeptical conclusion derives that errors are unavoidable inasmuch their very nature lies embedded in their concealment: that is, in an epistemic bubble, "any act of error-detection and error-correction is subject in its own right to the concealedness of error" (2012).

Furthermore, it can be argued that the characteristic of morality defined as "viscosity" is a converging point between our *bottom up* perspective and that of sociobiology and evolutionary psychology, which is more of a *top down* one. This concept clarifies some aspects of morality that further explain the insurgence of the moral bubble. Lahti and Weinstein (2005) introduce the concept of viscosity to provide an explanation of the gap between the absolutism of morality and the empirical evidence that moral regulations are often infringed with no major consequences neither for the whole moral system, nor for the very individual who performs the infractions. As a word borrowed from physics, viscosity refers to "the state of being thick, sticky, and semifluid in consistency, due to internal friction". To say that morality is *viscous* hints at its thickness and being glue-like, thus meaning its capability to be deformed, stressed, pulled apart and reassembled without showing decisive harm to its own stability and reproducibility. This is a trait displayed both at an inter-subjective and intra-subjective level, so that even major infringements of moral axiologies do not cause the moral to become self-defeating: such rapid regeneration of morality makes groups endorsing a morality "less likely to engage in rapid changes of commitment level that can compromise the efficacy of indirect reciprocity and ultimately threaten group stability. [Individuals belonging to those groups] will also be less likely to track drastic fluctuations in perceived group stability, decreasing susceptibility to sudden extrinsic threats" (2005, p. 58).

From another perspective, though, this viscosity is a decisive trait of the moral disposition at the singular agent level: it allows a person to steal once and yet still deem robbery to be morally wrong, to be unfaithful now and then and yet consider faithfulness as a fundamental moral virtue, to be violent and yet preach non-violence as the way to happiness and salvation. That is not a matter of hypocrisy, as Wilson (2002a) rightly points out in his analysis of Judaism in a group-serving perspective, but rather a necessary condition to the survival of morality itself by means of the moral bubble, whose scope is to avoid the cognitive breakdown that would be triggered by the constant appraisal of the major or minor inconsistency of our conduct with respect to our convictions. In a previous book I have already remarked how, since the eighteenth century, an ever-growing number of objects has been admitted to the category of those deserving moral treatment: women,[15] animals, local and global environment, political entities, cultural artifacts and so on. This can be argued to further stress moral viscosity, insofar as our actions are susceptible to be evaluated as *good, right and righteous* or *bad, wrong and evil* in ever more domains, which seems to encourage an hypertrophic growth of the *moral bubble*. The last consideration leads to the hypothesis, I elaborate in chapter six[16] that a kind of

[15] (Magnani, 2007b). By this I do not mean to endorse any misogyny but merely state a truth in the history of thought.

[16] Cf. section 6.5.

overmorality (that is the presence of too many moral values attributed to too many human features, things, events, and entities), can be dangerous, because it sets ready more opportunities to trigger fresh violence by promoting plenty of unresolvable conflicts.

Burton contends that beliefs in general (and so also moral beliefs) are endowed with a kind of *inertia* (which further relates to the shift between the mere belief and real knowledge) that is typical of the moral bubble: "The more committed we are to a belief, the harder it is to relinquish, even in the face of overwhelming contradictory evidence. Instead of acknowledging an error in judgment and abandoning the opinion, we tend to develop a new attitude or belief that will justify retaining it" (2008, p. 12). Furthermore, we consciously choose a false belief because it feels correct even when we know better: it seems that there is a gap between the specific sensation of feeling something to be right and knowing it to be right. Paradoxically, we keep and trust a wrong belief even if we have fresh knowledge which contradicts it. Similarly – within our bubbles – we adopt an axiological conviction (and we act according to it) even if we know its possible or actual violent outcomes.[17]

I have said above, following Woods, that a skeptical conclusion derives that errors are unavoidable, their very nature lies embedded in their concealment; that is, in an epistemic bubble. Indeed, in a sense, there is nothing to correct, even if we are aware of the error in reasoning we are performing. Analogously, we cannot complain about the way we think and change it, even if we are aware of the possible deceptive character of the reasoning we are performing. What is at stake here is the awareness that something *has a priority* and that it is also because it is *stable*: firm beliefs are endowed with an intellectual value because they are and *remain* ours (over a lifespan, maybe!) and they have an intrinsic moral value because they fit some moral perspective that *we* (as individuals) agree with. We usually share such perspectives with different groups we belong to but they can occasionally also be extremely subjective and – obviously – they can be seen as perverse from the perspective of others. Errors, and so deception and aggressiveness, are a constitutive "occluding edge" of agent-based linguistic acts and conducts. It is from this perspective that we can also grasp the effective importance that human beings placed on the so-called "intuition", where they simply reason in ways that are *typical*, and *typically* justified for them (Woods, 2012, chapter four).[18]

3.2.3 Gossip, Morality, and Violence

If humans are so inclined to disregard errors it is natural to devolve to "fallacious" reasoning a kind of natural, *light* military activity which becomes evident when vital interests of various kinds are at stake. In this case arguments that embed fallacies

[17] See chapter five, subsection 5.1.1, about some Kantian excerpts shedding further philosophical light on the concept of "moral bubble".

[18] It is interesting to note the recent attention to fraud, deception and their recognition in the area of multi-agent systems (MAS), e-commerce, and agent societies, but also in logic (cf. (Sun and Finnie, 2004)).

can nevertheless aim at 1) defending and protecting ourselves and/or our group(s) – normally, human beings belong to various groups, as citizens, workers, believers, family-members, friends and so on; 2) attacking, offending and harming other individuals and groups (one must bear in mind that groups are always potentially aggressive coalitions). Here, by way of example, it is well-known that gossip, a typical tool for coalition management, takes advantage of many fallacies, especially *ad hominem, ad baculum, ad misericordian, ad verecundiam, ad populum, straw man,* and *begging the question.*[19]

From this perspective, gossip (full of guessed hypotheses about everyone and everything, exploiting several fallacies) contemplates

- telling narratives that exemplify moral characters and situations and so inform and disseminate the moral dominant knowledge of a group (or subgroup) (a teaching and learning role which enforces the group as a coalition) *possibly* favoring its adaptivity, while *at the same time* facilitating some individuation, persecution, and punishment

   1. of free riders inside the group (or inside the same subgroup as the arguer, or inside other subgroups of the same group as the arguer),
   and
   2. of alien individuals and groups presenting different moral and other conflicting perspectives.

Even if Wilson's evolutionary perspective in terms of *group selection* is questioned because of its strict Darwinist view on the development of human culture, a suggestion can be derived, especially if we reframe it in the theory of "cognitive niches" that I will present in the following chapter of this book: schemas of gossiping establish cognitive niches that can be adaptive but of course also *maladaptive* for the group (2002a). Human coalitions produce various standard gossip-templates which exploit fallacies and that can be interpreted in terms of conflict-enabling cognitive niches. However it is unlikely that these "military" schemas, embedding both moral and violent aspects, can directly establish appreciable selective pressures.[20]

The disseminating process of gossip shown above is *moral* and at the same time it secures the more or less *violent* persecution of free riders inside the group. Furthermore, in the parallel process of protecting and defending ourselves and our groups, gossip is *moral* and at the same time it secures aggression and *violence* against other,

[19] Some considerations on the moral role of gossip are given in (Bardone and Magnani, 2010). On the efficacious and presumptive "adaptive" role of language and theory of mind to justify remote punishment and bullying of weaker and/or low-ranking individuals and groups by third parties cf. (Ingram et al., 2009). The study also presents a description of the linguistic manipulation of the reputation of others (and of the individual capacity to manipulate their own reputation in the mind of other individuals or groups) – through gossiping for example – and its role in recruiting the support of others. On the role of language to achieve some forms of symbolic status for a group, rather than to fundamentally achieve symmetrical cooperation, cf. also (Dessalles, 2000).

[20] I have deepened this eco-cognitive perspective concerning the intertwining between morality and violence in the first chapter (section 1.3.).

different, groups. In other words exploiting the so-called fallacies in gossip can be seen as *cooperative* because it carries moral knowledge, but it is at the same time *uncooperative* or even *conspiratorial* because is triggers the violence necessary to harming, punishing and mobbing. It has to be said that the type of violence perpetrated through fallacies in these cases is situated at an intermediate level – in between sanguinary violence and indifference. If the "supposed to be" – from our intellectual viewpoint – fallacies are eco-cognitively fruitful for the individual or the coalition, can we still call them fallacies, if we take into account their own perspective?

Gossip proves to be an effective coalition-management tool also at a higher cognitive level, to limit hostilities among individuals and cliques (competitively aligned against each other) and reassert once again the general values of the group. Gluckman states that gossip and scandal: "[...] control disputation by allowing each individual or clique to fight fellow-members of the larger group with an acceptable, socially instituted customary weapon, which blows back on excessively explosive users" (1963, p. 13). To make this point clearer, we might suggest a reasonable analogy: over the few past centuries, duels were a common method for settling arguments. Weaponry was not at the contenders' total discretion but was subject to a long-established regulation. Similarly, gossip can indeed be used as a social weapon; saying social, though, we do not only mean that it can be obviously used to damage other peers in the same coalition, but also that its very way of operating is determined by the coalition itself. Since the content of gossip is bounded by the evoked morals of the community, struggles are kept within the common set of shared norms, and any attempt to trespass these norms out of rage or indignation is discouraged by the very same mechanism. Thus, even internal struggles achieve the result of maintaining the group structured at a convenient degree.

The hypotheses generated by those so-called fallacies in a dialectic interplay (but also when addressing a non-interactive audience) are certainly conflict-makers[21] but they do not have to be necessarily conceived *a priori* as "deal-breakers" and "dialogue-breakers", as very usefully noted by Woods.[22] I would contend that the potential deceptive and uncooperative aim of fallacies can be intertwined with pieces of both information and disinformation, logical valid and invalid inferences, plus other typical mistakes of reasoning like perceptual errors, faulty memories and misinterpreted or wrongly transmitted evidence, but fallacious argumentations still can be – at first sight paradoxically – "civil" conventional ways for negotiating. That is, sending a so-called deceptive fallacy to a listener is anyway far less violent than sending a bullet, even if it *can* enter – as a violent linguistic behavior – a further causal chain of possible "more" violent results. Also in the case of potentially deceptive/uncooperative fallacious argumentation addressed to a non-interactive audience, the listener is, *in principle* (even if maybe not actually), in the condition to disagree and reject what is proposed (like in the case of deceptive and fallacious

[21] The relationship between arguer and audience is analyzed by (Tindale, 2005) in a contextual-based approach to fallacious reasoning.

[22] Several chapters of (Woods, 2012) are devoted to an analysis of the *ad hominem* argument and contain a rich description of various cases, examples, and problems, which broaden my point.

advertising or political and religious propaganda). The case of a mobbed person is more problematic, as it is often impossible to prevent the violent effects of mobbing performed through gossip: any reaction of the mobbed ineluctably tends to confirm the reason adduced by the mobbing agencies as right. This is one of the reasons that explains how mobbing is considered a very violent behavior.

In sum, when an argument (related or not to the so-called fallacies) "perpetrated" by the proposer(s) is accepted by the receiver (for example when gossip bearing plenty of *ad hominem* and *ad populum* arguments is fortunate and absorbed inside a group), it is proved successful, which makes it – simply and necessarily – a good argument. When the argument is rejected, often, but not necessarily, it happens because it has been recognized as a fallacy and an error of reasoning: *Proposition 12* (*Error relativity*), clearly states: "Something may be an error in relation to the standards required for target attainment, in relation to the legitimacy of the target itself, or in relation to the agent's cognitive wherewithal for attaining it" (Woods, 2012).

3.2.4 Judging Violence: Abductive Fallacy Evaluation and Assessment

I recently watched a talk show on television devoted to the case of a Catholic priest, Don Gelmini, accused of sexual abuse by nine young men hosted in a rehab facility belonging to Comunità Incontro in Italy, a charitable organization now present worldwide, which he had founded many years before. Four journalists argued in favor and against Don Gelmini, using many of the so-called fallacious arguments (mainly *ad hominem*) centered on the past of both the accused and the witnesses. I think the description of this television program is useful to illustrate the role of abduction in the filtration and evaluation of arguments, when seen as *distributed* in real-life dialectical and rhetorical contexts. As an individual belonging to the audience, at the end of the program I concluded in favor of the *ad hominem* arguments (that I also "recognized" as fallacies) used by the journalists who argued against Don Gelmini and so I accepted the hypotheses put forward against him. Hence, I considered the data and gossip embedded in the fallacious reasoning describing the "immoral" and "judiciary" past of Don Gelmini more relevant than the ones which described the bad past of the witnesses. I was aware of being in the midst of a riddle of hypotheses generated by various arguments, and of course this was probably not the case of the average viewer, who may not have been trained in logic and critical thinking, but it is easy to see that this fact does not actually affect the rhetorical success or failure of arguments in real-time contexts, as it was also occurring in my case. As listeners, we are all in an "epistemic bubble", compelled to think we know things even if we do not know them. In this case, the bubble forces you to quickly evaluate and pick up what you consider the best choice. I would like to put forward the idea that, at least in this case, the evaluation of the *ad hominem* arguments has to be seen as the fruit of an abductive appraisal, and that this abductive process is not rare in argumentative settings.

An analogy to the situation of trials – in the common law tradition as described by Woods (2010) – can be of help. Like in the case of the judge in the trial, in our case the audience (and myself, as part of the audience)[23] were basically faced with the *circumstantial* evidence carried by the two clusters of *ad hominen* arguments, that is, faced with evidence from which a *fact* can be reasonably inferred but not directly proven.

In a situation of lack of information and knowledge, and thus of constitutive "ignorance", abduction is usually the best cognitive tool human beings can adopt to relatively quickly reach explanatory, non-explanatory, and instrumental hypotheses/conjectures.[24] Moreover, it is noteworthy that evidence – embedded in the *ad hominen* arguments – concerning the "past" (supposedly) reprehensible behavior of the priest and of the witnesses were far from being reliable, probably chosen *ad hoc*, and deceptively supplied, that is, so to say, highly circumstantial for the judge/audience (and for me, in this case). I had to base my process of abductive evaluation regarding the fallacious dialectics between the two groups of journalists on a kind of sub-abductive process of *filtration* of the evidence provided, choosing what seemed to me the most reliable evidence in a more or less intuitive way, then I performed an abductive appraisal of all the data. The filtration strategy is of course abductively performed and guided by various "reasons", the conceptual ones, for example being based on conceptual judgments of credibility. However, these reasons were intertwined with other reasons such as variously conscious emotional reactions, based on feelings triggered by the entire distributed visual and auditory interplay between the audience and the scene, in itself full of body gestures, voices, and images (also variably and smartly mediated by the director of the program and the cameramen). Along these lines, I might add that the journalists "fallaciously" discussing the case were concerned with the accounts they could trust and certainly emotions played a role in their inferences as well.

In summary, I was able to abductively make a selection (selective abduction) between the two non-rival and incommensurable hypothetical narratives about the priest, forming a kind of explanatory theory – of course I could have avoided the choice, privileging indifference, thus stopping any abductive engagement. The guessed – and quickly accepted without further testing – theory of what was happening in the dialectic exchange further implied the hypothesis of guilt with respect to the priest. That is, the *ad hominen* of the journalists that were speaking about the priest's past (he was for example convicted for four years because of bankruptcy fraud and some acceptable evidence – data, trials documents – was immediately provided by the staff of the television program) appeared to me convincing, that is, no more negatively biased, but a plausible acceptable argument. Was it still a fallacy from my own actual eco-cognitive perspective? I do not think so: it still was and is a fallacy only from a special subset of my own eco-cognitive perspective, the logical/intellectual/academic one! The "military" nature of the above interplay between contrasting *ad hominem* arguments is patent. Indeed, they were armed

[23] Furthermore, like the jury in trials, an audience is on average composed of individuals who are not experts capable of "overt calibration of performance to criteria", but instead ordinary – untutored – people, *reasonably* used to reasoning in the way ordinary people do, performing a kind of "intuitive and unreflective reasoning" (Woods, 2010).

[24] On these various aspects of abductive reasoning cf. (Magnani, 2009, chapter two).

linguistic forces involving "military machines for guessing hypotheses" clearly aiming at forming an opinion in my mind (and in that of the audience) which I reached through an abductive appraisal, quickly able to explain one of the two narratives as more plausible. In the meantime I became part of the wide coalition of individuals who strongly suspect that Don Gelmini is guilty and we can potentially be engaged in further "armed" gossiping.[25]

1. In special contexts where the so-called fallacies and various kinds of hypothetical reasoning are at play, both at the rhetorical and the dialectic level, their assessment can be established in a more general way, beyond specific cases.[26] An example is the case of a fallacy embedding patently false empirical data, which can easily be recognized as false at least by the standard intended audience; another example is when a fallacy is structured, from the argumentative point of view, in a way that renders it impossible to address the intended audience and in these cases the fallacy can be referred to as "always committed".
2. Not only abduction, but also other kinds of (supposed) fallacious argumentation can be further employed to evaluate arguments in dialectic situations like the ones I have mentioned above, such as *ad hominem*, *ad populum*, *ad ignorantiam*, etc., but also hasty generalization and deductive schemes.
3. The success of a "fallacy" and of the inherent "fallacious" hypothetical argument can also be seen from the arguer's perspective insofar as she is able to guess an accurate abductive assessment about the character of her actual or possible audience. From this perspective an argument is put forward and "shaped" according to an abductive hypothesis about the audience, which the arguer guesses on the basis of available data (internal, already stored in memory, external – useful cues derived from audience features and suitably picked up, and other intentionally sought information). Misjudging the audience would jeopardize the efficacy of the argument, which would consequently be a simple error of reasoning/argumentation. These strategic appraisals and assessments can be considered as the basis of *preaching*: making appropriate judgements about the audience was the main objects of teachings in rhetoric.
4. As clearly shown by the example of the TV priest, arguments are not only distributed, as I have contended in the subsection 3.2.1 above, but they are also embedded, nested, and intertwined in self-sustaining clusters, which individuate peculiar global "military" effects.

Thom (1980, pp. 281–282) usefully observes that there is, underlying the idea of information,

[25] A vivid example of the aggressive "military" use of language is the so-called "poisoning the well" argument: "[...] a tactic to silence an opponent, violating her right to put forward arguments on an issue both parties have agreed to discuss at the confrontation stage of a critical discussion" (Walton, 2006, p. 273). "Poisoning the well" is often considered as a species of *ad hominem* fallacy. A tactic to silence the opponent is also typical of the so-called "Fascist state of the mind", I will describe in chapter five, section 5.3.

[26] Examples are provided in (Tindale, 2005).

> [...] the existence of a seeker for whom it is profitable to learn this information, together with a giver, who gives, of his own free will in general, the information to the seeker. In all uses of the word information one is unable to identify the four elements – seeker, request, giver, advantage accruing to the seeker through knowledge of the information – one must suspect dishonesty in the use of the word.

Fallacies play a fundamental role in these cases, also when there was no request and when information has little or no interest for the recipient, like in the case of overwhelming advertisement, which pretends to offer only information. As far as this constitutive hypocrisy is concerned, Thom concludes how "in propagating the faith (*de propaganda fide*) at least the Church is aware of its aim and proclaims them".

3.3 The Semiotic Emergence of Violence through Deception

As I will attempt to show in the following two subsections, violence is often triggered – even unawarely – when agents' expectations are *semiotically* disrupted by the behavior of their counterpart. That is to say, a violent outcome, that can be experienced by one or both parties, should be expected whenever one's signaling does not conform with the "standard" implications meant by the signal deployed. To this end, I will analyze two clarifying situations: the first illuminates a "natural" emergence of violence in the predatory behavior of a species of *fireflies*, whose female specimens prey on other male individuals by mimicking a reproductive luminous signaling; the second, instead, will deal with a further evolved, social, practice defined as "moral hazard", in which a chance of cooperation is signaled, but later it is perverted to serve one's own interest within an apparent but deceitful cooperative context.

3.3.1 Fireflies and Luminescent "Femmes Fatales"

A recent study on animal communication from a semiotic perspective is related to the reliability of animal signs and the problem of the benefits of deceiving in sign exchanges (El-Hani et al., 2009): deception in the case of aggressive signal mimicry in fireflies is investigated. This case study is appropriate to confirm the hypothesis of the previous sections about the intrinsic violent character of information. Of course we can see these animal deceptive behaviors as violent only thanks to a kind of anthropomorphic approach, and at the same time it is obviously controversial to ascribe to non-human animals intentional states. Nevertheless the clear explanation of the semiotic emergence of cheating in fireflies, that can be extended without much modification to the semiotic emergence of cheating in humans, simply explains – so to say – "the banality of the emergence of evil", to take advantage of Hannah Arendt's famous expression.

Firefly "femmes fatales" are specialized in mimicking the mating signals of other species of fireflies with the purpose of attracting responding males to become their prey. These aggressive mimics are a major factor in the survival and reproduction of both prey and predator. It is a case of deception through active falsification of information that leads to efficient predation by *femmes fatales* fireflies and to the triggering of further evolutionary processes in their preys' communicative behaviors.

We have to say that there is evidence about the animal "plastic" capacity of deception: individuals can learn/create special method of deception to attack preys, depending on the particular niche they live in. The case of some jumping hunting spiders illustrated by Wilcox and Jackson is striking. By stalking their prey across the web, they manage cheat on it, through highly specialized signals, also suitably exploiting aggressive mimicry. The interesting thing is that they plastically adapt their cheating and aggressive behavior to the particular prey species at stake, all this by using a kind of trial and error tactic of learning, also reverting to old strategies when they fail (2002).[27]

A semiotic model of the emergence of firefly deception is provided where deception is seen as part of the explanation of why communication evolves towards increasing complexity in sign processes. Of course the semiotic processes involved in such animal efficacious deceptions do not usually result from learning processes taking place in the individual's cognitive system, but from the fine-tuning of capacities (inherited by natural selection among variants over thousands or millions of generations) faced with selective pressures. In the case of human beings, who make common use of natural language to cheat, plastic cognitive capacities are at play, permitting the use of learning and reasoning processes.

Male fireflies of several species pertaining to the genera *Photinus*, *Photuris*, *Pyractonema*, and *Robopus* are predated by females of several species of *Photuris* fireflies. Here the cognitive-semiotic situation as described by the authors:

> Aggressive and related mimicries in fireflies can be modeled on the grounds of the modified version of Mitchell's (1986) account of deception [...]:
>
> (i) An organism R registers a sign Y emitted by organism E, and E can be described as benefiting when
>
> (iia) R behaves toward Y, as if
>
> (iib) Y means that X is the case; but
>
> (iii) it is untrue that X is the case.
>
> If we consider, for instance, the attraction of *Photinus macdermotti* males by *Photuris* females, the latter will be modeled as emitters of a sign Y, which the former registers and then behaves toward Y as if it means or indicates an opportunity for mating, X. To understand the reason why *Photinus* males behave like this in the face of Y, there is no need of introducing any conscious intentional state in the picture. The reason simply lies in the fact that during evolution *Photinus* experienced several instances of correlation between the sign type Y and the act of mating, and such a correlation had an effect on the chances of successful reproduction of individual organisms in such a

[27] On the role of abductive cognition in animals cf. (Magnani, 2007a). On animal signals, especially the honest and quality ones, cf. (Maynard Smith and Harper, 2003); the puzzling problem of "animal culture" is dealt with in (van Schaik, 2004).

> manner that those answering to *Y* were more likely to obtain the advantageous result *X*. Nevertheless, when a *Photuris* rather than a *Photinus* female is emitting *Y*, it is not really the case that *Y* means *X*. It rather means another result, *Z*, namely that the male will be probably eaten. It is not that Photuris femmes fatales as individuals intend to deceive *Photinus* males, but simply that there is a correlation between the event at stake and past events in which *Photuris* females benefited from emitting that sign *Y* to *Photinus* males, and natural selection favored individuals showing that communicative behavior of emitting *Y* in the presence of flying *Photinus* males. *Photuris* females are just trying to benefit again from emitting *Y* to *Photinus* males (El-Hani et al., 2009).

It is well-known that, following Peirce, signs are something that determines something else (its interpretant) to refer to an object to which it itself refers (its object) in the same way. This is a process of semiosis. The interpretant will become in turn a sign, and so on *ad infinitum* (CP, 2.303), or, in some instances, a chain of sign processes can lead to a final action, which amounts to a termination of the semiotic process. Signs can mediate the communications of habits embodied in the object of the interpretant. In Thom's words the *Y* emitted by the *Photuris* fatale females is a pregnance for *Photinus* males, which are compelled to an "interpretive" behavior, wired by evolution. The sign emitted by a *Photinus* male in turn indicates the male itself as an object to a female having the effect (interpretant) of predisposing her to mate with that particular male. Unfortunately, the sign emitted by the *Photuris* females constitutes a kind of *misinformation* for the males, that are deceived and then killed instead of invited to have sex and reproduce: "Deception relies upon the fact that this is the usual interpretant following the emitted sign. That is, a habitual relationship between sign and object must exist in order for an organism to act in such a way that it is deceived when a benefiting organism sends the same sign for indicating, however, a different object" (El-Hani et al., 2009).

The whole situation of deception indicates an example of increasing sign complexity during evolution because we face a symbolic habit that is taken as the basis for the generation of a more complicated habit at a higher level, which in turn causes a possible fallacious interpretation. The authors obviously note that we have to consider such signs as kinds of "proto-propositions". The example also shows that the development of nested deception strategies is really basic in the biological evolution of organisms, between species that compete for suitable ecological niches, but also between individuals or groups within the same species.

Of course non human animals do not lie, at least in the moral sense we humans attribute to the word "lie" but, as we have seen in chapters one and two and in the previous sections of this chapter, this deceptive potential is extremely developed in humans thanks to linguistic communication and its more or less aggressive military and fallacious character. Of course deceptive behavior is highly prone to cause or engage in violent outcomes.

Further insight on the problem of deception in an evolutionary perspective is provided by Smith (2004). The book sets out to achieve an adaptationist decomposition of at least one aspect of our unconscious psychology: that which delivers/underlies social manipulation. We tell stories about our lives and these narratives are for our own private consumption, to comprehend events as well as to predict future, but they also can become public. However, many narratives are not under personal control, but derive from the unconscious or sub-personal-level control of a social module. A module which is domain-specific, and which performs military insights, exactly as it is maintained by the Machiavellian Intelligence Hypothesis,[28] and which is not only related to social scanning. In sum, we also convey unintended messages, we are always constrained by our sensitivity to other people. A great part of knowledge about the world is socially constructed in a language – a kind of sub-textual signaling – that does not directly represent reality. In this perspective it is also clear that seemingly normal conversations between people, in various kinds of relationship, embed and signal aggressions, discontent, frustrations, emotions, etc.

It is worth noting the fact the human deception and its possible violent outcomes are often performed by taking advantage of technological artifacts. For example, in the area of the analysis of adversarial reasoning in computer mediated communication – which also addresses the technologies for opponent strategy prediction, plan recognition, deception discovery and planning, and strategy formulation that not only applies to security issues but also to game industry and business transactions – deception is usefully described as a violent "semantic attack". It is for example active in cognitive hacking, which is specifically against the users' mind and behavior, "by manipulating their perception of reality", in contrast to physical and syntactic attacks, which operate against hardware or software systems.[29] In this case information is used in a violent and sophisticated way – beyond the well-known violent effects of traditional mass-media propaganda – to influence and to affect the behavior of humans through computational tools: a full process of *information warfare*, as it is called.

3.3.2 *Moral Hazard as the Violent Perverting of Cooperation*

Moral hazard is a recurrent feature of human morality. Dowd (2009) defines moral hazard as the situation in which one party is responsible for the interests of another having an incentive to let his interests come first. One of the most topical examples of moral hazard found in literature is concerning the subprime scandal and the related financial crisis of fall 2008. Banks started off by selling their mortgages regardless the probabilities they would default or not, instead of having a view to

[28] Posed in the late 1980s (Whiten and Byrne, 1988, 1997; Byrne and Whiten, 1988), the "social brain hypothesis" (also called "Machiavellian intelligence hypothesis") holds that the relatively large brains of human beings and other primates reflect the computational demands of complex social systems and not only the need of processing information of ecological relevance: the ability to manipulate information and not simply to remember it, to recognize visual signals to identify other individuals, sufficient memory for face recognition and to remember who has a relationship with whom, use of tactical deception, coalition, ability to understand intentions, to hold false beliefs, and "mind-reading", known as "theory of mind" etc. Language itself would have evolved from physical grooming as a way of creating social cohesion as the size and complexity of the social group increased (cf. also (Dunbar, 1998, 2003)).

[29] (Kott and McEneaney, 2007).

holding the mortgage to maturity. Banks happened to have an incentive to undermine or misperceive all the risks associated to lending money to those who would hardly be able to repay the loan. The main consequence was that the financial market was intoxicated by the so-called *subprimes* leading some banks to bankruptcy (and the rest of global capitalism to a credit crisis).

Moral hazard can be also applied to several other domains, for instance, the insurance markets. In this case the moral hazard regards the insured party that may have an incentive to misbehave, as he knows that the negative consequences of his actions will be covered by the insurance. One of the side effects caused by moral hazard in the insurance markets is that the costs of being insured will tend to raise (Shavell, 1979). Moral hazard may also regard a worker having an incentive to shirk on the job. For instance, once a person acquires a secure or permanent position, he will be less motivated to put his company's interests first, as he will not bear the consequences of poor performance at work.

My main take is that moral hazard points to a deceptive/faking dimension in which cooperation may end up: people might *think* that they are cooperating when they are not. In this sense, moral hazard is not to be considered a free riding behavior. In the case of free riding a person simply refuses to cooperate in order to gain a private profit (a ride on a bus, for instance). Conversely, moral hazard emerges from a network of mutual interests in which a person offers his cooperation. In fact, as I mentioned above, one of the central aspects of moral hazard is that a person is responsible for the interests of another party. From an analytical perspective this relational dimension of moral hazard can be further specified as the result of four main elements that I will now consider:

1. asymmetric information and its docile dimension;
2. an incentive to put one's own interests first;
3. poor risk-assessment;
4. responsibility avoidance by attributing negative outcomes to bad luck (or the *invisible hand*).

Let us start with the first element, asymmetric information. We have asymmetric information when in a contract one of the parties has more or better information than the other. For instance, when a person sells a car, he certainly knows more about the car than the potential buyer. Indeed, asymmetric information may open up a deceptive dimension, as the seller, for instance, can always hide some information from the potential buyer that might change the buyer's mind. However, no matter what the seller's intentions are, asymmetric information does not necessarily lead to a deceptive dimension. Actually, asymmetric information may give cooperation a chance. In fact, it is worth noting that the buyer should rely to some extent on what the seller tells him about the car. He could not get this piece of information from anybody else except him. The seller himself is committed to disclosing some information he has and that the buyer does not have. Plain and simple, both the seller and the buyer should be docile, as they are involved in a relation in which they basically exchange information and rely on each other. So, basically, a moral hazard is instantiated because the two parties involved decide to cooperate disclosing some information that might benefit both of them.

When a person faces a moral hazard, then it is because he has an incentive to put his own interests first. And this is the second element. It is worth noting that his interests are not necessarily contrasting with the ones of the other party. It might happen that pursuing a certain course of action may benefit both parties, and so potentially granting the best result to all.

The third element is related to risk-taking. When facing a moral hazard a person may overlook the risks attached to the available options. Why? Because he is not going to be the one bearing the consequences of his choice. It is worth noting here that failing to adopt the perspective of another person is not necessarily due to cheating. *Perspective taking*, namely, putting oneself in somebody else's shoes, might simply be *biased*. For instance, people seem to be more concerned about losses. As Kahneman and Tversky (1979) claimed, losses tend to appear larger than equivalent gains. This propensity to look more carefully at losses might be a cognitive aid to weigh the risks attached to a certain decision. But that aid is *de facto* unavailable when a person will not bear the consequences of his decisions just like in the case of moral hazard. And so, his risk assessment might be eventually impoverished.

This third element is closely related to the fourth and last element I want to stress now. When a person opts for a decision – overlooking the risks attached – he will tend to blame "bad luck" more than himself. That is, he will tend to over-attribute any bad outcome to misfortune, thus avoiding to acknowledge his own responsibilities. For instance, in the case of the financial crisis operators simply blamed the so-called *invisible hand*, which transformed individual choices into public outcomes without any possibility to intervene. Plus, it is worth noting that those who had a moral hazard could insists they were simply aiming at a higher gain for both, so that they were actually cooperating, and it was just bad luck who struck both parties. This is connected to the second element: the two parties' interests are not necessarily contrasting.

Now, let me turn to an interesting case that will immediately shed light on the violent dimension of moral hazard as the perverting of cooperation. The case I am going to briefly treat is the case of humanitarian intervention as an example of the perverting of cooperation.

Humanitarian intervention, that is, the prevention of genocide or ethnic cleansing, is considered by the Western countries as something good. Emerging after the collapse of the Soviet Union, humanitarian intervention is now deemed as a standard norm shaping international relations, especially the ones involving Western rich countries and the underdeveloped ones, usually characterized by violent and despotic governments. Humanitarian intervention is also codified by the UN General assembly in a general form that should be pursued by the international community as a whole. So, whenever a minority or a population is engendered by the act of a state, the international community has the right and the duty to intervene to protect that population from genocide or ethnic cleansing.

Recently, Kuperman (2009) called into doubt not the morality underlying human intervention, but the impact that it may have on some particular occasion. In brief, he argued that humanitarian intervention may lead the rebels or the members of a minority to have a moral hazard and thus, in my terminology, to the perverting of co-operation. Basically, the main point he makes is that moral hazard creates a situation in which the humanitarian intervention bolsters rebellion by promoting irresponsibility among minority leaders. As Kuperman contended, minority leaders may have an incentive to act irresponsibly, for instance, by openly provoking state retaliation against inert civilians. Paradoxically, a violent retaliation would be a good opportunity for them to ask the international community for a military intervention on their side. Indeed, this move would be successful, if the intervention could stop the genocide. However, as Kuperman noted, when the international community takes appropriate action to prevent it, it is usually too late, and the costs of this strategy are too high.

As it is clear in the case of humanitarian intervention, the perverting of cooperation, essential to moral hazard, has a major consequence: good intentions may be perverted and literally turned upside down encouraging free-riding behaviors. What I find interesting to stress here is that, paradoxically, the free-riding dimension of the moral hazard remains hidden to those who have it. Just like in the case of the minority leaders, they think they are cooperating when they are not, actually. The perverting of cooperation masks the whole dynamics favoring that moral embubblement I have illustrated above in subsection 3.2.2.

To summarize, the perverting of cooperation affects the four elements I have mentioned above that are turned upside down. First, the docile dimension for which the two parties are committed to disclose some information is perverted into a violent and aggressive dimension. One of the two agents exerts influence over the other by taking advantage of his bargaining power as he has more information. Secondly, the fact that one is responsible for another party's interests is turned into a free-riding dimension in which he puts his own interests first ignoring the other party's. Third, risk-taking propensity is certainly essential to capitalism for gaining a competitive advantage over others. To put it roughly, *no guts, no glory*. This potentially benefiting attitude is however transformed into a careless one leading to an impoverished cognitive performance resulting in a poor risk assessment damaging other people's interests. Fourth and finally, the same risk-taking propensity is also perverted by over-attributing actual failures and damages to bad luck so as to allow the person having a moral hazard to avoid blame and sense of guilt.

3.4 The Violent Force of Law and of Morality

Morality and law always have a potentially violent character. They aim at teaching how to behave, what is forbidden, what are the obligations and the expected punishments. Imagine the effect of moral relativism and moral dogmatism in the interplay between non-religious and mostly-religious people: religious people sometimes perceive relativism as an offense and a kind of violence to their beliefs, non religious people in turn think relativism very just and proper, and feel themselves a bit offended by the dogmatism of religions and/or the way in which they are looked down on by believers; of course the dogmatisms of two different religions often conflict in a violent way, just because they embed different moral views of life and behavior, but often they trust each other with more respect than when confronted with relativism. Moreover, people that are compelled to follow a certain moral or legal rule often follow a different morality or respect a different law (their morality can easily clash with the law system in question): the result is that they usually perceive the morality imposed on them as extremely violent. This is what we see everyday in simple situations in families, workplaces, collectives, where the conflict between moral issues is continuous and widespread. Furthermore, human beings usually possess various moral frameworks, that can be willingly and easily interchanged in the span of one day, as I will describe in the following chapter, so the likelihood of moral conflict is very high.

Walter Benjamin (1978) wondered about the possibility of distinguishing between legitimized or justified force and violence and the other remaining forms of them, for example the ones that are either prior, excessive of, or not obedient to law, rights and morality. Derrida's answer to this question seems clear and appropriate to me: we can paraphrase his thoughts and state that there cannot be a separation between legitimated and illegitimated violence, there is no separation between good and evil, because one is the *other face* of the other. For example, morality and law are always fundamentally intertwined with violence, as I have already illustrated above in section 3.1. Writing, if one considers its character of posing a foundation, producing, judging, or knowing, is violence in itself: with a kind of violence that at the same time manifests and dissimulates itself as violence, and so it is a space of continuous (potentially irritating) equivocation. Grosz – while interpreting Derrida – observes: "Justice, law, right are those systems, intimately bound up with writing (the law is writing par excellence, and the history of legal institutions is the history of the reading and rewriting of law, not just because the law is written, and must be to have its force, but also because law, and justice serve to order, to divide, to cut). 'Justice, as law, is never exercised without a decision that cuts, that divides' ":[30] "what reverberates here is the topic of an 'illegal' violence that founds the rule of the law itself" (Žižek, 2009, p. 59).

Of course ethics and morality share with the law the same problem, but also with knowledge and language, as we have seen above in the previous chapters, when also speaking of aggressive "military intelligence". Derrida's "deconstruction" is not merely the denunciation of the violence of law, morality, and so on, but – ineluctably – an engagement *with*, a participation *in* this violence. A simple result that is a "gift" from philosophy we cannot disregard. It is in this perspective that violence can be seen as clearly constitutive of religion, politics, thought, morality, knowledge, sociability and most everything typical of *homo sapiens sapiens*.

In summary, morality and law are not the barriers that divide violence from civilization, for example "sectionalizing" the violent, as something chronologically before them or left outside them, and so supposed to be captive under their sagaciousness: establishing morality and law as the space of a regulated violence just dissimulates their own violent office, as is evident in the structural violence, I

[30] (Derrida, 1990, p. 963), quoted in (Grosz, 1999, p. 11).

have already quoted many times: "[...] structural and symbolic domains are not metaphors for 'real' violence but enact their own forms of violence" (Catley, 2003). When looking at verbal, linguistic abuse and aggressiveness, we often say, partly to reassure ourselves "we are just dealing with verbal violence", and by adding the adjective "verbal" the term violence is in some sense weakened or even erased, certainly diminished in its intensity, in such a way that violence seems to disappear. The thought that very often accompanies such an utterance is something like "ok, guys, verbal violence is not exactly 'real' violence", it is a minor form of violence that we can hardly recognize as sharing something relevant with bloody violence. In such a way we certainly stress the importance of verbal (and of structural) violence, but at the same time we forget "the violence of every violence", disregarding verbal and written abuse as proper and effectual occurrences of violence.

3.4.1 Lawmaking and Law-Preserving Violence

Let us come back to Benjamin (1978) and his seminal ideas about violence that I have already mentioned above. Benjamin acknowledges that violence is strictly linked to the concept of law. For him, law presents two violent aspects: on the one hand, in the case of a state for instance, lawmaking violence is foundational and concerns the affirmation of a new constitution or a new declaration of independence: it is violence that sees itself legitimized. On the other hand, law-preserving violence is the everyday violence perpetrated by an already founded state, it is conservative and protective.[31] It is obvious that morality shares with law the same destiny, because every moral framework has to be aggressively imposed and then preserved, through punishment. The two aspects of law are intertwined: in this sense the intertwining between morality/law and violence in capital punishment is particularly illuminating. Capital punishment seems to express and "mirror" a metaphor of "law itself in its origin", as a brute example, Benjamin says, of mythical violence, being immediately moral and immediately bloody. Consequently, there is an "ungroundedness" of law that, paradoxically, lacks "legal" foundation and "legitimation": at the same time "Human life itself is forced in modernity to bear the burden of the law's own ungroundedness" (Abbott, 2008, p. 86).

We have to note that in Benjamin the elimination of violence is seen in the perspective of a theological framework. For Benjamin humans could go beyond their constitutive link to law/violence thanks to the bloodless and expiatory powers of "divine violence" and its capacity to disrupt the workings of mythical violence with a kind of horror. Mythical violence is law-making but divine violence is law-destroying and would lead to a non-violent relation to law. Some interpreters see in this conclusion a kind of modernist messianism where the coming of the Messiah is at the same time the human remembering of the forgotten animal substratum (the denial of the *animal-in-us* is indeed the condition that supports the law's exceptionality). The Messiah would redeem humans "to" their animality. Violence generated by the law will disappear with the coming of the divine violence (that is to say that

[31] (Abbott, 2008, p. 83).

the Messiah only "expiates" and does not cause more law and violence). It would seem that Benjamin's divine violence does not "[...] represent the arrival of the divine on earth, but rather the earth's abandonment by the divine. Basically Benjamin's messianism is characterized by a kind of post-secular atheism: [...] Divine violence represents a kind of cut whereby the profane world finally separates from the transcendent" (Abbott, 2008, p. 92), or, in Agamben's words, it is "not an event in which what was profane becomes sacred and what was lost is found again" but instead "the irreparable loss of the lost, the definitive profanity of the profane" (1993, p. 102).

Beyond their abstract and intellectual aim at going "beyond violence" I have already criticized above, these thoughtful speculations gracefully reverberate something more realist than expected, if we consider that recent rich interdisciplinary research stressed the role of violence in creating, maintaining and restoring order but also in challenging, transforming and destroying order and morality. Both in peaceful and in war contexts, from Iraq to Palestine, from urban settings like Chicago and Stockholm to South Africa and Northern Ireland, from India to Rwanda to Imperial Germany (Kalyvas et al., 2008).

3.5 Ochlocracy: Violently Perverting Democracy and Its Values

In the previous section, I contended that morality and law always have a violent character insofar as – and we should not forget it – they are supposed to teach people how to behave, what is allowed, what is not allowed, when to administrate punishment, and so on. The distinction between *law-making violence* and *law preserving violence* has proved useful to stress the paradoxical role of violence in producing and/or maintaining a legal order, such as the democratic one: obviously, also democracies are founded on a more or less violent process (sometimes sanguinary) of "inauguration". In this section my aim is to look into the potentially violent character of the crowds. In section 3.5.1, I will begin by dealing with the issue related the so called "wisdom of crowds". I will set out to explore the different levels of violence characterizing the various collective forms of intelligence and, most of all, its degenerating forms. In section 3.5.2, I will broaden my inquiry on the various problems related to the degeneration of democracy pointing to a specific phenomenon related to ochlocracy that I will call the "perverting of sympathy", still a case of people's moral embubblement.

3.5.1 Wisdom of the Crowd or Violent Tyranny of the Crowd?

The most fascinating demonstration of collective intelligence or wisdom of crowds is offered by the so-called *Condorcet Jury Theorem* (Sunstein, 2005a). Basically, the theorem predicts – by a mathematical demonstration – that, when making a decision or guessing an answer, groups will do better than individuals (and large groups will do better than small ones). Probably, the most classical example of this phenomenon is the jelly-in-the-jar experiment: suppose that a jar contains a number of belly-beans and it is asked to a number of people to guess how many beans it contains. If any single guesser has more than a 50 percent probability of guessing the correct answer, then it mathematically follows that the probability of a group of guessers will necessarily increase as the group size gets larger.

The wisdom of crowds is a wonderful example of *knowledge without knowledge* that apparently goes beyond a *just luck* phenomenon.[32] The case of Thiokol[33] is certainly one of the most thrilling and surprising cases to mention in relation to group intelligence. Thiokol was one of the contractors participating in the catastrophic launch of the Challenger space shuttle, that took place at the end of January 1986. The Challenger exploded ten miles above the heads of hundreds of incredulous spectators. Immediately after the accident, investors started selling the stocks of those contractors involved in the mission. Predictably, all five contractors' stock went down from 3 to 6 percent. Something strange happened the day after the disaster: The Thiokol stock collapse pushed towards a trading halt, as Thiokol's stock was down nearly 12 percent. It doubled the loss of the day before, because Investors were blaming Thiokol for the Challenger disaster. Surprisingly, six months after, a commission instructed by the US President – headed by the physicist and Nobel Prize Richard Feynman – reported that the cause of the collapse was due to the O-ring seals produced by Thiokol. Unexpectedly less resistant in cold weather, the O-ring seals allowed the gasses to leak out, producing the explosion shortly after lift-off.

How did the investors know about the O-ring seals? The Commission reasonably took some months to give a response to the causes of the disaster. How could the stock market be so accurate and so fast in indicating the party responsible? Surowiecki pointed to the wisdom of crowds (2004).

In spelling out his claim, he contended that it was not so surprising that the stock market made the right guess; it is just that quite often we do not realize how much information a group of individuals may contain and use. Building on this intuitive remark, Surowiecki indicated four conditions underlying the emergence of collective or group intelligence: diversity of opinion, independence, decentralization, and aggregation (2004, p. 10).

1. *Diversity of opinion* is basically related to the fact that everybody should be allowed to hold any belief or any piece of information, no matter how eccentric it might look like. In fact, that particular piece of information might be the missing piece of a puzzle and so lead the group to think differently.
2. *Independence* means that one should be free to develop his ideas independently from others. Suppressing private information, simply because they contrast with what is believed by the majority, might be blocking a fundamental piece of the puzzle to emerge publicly.

[32] A recent review of collective behaviors is provided in (Goldstone and Gureckis, 2009). The emergence of collective behavior in spontaneous organizations is further detailed in (Moussaid et al., 2009).

[33] Reported in (Surowiecki, 2004).

3. *Decentralization* deals with the possibility to draw on local knowledge. This means that a person should be in the position to have valuable first-hand knowledge sources.
4. The last condition – *aggregation* – is related to having some mechanism able to turn private judgments or beliefs into a collective decision. In the case of Condorcet Jury Theorem, voting is such a mechanism, in the case of the Challenger it is the stock market.

More generally, the four conditions I have briefly surveyed feature a sort of *institutional agency* characterizing an *institutional agent*. The term institutional agent has been introduced by Gabbay and Woods[34] to indicate a kind of agent which differs from the individual (or practical) one in terms of better access to cognitive resources, better cognitive assets, better computational capacity, higher standards of performance. As I stressed in my *Abductive Cognition* (2009), one essential feature of institutional agency is the role played by the externalization of the mind.

Basically, human beings overcome those limitations related to their individual cognitive assets by distributing various functions to external objects. The products of such a process are then re-internalized and, in the emergent interaction, they grant humans additional cognitive capabilities. The institutional agency emerges as the various externalizations are made stable so that people can rely on them, but they also have to comply with them. For instance, the role played by the stock market is clearly resulting from an institutional agency: it is not simply the unreflective act of a blind crowd. The artifactual dimension of the stock market is often overlooked. In this sense, I would suggest that the *invisible* hand should rather be called the *technological* hand, considering for instance the extensive use of mathematical tools in managing the various financial products sold on the stockmarket. Investors do not just act randomly, they have received a specific training in which decades of knowledge were condensed and transferred. Universities and various institutions devoted to knowledge transmission and innovation took an essential part in the process.

Science is another example that can be described as a wise crowd, as discoveries and advancements can hardly be reduced to the work of a single scientist, especially in contemporary science. The first three conditions listed by Surowiecki (diversity of opinion, independence, and decentralization) are usually met. Ultimately, scientists are encouraged to express their ideas, no matter how eccentric they could be, as they might turn out to be correct, even destructive for the current paradigm (diversity of opinion). They are urged not to take for granted the results obtained by other scientists if it is the case (decentralization), and they are also called to develop a critical attitude towards any theory or explanation presented to the scientific community, thus avoiding the risk of blind conformity (independence). The last condition – aggregation – is of particular interest to stress the institutionalized dimension of science as a wise crowd. Science has several aggregating mechanisms, some of which are *qualitative* and some others are *quantitative*. As concerns the quantitative mechanisms, probably the most well-known in science is the so-called *impact factor*. When a scholar or scientist cites another source as a reference in an article, it

[34] (Gabbay and Woods, 2005).

is likely he implicitly *votes* for that source, meaning that it is an implicit statement he makes about its scientific reliability and relevancy. When appropriately aggregated by using particular statistical tools, all these votes may give us a quantitative measure of the impact a certain article or journal has had over a certain research community. And so, when discriminating among various publications, a researcher might give more weight to the most cited ones.

Peer-review is another aggregating mechanism characterizing science, which is, however, a little bit different from that of impact factor. As is well-known, peer-review is a self-regulation process variously employed by academic and scientific journals. As the editor of a journal receives a manuscript, it is sent to two or more reviewers for evaluation. Peer-review is both a qualitative and quantitative aggregative mechanism. It is quantitative, as the reviewer's recommendation is basically a vote in favor or not of an article being published. In case of disagreement, the editor of the journal can decide to accept or refuse an article on the basis of the votes expressed by the reviewers. It is also qualitative, as the reviewers usually provide the author(s) with suggestions about how to improve the manuscript. Indeed, the contribution furnished by peer-review as an aggregating mechanism is relevant insofar as it is a systematic procedure which is not restricted to a closed and small group of researchers, but it is a wide-spread practice among researchers, scholars, and practitioners all around the globe.

Among the qualitative aggregating mechanisms, international conferences and workshops certainly play a major part in science and academia. Although international conferences are not aimed at solving a particular problem, they certainly contribute to making scientific communities grow and develop. International conferences play a pivotal role in distributing and disseminating more or less provisional results and give the researchers belonging to a certain community the chance to expand their views and perspectives not only as individuals, but also as part of a whole.

Finally, also a new justification of democracy can take advantage of the idea of "collective intelligence" illustrated above, with the help of the concept of folk epistemology, now much studied in the cognitive science community.[35] Acknowledging that "Contemporary democratic societies are plagued with controversies that emerge from the need for a democratic political order to justify itself to a morally conflicted citizenry", Talisse asks himself about the principles justifiable to all citizens – beyond the moral conflicts at stake – that in turn would provide the justification for democratic government (2009, p. 42). The democratic legitimacy in terms of strict moral principles appears impossible when citizens are divided because of their basic, inviolable moral commitments: for example, Rawlsian "overlapping consensus" is too fragile, because the moral controversies of current democracies are too strong. To solve the problem, Talisse proposes to consider *epistemic* principles, rather than proper moral principles:

> In particular, I shall argue that there is a set of epistemic commitments that we hold in common, no matter how deeply we are divided over our moral doctrines. I refer to

[35] Cf., for example, (Mercier, 2010).

> these commitments as *folk epistemology*. [...] the term *folk epistemology* is intended to capture the epistemic practices of the man-on-the-street, the pre-theoretical and intuitive epistemic commitments that are so deeply embedded in our cognitive lives that it is the task of professional epistemologists to explain them and render them systematic (pp. 44–45).

Citizens can converge to a democratic consensus only through the systematic "epistemic" practice of a negotiation performed thanks to the explicitness of best reasons, evidence, and argument, in the framework of an individual commitment to the truth. I have to note that this of course is still a kind of *moral* commitment, a kind of preference for the truth, in the light of the tradition of science and Enlightenment, but Talisse thinks it is less likely to trigger conflicts – and therefore violence – than other more passionate and self-identifying moral commitments. This practice of negotiation is in itself an enterprise of democratic "justification", where the opponents' reasons have to be taken seriously, even if:

> [...] the folk epistemic justification of democracy does not provide a theory of legitimacy. It shows only that despite our deep moral differences we each have a reason - the same reason - for upholding democratic commitments even in the light of democratic outcomes that strike us as morally unacceptable. [...] According to Sunstein, then, the democratic state should aspire to create a "republic of reasons". He acknowledges that proper deliberation requires that citizens embody certain epistemic character-traits (p. 51 and p. 53).

The democratic *epistemically perfectionist* state is committed to reinforcing a *proper epistemic* community, as the fruit of a collective intelligence.

3.5.1.1 How a Freely Choosing Being Ends Up Being Violently Chosen

I contend that the four conditions indicated by Surowiecki more or less explicitly stress the essential tie between democracy and liberalism, on the one hand, and the emergence of particular institutional agents like market and science, on the other hand. In fact, all the four conditions are met as a society praises certain basic values characterizing liberal democracy: for instance, freedom of speech, the respect of minorities, tolerance, and some other attitudes related to the Popperian "open society" like being open-minded, resilient to change, sensitive to anomalies and willing to revise one's own beliefs.

From this perspective, I shall turn to a problem affecting our democratic societies today, that is what the democrats[36] see as the violent emergence of a degenerated and perverted form of democracy – the so-called *ochlocracy* or *mob rule*. In my view, the degeneration of democracy basically transforms the wisdom of crowds – still capable of violent acts – into the *tyranny* of crowds, which basically perverts the various institutional agents emerging and/or developing in democracy. In the next

[36] By using the word "democrats" I refer to supporters and advocates of *democracy*, and not to members or supporters of the Democratic Party in American politics, who would be indicated as "Democrats".

section I will deal more in detail with a particular phenomenon related to ochlocracy, what I will call the *perverting of sympathy*. As for now, I will focus my attention on illustrating the kind of violence emerging in ochlocracy which is basically due to the bandwagon and conformity effects.[37]

As I have described so far, wise crowds are not simple crowds. In my terminology, wise crowds are related to the emergence of a new form of agency that I called institutional agency. That is, in order to have an agency as such, some requirements should be met: diversity of opinion, independence, decentralization and aggregation. My claim is that when the four conditions are loosened, then the wisdom of crowds progressively slides into the violent *tyranny* of crowds.

According to the Condorcet Jury Theorem I mentioned above, groups do better than individuals, when any single individual belonging to the group has more than 50 percent of probability to guess the correct answer. Mathematically speaking, the larger the group is, the more it will approximate the correct answer. However, as one may easily note, the probability does not increase if the first chooses and the others simply *conform* to what the first did. Indeed, he might be right, but he might be wrong. As brilliantly put it by Sunstein "conformity is often a rational course of action, but when all or most of us conform, society can end up making large mistakes" (2005a, p. 3).

As people start conforming, the main consequence is that the total level of information available to the group simply, diminishes[38] and so does the chance to have that piece of information vital to solving the problem at stake. A scientist, for instance, may opt to support a particular theory, although he may have good reason not to, just because his community strongly supports it. He may deliberately choose to reject as erroneous the results he got from an experiment, because they do not conform to the view held by the majority of his colleagues. Indeed, lacking confidence may promote conformity, but, as the bandwagon effect becomes more influential, the price one has to pay for dissent may be very high.

Diversity of opinion and independence are clearly threatened, as conformity and bandwagon become stronger in society. However, some distinctions should be made. As Cialdini and Goldstein pointed out, the reason why one adjusts to the group may vary (2004). Even though the result is basically the same, one may conform because he is afraid of being more or less violently laughed at, for instance. Or, he takes what the majority of people think as an indicator to increase the accuracy of his decision. That can happen especially when one has poor information, and what the group thinks may be cognitively helpful. This helps us distinguish mindless conformity from other forms of conforming that are more *strategic* – so to speak – meaning that conformity is part of a strategy in which conformity is one of those elements that are pondered and taken into account to reach a final decision. In this sense, a person might eventually assume the same position exhibited by the majority of

[37] The bandwagon effect is that process in which a person follows what the majority of people does (Leibenstein, 1950). The conformity effect is that process enhancing the probability that a given behavior or trait will become common in a group or population (Efferson et al., 2008).

[38] (Sunstein, 2005b, 2007).

people. However, the process through which he arrives at that does not necessary imply the dismissing of independence and diversity of opinion.

Another interesting case worth mentioning here is when people conform to standards of behavior. I maintain that a line should be drawn to distinguish between mindless conformity and *compliance* that can be viewed as a more sophisticated form of conformity. I contend that compliance can be viewed as a disposition underlying the aggregation of information and knowledge, not merely a poor pattern of behavior as usually mindless conformity is considered. Accordingly, compliance may even promote diversity and independence of opinion. For instance, a mathematician follows – and thus conforms to – certain rules in demonstrating a theorem, not because he conforms to what the majority of mathematicians do, but because he wants to make his contribution intelligible for his group of peers. In this sense, the emergence of standards or norms of behavior is more related to compliance than conformity. This is what emerged from a recent neurological study: Klucharev and colleagues (2009) posit that compliance[39] is an error-related response connected to reinforcement learning. Therefore, compliance can be considered as a monitoring mechanism that tells us when we deviate from group norms. Such a mechanism does not simply aim at forcing us to dismiss what we think. Conversely, it facilitates communication and supports the collaborative dimension of knowledge.

I mentioned the fact that conformity and bandwagon weaken the various institutional agents emerging when diversity of opinion and independence – above all – are granted and successfully implemented in society. I maintained that, as conformity gets stronger and stronger, people dismiss their own positions and choose not to share what they know. The main effect is a general impoverishment affecting the society as a whole. Now, I will mention another relevant phenomenon emerging when conformity spreads in society, the so-called *paradox of choice* (Schwartz, 2004) which I will refer to as the perverting of free choice.

During the last decade a number of studies have been published and discussed about a problem that clearly sounds like a paradox: it is not true that the more choice one has, the better the choice is likely to be.[40] Contrary to popular belief, a limited-choice set may eventually lead customers to be satisfied, whereas extensive-choice set may even make them feel frustrated or at least less satisfied about their choice. It is worth noting that the paradox of choice usually does not affect those who enter a shop having in mind a particular product or service they want to have (Iyengar and Lepper, 2000). In that case a larger range of alternatives is praised. Why? The different perception of the opportunities one might have points to the simple fact that the choice among a larger range of alternatives requires a customer to be better informed about a number of variables. As Lurie (2004) put it, "more alternatives mean more information". That is, the experience of *choice overload* is clearly demotivating, since one lacks the information that he would need not only

[39] (Klucharev et al., 2009) actually refer to "conformity", but their description makes it much more akin to my conception of compliance.

[40] Cf. (Iyengar and Lepper, 2000; Schwartz, 2004; Jessup et al., 2009; Haynes, 2009; Fasolo et al., 2009).

to make sense of the various alternatives, but also to be able so match the available options with his preferences (Jessup et al., 2009; Haynes, 2009). In fact, usually customers report a feeling of frustration as they did not buy what they really wanted. The interesting aspect of the paradox of choice is that customers continue to praise and appreciate extensive-choice sets. Apparently, they still perceive abundance as something positive, even when they will later question the choice they made.

Frustration and dissatisfaction due to extensive-choice sets may point to a general – more or less perceived as violent – phenomenon that I call *the perverting of choice*. I contend that such a perverting mechanism may be due to mindless conformity. As mentioned above, one of the main effects brought about by mindless conformity and bandwagon effect is that people do not disclose their private information and beliefs. Conversely, they conform to what other people say or believe. Moreover, it is worth noting that at a certain point people may systematically adopt conformity not only as one of the indicators to ponder in case of a decision, but as a *habit* that is automatically picked just like any other. That is, they do not even need to make an effort to adjust their own view to that of the group or the majority, simply because they have already given up on any chance to hold a view. Such a passive attitude can be considered as a sort of adaptation to the violence and tyranny of the crowd.

As the perverting of choice takes place, one the one hand, people are still willing to have a large range of alternatives among which to choose but, on the other hand, their chance to buy something they are not satisfied with exponentially increases. As it happens, the adoption of conformity as a habit strips them of the information that would have guided them to compare the different products and formulate their own preferences. The main result of the perverting of choice is that people are decreasingly *choosers*, and increasingly *chosen*. That is, the violence perpetrated by the crowd is resolved so as to turn subjects of their own choices into objects of the choices made by somebody else.

3.5.2 Ochlocracy as the Perverting of Sympathy

Ochlocracy is defined by Polybius as the degenerated and pejorative form of democracy (1979). The thesis I will elaborate is that ochlocracy is established as a number of democratic values are simply caricatured and eventually violently dismissed. In this section I will illustrate what I call *the perverting of sympathy* as an example of deformation of democratic value.

In his *The Rise of the Roman Empire* Polybius listed some conditions under which ochlocracy arises. A key point in his analysis is that at a certain moment people simply cease to value equality and freedom that are indeed two pillars in democracy. He argued that, as they get accustomed to equality and freedom, such values simply become *transparent*, and the process of caricaturing democracy may take place. Polybius mentioned a fundamental aspect connected to this element: efforts and merits (but I would also add expertise and knowledge) are no longer preconditions to hanker after office. Conversely, it is the systematic corruption of the people – by breeding their senseless cravings – that grant some to achieve positions that would not have been possible to achieve before.

As Polybius brilliantly noted, people seek to raise themselves above their fellow citizens. Indeed, the major consequence is that the society as a whole is progressively weakened, and so the social bonds cementing the social contract. Polybius did not dig into the possible cause of such a collapse: he simply argued on a philosophy of history characterized by the cyclical nature of systems of government. However, building on his insight, I contend that there is a key element that is often overlooked, and that might help us explain the emergence of ochlocratic phenomena: in democracy people's quality of life is drastically improved. On the contrary, what has been propagandized during the last decades is a sort of distortion or a caricature of democracy. Probably, the best example comes from the former US President Ronald Reagan who once provided this moral virtue of capitalism and democracy: he said that everybody should have the chance to become a millionaire (here it is worth noting that only 1% of the population has the statistical likelihood of becoming a millionaire). In my perspective this is clearly a violent misrepresentation of democracy (and capitalism). Indeed, giving everybody the chance to become a millionaire is not a realistic goal for a democratic society to set. It just creates expectations that cannot be fulfilled. Let us see why.

I contend that, generally speaking, ochlocracy favors obliteration of the causal chain that is responsible for creating the wealth in a democratic society, and so the idea of democracy itself. It gives the illusion – bred and fed on ignorance perpetrated by aggressive and populist minorities – that it is not the democratic society as a whole that promotes the conditions for wealth to be created, but the single individual regardless of the societal context. Here I am talking about a caricature of individualism, the one according to which a person aims at raising himself above his fellow citizens, as Polybius noted. This caricature of individualism widens the gap between the extent of formal rights and the actual ability to fulfill them. According to Zygmunt Bauman, this is precisely the major cause for the emergence of resentment as a psychological and social phenomenon. The excitement and craving for success and power is soon replaced by the humiliation to see that what has been desired for a long time is simply unreachable (Reagan's dream regards very few). Those who are humiliated develop a strong cognitive dissonance[41] that, as Bauman put it, casts what one wants "as simultaneously desired and resented, elevating and degrading" (2009, p. 5).

Cognitive dissonance is what best characterizes the attitude of people who hold resentment towards democracy. Let me develop this point: democracy is the kind of government that also grants everybody the chance to be part of a higher level of the hierarchy in comparison with their parents. That is, the place one holds in society is not legally determined by birth status, but by some form of competitive process in which everybody has the right to compete *on equal terms*. Usually, high social

[41] On the concept of cognitive dissonance see also above subsection 3.2.1 and subsection 5.1.1 in chapter five. See also above, subsection 3.2.2, my analysis of (Burton, 2008) and some observations about beliefs in general.

mobility is taken as a sign attesting to the prosperity of a society, so it is commonly regarded as a value to pursue at every cost. Interestingly, some people seem to develop dissonant feelings towards this particular ideal. As I have already said, thanks to a related demagogic overemphasis on "individuality", ochlocracy favors the obliteration of the causal chain that is responsible for creating the citizens' wealth in a democratic society, and so the idea of democracy itself. Such a "morality of the centrality of the individual" (presented as a new, good, introduction) generates the following various effects. On the one hand, those people tacitly know that they were given a chance to choose their own way in life thanks to democratic virtues like the equality of opportunity. They owe so much to democracy. On the other hand, they narcissistically and presumptuously overvalue themselves and their contribution to society.[42] That is, they simply obliterate the premise according to which it was *democracy* that gave them a chance to improve their social status, and this should not be taken for granted. Various caricaturing forms of individualism emerge and eventually lead these resentful people to think that they are basically *better* than most everyone else, and that they did not get what they really deserved.

Cognitive dissonance and resentment trigger envy – exactly in the sense I already illustrated in this book, referring to Girard – towards their fellow citizens. Interestingly, this process is activated no matter what the envied individual's status is, whether it is lower or higher than the subject's. This is indeed quite paradoxical, because usually the feeling of envy arises when a person, in comparison with another, lacks something that she desires, be it power, money, beauty, etc. To put it brutally, they may even come to envy their maids. My contention is that such envy results from what I call *the perverting of sympathy*. I am now going to illustrate how the violent perverting of sympathy is connected to cognitive dissonance.

It is well-known that one of the moral principles describing the very essence of liberal democracies is that one is to imagine himself in the position of everybody else in society. Also known as the "veil of ignorance" (Rawls, 1971), this principle is supposed to preserve fairness and impartiality when discussing how to interpret what the common good is. Adam Smith in his *The Theory of Moral Sentiments* referred to sympathy as the ability to imagine oneself as another person. As he put it "we can form no idea of the manner in which they are affected, but by conceiving what we ourselves should feel in the like situation" (2010a, p. 8). In my view, the

[42] This is connected to overconfidence: people simply fail to recognize their incompetence and limitations. Overconfidence is usually triggered by ignorance (Dunning et al., 2003). That is, it is very likely that those who are not particularly good at performing a certain task (intellectual or not) will overestimate their accuracy in doing it. Accuracy does not necessarily follow from confidence. Quite the contrary, I contend that one's ignorance is immediately transformed in *ignorance of ignorance*. This is more likely to happen when one is *utterly* ignorant: utter ignorance *embubbles* a person making him immune not only to a particular piece of knowledge, but any. In doing so, it is his ability to know in general that is blocked or, at least, weakened. For a formal analysis of ignorance of ignorance, see for instance (Kamitake, 2007).

perverting of sympathy is a process of perspective-taking that turns out to be a means to separate people, not to bond them together.[43]

It is worth noting that the perverting of sympathy is not to be viewed as a failure in perspective-taking, properly speaking. Usually it is said that the main obstacle to sympathy is to overcome egocentrism and one's own current state (Epley and Caruso, 2009). I contend that resentful people do not fail in doing that: they fail in another respect. By definition, adopting another's perspective requires a person to get over his own in order to infer how the other person looks at the world. That is, a person must imagine first how he would feel, and then try to imagine himself as another person. But what if one fails to get over his own perspective? I posit that this is precisely the case of perverted sympathy. Resentful people – caught in the trap of cognitive dissonance – give up on their perspective and, of course, identity. Consequently, they start looking at the world through the enemy's eyes – the other's – not their own (I will describe this process as the "paradox of the insider enemy", cf. the following section). The main consequence of such a process is that resentful people simply fail to acknowledge the virtues of democracy, as the perverting of sympathy makes them only perceive its weakness just like an enemy of democracy would see it. Looking at democracy through its enemies' eyes does not mean developing a critical attitude that might expose some negative aspects that democratic society certainly has. Conversely, it is meant to give a grotesquely deformed representation of what democracy is. Even though resentful people blame, for instance, the so-called *rogue states*, which are depicted as a threat for our Western values, they usually call for a stronger government: they want a government no longer restricted by those democratic procedures that are viewed as pointless, futile, and baroque – just an obstacle to justice. In sum, the perverting of sympathy contributes to spreading throughout the population a caricature of democracy that can only be suitable for *chauvinist buffoons* and *ochlocratic clowns*.

I contend that this is a response to cognitive dissonance. Let me illustrate this point: by definition cognitive dissonance arises whenever a person simultaneously holds two contrasting cognitions. This is the case of the aforementioned people. As

[43] A terminological note is needed here. I use the term sympathy rather than the term empathy, as the former puts more emphasis on the *imaginative* role involved in perspective-taking. Whereas the latter stresses more the embodied and simulative dimension of the social mind. In fact, it is worth noting that during the last decade or so an exponentially growing body of research has appeared on the neural mechanisms underlying empathy, namely, the so-called mirror neurons. Originally discovered in monkeys (Rizzolatti et al., 1988), mirror neurons are a system of neurons firing *both* when a person observes an action performed by another individual and when he executes the same or a similar action, like grasping, tearing, holding or manipulating objects. Such a neuronal structure provides a clear example of sensorimotor brain parts performing a resonance task devoted to both performing and *understanding* actions (Tummolini et al., 2006; Jeannerod and Pacherie, 2004; Anquetil and Jeannerod, 2007). The discovery of mirror neurons has given rise to speculation not only about empathy but also on various aspects of social cognition: intentions, joint action, mind-reading, emergence of language (Gallese, 2005, 2006).

I have already said, some people experience a strong cognitive dissonance, as they could not reach what they wanted to have but thought they deserved to have. The desired value or object (whether it is money, power, etc., does not matter) is at the same time elevated and degraded. The same attitude is held towards democracy, as described above: on the one hand, democracy is what gives them the chance to acquire a better position in society – better than their parents' or grandparents'; on the other hand, they regret that they could not raise themselves above their fellow citizens, as democracy commands that everybody can compete on equal terms.

As cognitive dissonance appears in one's mind, it tries to reduce the tension between the two views. Usually cognitive dissonance is reduced by means of oppressing one of the two views. This objective can be reached, for instance, by obliterating facts supporting one view and fabricating appropriate evidence justifying that suppression in favor of the other. My take is that the perverting of sympathy described so far is a mechanism meant for resolving the tension generated by cognitive dissonance. More precisely, looking at democracy through the eyes of its enemies enables the resentful to obliterate the virtues of a democracy that is, in turn, represented as weak and corrupted, which is usually the way Western democracies are depicted by some anti-Western radicals. From a cognitive perspective, the perverting of sympathy functions like a *moral* and *cognitive mediator*, which should apparently help broaden one's perspective. In reality, it simply distorts the reality rendering both the moral and cognitive character of the resentful individual like that of a *tragic clown*.

3.6 Beyond 9/11. Democrats, Speak Out! Would You Like to Be Terrified or Deterred?

Finally, given the fact I live in a world where democracy is perceived as threatened, this last section refers to various violent processes that affect its prosperity or even survival. Various processes of moral embubblement are involved, where dissimulation of violence or unawareness of being violent are still at play: the case of 'the paradox of the insider enemy" (of democracy) is an illuminating example.

3.6.1 The Widespread Violence against **Rechtsstaat** *and Democracy*

Among the first betrayers of democracy there certainly are those addicted to the "free market", who have been raging against every kind of welfare in the entire Western world over the last decades. Before coming back to other more basic problems which affect democracies and incline them to ochlocracy I need to say a few words about this violent "free market" obsession, which has become a true challenge to the survival of modern constitutional democracies.

On the violent attacks against democracy, perpetrated by the stupidity – more or less gravid of violent outcomes for everyone in our globalized world – of economists, politicians, business men, journalists, etc., rude, aggressive, and tireless upholders of the "free market", it is interesting to mention the recent book by Smith, which is eloquently subtitled "How Unenlightened Self Interest Undermined Democracy and Corrupted Capitalism" (2010b). For example the book stresses that mainstream economists assumed what I would define as an *idealizing* and *abstracting* attitude towards the free market. Even though such an attitude is characteristic of the work of scientists in devising their models,[44] it turned out to be a specific form of what researchers in informal logic started to technically call "bullshitting" (Carnielli, 2010),[45] related to a galaxy of biases (i.e., confirmation bias, bandwagon, etc.), which only the recent financial crisis has blatantly put on display. This form of bullshitting resulted even more violent, as it was actively supported and encouraged not only by politicians, but also by intellectual elites. They successfully managed to dupe the public opinion taking advantage of the influence economists (as experts) usually exert on the population.

Smith brilliantly illustrates how the "fundamentalists" of the free market were, and still are obsessively and tirelessly more attracted by their blind trust in that sacred myth (which, to be clear, for example considers "government action of any sort beyond bare minimum as 'interference'. That vision is in fact close to anarchy" (Smith, 2010b, p. 106)) than to the protection of the suitable and contemporary political conditions that have allowed the emergence of the free market: chiefly, they displayed utter disregard for the principles permitting the market as we know it, that can be traced down – I would say – to *democracy*. Because of this attitude, the free market was sold and democracy undermined, as the current economical crisis and the democratic quality of political governments of the Western world vividly show. Unfortunately it is going to take years before the masses realize this, maybe decades, as for now in our current – more or less "ochlocratized" – countries it is just the sophisticated (and so highly fastidious) awareness of marginalized elites.

Let us come back to another, more explicit, challenge to democracy, of course intertwined with the effect of "free market" fundamentalists that I have just outlined. The "perverting of sympathy" I have illustrated in the previous section can be broadened thanks to new considerations that take advantage of what I call the "The Paradox of the Insider Enemy"[46]

We can start by dealing with the problem of xenophobia in Western democracies. Do we have to prefer the violence embedded in anti-xenophobic attitudes or the violence of being xenophobic? Some doubts can arise, but I confess that I finally choose the first as a lesser evil: that is I prefer the type of violence embedded in a democratic attitude and in tolerant liberalism. Currently the "political mind" of an increasing number of citizens of liberal democracies is – unfortunately – becoming xenophobic, as a reaction to the effects of globalization such as the free circulation of people and workers all over the world and the consumerism of the planetary free

[44] (Woods and Rosales, 2010).

[45] Cf. chapter four, section 4.6.

[46] Cf. below subsection 3.6.3.

market.[47] Furthermore, they consider democracy weak in its ways of self-defense and self-protection. It seems that citizens who possess that "political mind" have forgotten that

1. globalization is a product of post-capitalism, a recent development following the attitudes of liberal tolerant constitutional democracies and certainly not due to those political regimes affected by rudimentary more or less violent xenophobic policies;
2. xenophobia is a cardinal aspect of archaic social groups, collectives, and states, of economically undeveloped social organizations, of fascist or despotic regimes, and so on;
3. the tolerant liberalism of constitutional democracies "is not weak". It is not at all exempt from violent aspects and certainly it is not so vulnerable as they happen to think: it exerts punishment against out-group (often themselves xenophobic) people and condemns and punishes inside xenophobia as illegitimate both from a moral and legal point of view, together with many other archaic violent behaviors; democracies make terrible wars against other states showing all their might; globalized corporations which are present in democracies are extremely aggressive, etc.[48]

In the previous subsection 3.5.2 I already observed that citizens start looking at the world through their enemies's eyes – the other's – not their own. Let us further explore this process taking advantage of some considerations about the concepts of *Rechtsstaat* and of constitutional democracy. Citizens of current democracies who adopt xenophobic or racist attitudes against people that are already xenophobic in themselves, for example for fundamentalist religious reasons, *perversely* manifest an implicit identification with the targeted object, because they are triggered by a kind of mimetic envy, in Girard's sense (1977; 1986). Such a citizen seems to think: me too, I would enjoy being like them, I would like to avoid being bored and enchained by the constraints of *Rechtsstaat* and democracy. In this way, those citizens paradoxically renounce the *violence* of liberalism, which forbids xenophobia and punishes xenophobic attitudes, and they finally manifest a preference for *another violence*, the one embedded in the xenophobic morality/mentality of the targeted group. From the perspective of a defendant of democracy like me this is a very sad transformation, bearing a number of violent outcomes.

---

[47] On violence promoted by the new forms of crime typical of the dark side of globalization, and the related new legal challenges imposed by the need of global justice and innovative international or supra-national institutional arrangements, cf. (Letschert and van Dijk, 2011). Human trafficking for sexual purposes, organized crime/corruption, victims of terrorism and of the privatization of wars, global corporate crime, cyber piracy, and cross border environmental crimes are eloquently described.

[48] Balibar (2005) provides further sensible information on the following central problems of contemporary societies: the question of access to national and international rights, the current crisis of popular sovereignty and democracies in Europe, the politics of human rights and its complicity with forms of mass violence and the reaction of liberal democratic nation-states to the collapse of the Soviet Union and the consequent complications.

As for the citizens of current democracies, the ideological/political side of this identification imbues them with an implicit preference for vague forms of political regimes echoing a barbaric antiquity, where factions and hordes dominate, or – at the best – it lures them into favoring the superiority of dictatorships of any kind. Of course, we know perfectly well that these forms of social order are both characterized by the absence of *Rechtsstaat* and, all the more reason, of democracy. Moreover, in this perspective, the mental state of the xenophobic citizen inevitably also shades in the annihilation of every "internal" opposition, an annihilation which in turn characterizes what is called the "Fascist state of the mind", as I will describe in chapter five. In presence of these massive transformations of mentality, the result is that democracy quickly becomes infected by too much demagogy, and often tends to mutate in an obscene *ochlocracy* (cf. the previous section), more or less fascistically dominated by that kind of violence which is typical of a despotic group. The recent cases of G. W. Bush's presidency in the US and of the prime minister Silvio Berlusconi in Italy eloquently illustrate the propensity of modern democracies to degrade into ochlocracy: the former, with his insistence on the myth of the anti-intellectual cowboy president, delivered the liberty-suppressing Patriot Act and stuck the country in a long, costly and rather unjustified war, as it later proved to be based on false allegations. The latter does not take care to keep a low profile as regards his being a media tycoon – bearing by an embarrassingly self-evident conflict of interest – and teases (taunts?) all other institutional forces with a constant populist attitude, which much pleases his supporters.

3.6.2 What Westerners Want (When They Do Not Like Constitutional Democracies Anymore)

It is easy to answer: for example they prefer the violence of being *terrified* than the mere violence of being *deterred* through punishment. If I dislike democracy I must at least make the effort and imagine, for instance, what punishment becomes in a non-democratic regime. Hildebrandt (2010, p. 168) efficaciously notes that

> If I have no idea how my behaviour will be qualified in a court of law I will not be deterred from violating the criminal law for the simple reason that I cannot determine how it relates to my actions. Instead of deterred I may be terrified, due to the uncertainty of how I will be judged. This may even paralyse me into refraining from any decision on the exception that I may still feel is at stake. I may thus abstain from taking measures I actually deem necessary, or act out of a sense of necessity without having a clue as to whether this action will be evaluated as wrongful at a later point in time. If the only criterion of deterrence is its effectiveness, as in utilitarian theories of punishment, the punishment of "wrongful" interventions under such uncertainty must be rejected. As Schmitt would have it, a state of emergency is beyond "wrong" and "right" even if it may produce a new understanding of what is "wrong" and what is "right".

The case of the state of emergency, which has become a dominant paradigm for governing in contemporary politics of constitutional democracies, furnishes good teaching, marked by the potential disappearance of *legal certainty*, as well as of *justice* – equality before the law – and *purposiveness* of the legal norms. Agamben (2005) usefully notes that in Guantánamo (as an evidence of what he calls "Bare Life" production under the sovereign exception) the inmates did not enjoy the protection either of the current criminal law or of the laws referred to the prisoners of war: a kind of absurd legal void, perverse in the light of democratic mentality. To give another example, think also of the permanent and daily attack basically performed by current Italian Prime Minister, with the complicity of a strict despotic governmental pack, against the Republican Constitution, an attack canonically justified in terms of "necessity" and "emergency".[49]

Through the state of emergency the creation of an embarrassing and dangerous legal void is often also produced. These aspects are more or less present in both the state of emergency and in non-democratic regimes, where instead the democratic rule of law is not present: its role is played by the supremacy of a kind of "living law", which coincides with the king, the dictator, the oligarchy or simply to the authority at stake in certain political (but also bureaucratic) situations. The case of the state of emergency is of course a case of *law-making violence*, which contrasts the *law-preserving violence*[50] typical for example of the daily self-sustaining policing activity of a living democracy. Both states of emergency and dictatorships involve to various degrees an undetermined power of "police" on the part of government, which cannot be limited and so coincides with a self-legitimated, violent law-making enforcement. Hildebrandt (2010, p. 168) further comments that

> The suggestion that the powers of the sovereign are necessarily unlimited is also flawed in as far as this is concluded after comparing sovereign power with the force of law, suggesting that because law is a matter of strict definition it cannot limit a sovereign that defies definition. [...] However, just like "police power" the "law" is fundamentally underdetermined. Legalist understandings of law hide the intricate relationship between norm and decision (normativity and authority) that in fact indicates that in the end the sovereign depends as much on the "law" as the "law" depends on the sovereign. Whoever tries to determine the priority of chicken or egg here, misses the point.

Let us come back to the problem of the state of emergency. We know perfectly well that recent terrorist attacks and the various critical economical or social effects of globalization (such as the immigration towards the rich Western EU of the *Rom* people) have suggested to some "democratic" governments the promotion of various forms of state of emergency, which suspend aspects of the existing jurisdiction. In this case problems with legal punishment arise. I strongly agree with Hildebrandt: "In my opinion retributive theories of punishment would have a serious problem with the idea of an emergency that creates – or allows the creation of – new legal norms. As long as retribution depends on a pre-existing legal order it is difficult to hold government officials responsible for interventions undertaken when the legal order is breaking down" (p. 169).

---

[49] There is a vast amount of literature concerning the problem of the state of emergency, I will just refer the reader to the titles listed in (Hildebrandt, 2010) and to the book by (Agamben, 2005).

[50] See above subsection 3.4.1.

Punishment, in a democratic order, should communicate censure, so re-establishing the force of the violated legal norm. This is a regulatory standard that can be brought before a court of law to re-establish the relationships and mutual trust between citizens who share the jurisdiction (as potential victims and offenders). In the state of emergency the modern "monopoly of violence" on the part of the *Rechtsstaat* is – more than obviously – at stake and, like Schmitt (2005) notoriously contended, the juristic order tends to be reduced to a simple *decisionism*, which can jeopardize the pre-existing law, in our case that of a constitutional democracy.

It is also extremely obvious that, to defend the "law" of a constitutional democracy, a particular *moral* option is necessary. This is just the one which is lacking in the perverted case of our democratic citizens, so fragile and perplexed about democracy. They also appear to be more or less convinced of still being democratic but at the same time *de facto* sponsors of the suspension of democratic order. Paradoxically, they often say they just want to defend democracy. In sum, they create a possible failure of democracy, the very same failure of democracy they see as being perpetrated by some enemy. The new suspended emergency order is seen as being "appropriately" violent and powerful and thus the only one which can protect the democracy which is so weak and fearful: of course, they do not notice they are the ones making democracy weak by jeopardizing its force and distrusting its rules! They – to address attention to the psychic content – construct a psychic analogous of a state of emergency, a fearful "emergency mind" which constantly jeopardizes their "democratic mind".

The challenges to democracy and to *Rechtsstaat* I have just illustrated can also be usefully seen from the point of view of political philosophy, taking advantage of some suggestions provided by Balibar (2005). First of all Balibar contends that we do need to democratize the institution of national borders for reciprocal and multilateral negotiations for the regulation of migration. Indeed Balibar contends that the violent production of immigrants and refugees – closely related to an increasing obsession with identity – through wars, embargoes, incautious support of primitive despotic regimes, no less than through forceful "humanitarian" interventions, has instantiated a European advancement of violent quasi-apartheid structures. I think this consideration by Balibar can be indirectly related to those suspensions of democratic order I have just illustrated. Furthermore, greater collective security should be achieved by multiplying the modes of translation used to mediate inter-communal objectives, rather than by violently monopolizing otherwise diverse social forces through biopolitical, national, religious, and other strategic alignments. These social forces form a unitary mode of power that is compelled to continuously defend itself in often violent ways just as much from interior as from exterior pressures that threaten to disintegrate it. It is on the basis of this analysis that Balibar questions – in order to overcome present widespread multifarious violence – the urgency of reshaping the equilibrium among citizenship, European unification, nationalism and the politics of globalization, and the relationships between national and international law. In the last case the additional problem is to prevent transnational crises (which are still crises of proliferating nationalism – and of violent ethnical or racial nationalism – but are also related to transnational forms of mass violence, and to their mixture with the violent aspects of globalization which generated a pressure to denationalization). In sum, democracy should be enhanced and protected by a collective dynamic which is separate from the obsessive nationalist political framework, a dynamic he considers as a kind of political "civility".

3.6.3 The "Paradox of the Insider Enemy"

I have just said that believing in, defending, and preserving democracy is certainly an individual *moral* option we can undertake or not. Performing this moral task is always related to various violent actions due to conflicts, punishments, quarrels and so on, for example against xenophobic people. Let us just come back to our democratic citizen who became xenophobic, as a reaction to the effects and offenses (perceived as an intolerable violence) of globalization and terrorism and who ends up conceiving himself as the enemy he is targeting. I said that the mechanism is perverse from the psychological point of view: it is also *paradoxical* from a rational perspective. Let us call it the "The Paradox of the Insider Enemy", that we can see at work in many current constitutional democracies. Here are some notes about it.

1. The paradox involves an analogous one related to violence: I have said that those "democratic" citizens, who do not trust democratic liberalism's ability to perform *violence* anymore (both defensive and offensive), end up preferring *another* violence, the one embedded in the xenophobic morality/mentality of the targeted group. Very sad, for democracy and its civilizing project!
2. In the case of these disconsolate citizens of current democracies, the ideological/political side of the identification I described above leads them into accepting an explicitly degraded "moral" agreement with the culture of emergency and dictatorships (often trapped in the archaic morality of honor culture[51] and thus of resentment and revenge), contending that the power of the sovereign cannot be appropriately limited or confined. The power is no more seen as structurally intertwined with the rule of law and it – perversely – appears beneficial because is free from the blunt constraints of democracy, where instead legal norms and legal decisions cannot be separated. We can call those citizens *proto-enemies* of democracy: citizens who unawarely, aim at dismissing their democratic membership.
3. The urgency of a presupposed "necessity", which emerges from fearing the enemy (for example the religious fundamentalists, self-defined as *different* from us, or the Chinese enterprises and corporations, which inflict losses to Western capitalism) builds up their "emergency minds", inclined towards authoritarian and/or totalitarian ideologies [we said that a current practical effect is often a kind of ochlocracy, imbued with feelings of hatred. Hildebrandt say, analogously, that the practical effect could be a "reaction [which] implies revenge and could lead to a feud on a global scale, in the end annihilating the entire global order" (2010, p. 178)]. Here the paradox still consists in the fact that

[51] Cf. chapter one, section 1.2, this book.

the "emergency minds" contend – in a more or less sincere way – that they are "defending" democracy![52]

4. In the "emergency mind" the democratic culture and mentality is suspended, the adversary is considered as an enemy of constitutional democracy, who has to be treated in a *special* way. Furthermore, the new "emergency mind" of our former-democratic citizen also transforms itself in what I called a proto-enemy of democracy. Consequently, the alien *enemy* inside democracy can be violently tortured, or even killed, using rules that fall outside of the current democratic criminal liability (normally related to the state's monopoly of violence). The alien enemy is seen as not "included" in the democratic system of the fellow citizens "who share jurisdiction" (Hildebrandt, 2010, pp. 175–176). He is outside of the democratic social contract. Unfortunately, also the xenophobic citizen who treats him as an enemy has become similar to his foe, a "proto-enemy" of democracy.
5. Thus, the targeted people (for instance those who commit acts of terrorism) are no longer fellow citizens but only enemies, that is, they are *special* criminals, to whom the standard criminal jurisdiction cannot be applied. I repeat, unfortunately, that the new "emergency citizen" has become himself a proto-enemy of democracy. The reader should also note that his mental process has even led him to consider himself as a *special* person, a *special* and *unique* democratic person, vastly different from other *normal* fellow citizens, violently regarded as too weak, imperfect, and fearful! Paradoxically, he is utterly blind to his having become an individual who is potentially outside of the democratic social contract, who can contribute to obstructing the return to the ideal translucency of the constitutional democratic order. Sadly, we know he cannot do this, because he is trapped in the "paradox of the insider enemy"! The same happens in the case of "emergency minds" and "emergency authorities": "After all, if we wish to hold government authorities accountable for their interventions during the state of emergency, we need to acknowledge that in as far as they have willingly offended basic tenets of constitutional democracy they have caused harm to the social and legal infrastructure of communication that constitutes the democratic constitutional state" (Hildebrandt, 2010, p. 176).

[52] It is worth noting here that some aspects characterizing political activism might lead to the undermining of democracy and its virtues. Historically speaking, activism emerged from the so-called *counterculture*. The counterculture movement was a very heterogeneous social phenomenon that attempted to reject the status quo and openly criticized the mainstream culture which was seen as corrupting and alienating (Swartz, 2010). Activism spread questioning the traditional forms of political mediation characterizing democratic institutions. Parliaments and politicians were no longer considered effective or trustworthy in fulfilling people's ideals, as they just contributed to corrupting the system. Such an attitude still characterizes the activist mindset. Although motivated by a sincere commitment to justice, I maintain that activism should be reconsidered in the light of the insider enemy paradox. That is, in rejecting some very basic forms of democratic mediation, it seems that activists pervert their commitment failing to acknowledge that the kind of violence perpetrated by democratic institutions is still better than other undemocratic ones.

6. In the end, to avoid the "Paradox of the Insider Enemy", and its archaic violent consequences, we just need to prefer the legitimate violence of the rule of law of the democratic order. This turns into a firm *moral* option, a matter of personal preference and courage able to recognize various zealous dreams as illusions: "Dreaming about locking up communities is simply illusionary in a world that depends on mobility and transnational exchange" (p. 178).

3.7 Bad News from Evolutionary Game Theory

It has to be said that recent computational studies taking advantage of methodologies from *evolutionary* game theory suggest that ethnocentrism/xenophobia, usually thought to rely on complex social cognition, may arise through biological evolution in populations with minimal cognitive abilities (Kaznatcheev, 2010). Indeed, given the fact ethnocentric agents differentiate between in-group and out-group partners, and adjust their behavior accordingly, they are more cognitively complex than the so-called humanitarians and selfish agents, only capable of cooperating or defecting respectively. These studies connect a fitness cost to this complexity to examine and test the robustness of ethnocentrism. The research concludes that ethnocentrism is not robust against increases in cost of cognition. However, the model also confirms that humanitarians are suppressed largely by ethnocentrics. It results that the proportion of cooperation is higher in worlds dominated by ethnocentrics. These are more effective than humanitarians in suppressing free-riders.[53]

Moreover, of course we can see that both free-riders and humanitarians are considered by the ethnocentrics as kinds of betrayers and so both become their target: this leads to the conclusion that suppressing free-riders, such as selfish and traitorous agents, allows ethnocentrics to maintain higher levels of cooperative interactions. Sadly this abstract and interesting research seems to imply that xenophopic morality, with its violent character and the related punishments, turns out to be a guarantee of the morality of cooperation. Reciprocally, when humanitarians increase, so too do selfish agents, and the decrease in overall cooperation caused by higher levels of selfish agents exceeds the increase of cooperation caused by the huge cooperation humanitarians undertake across groups. To conclude, as I have already said in chapter one, it seems that in the light of these – fortunately – very abstract results, the xenophobic morality, with its obvious highly violent outcomes, is paradoxically a promotion and a guarantee of the morality of cooperation. Computational evolutionary models might be extremely powerful in predicting the emergence of certain dynamics – at a social and cultural level – and tell us a great deal about the conditions under which a certain behavioral pattern can be positively selected within a population. However, given the fact it is unlikely that human socio-cultural systems (and morality) could be entirely described as Darwinian ones (and also as loosely Darwinian)[54] what we actually see happening usually deviates from what is predicted by computational evolutionary models. That is, they basically apply more or

[53] This hypothesis is still the object of controversy, cf. (Shultz et al., 2010).

[54] On loosely Darwinian models cf. (Magnani, 2009, chapter six).

less Darwinian algorithms to a domain that is basically not Darwinian. It seems to me that computational evolutionary models suffer from the same epistemological vices I described in chapter one,[55] where I questioned the applicability of evolutionary models to various aspects of cognition and culture.

[55] Cf. subsection 1.3.4.

Chapter 4
Moral and Violent Mediators
Delegating Ourselves to External Things: Moral/Violent Niches

4.1 Cognitive Niches as Moral Niches

4.1.1 Human Beings as Chance Seekers

The concept of cognitive niche is useful to frame morality and violence in a naturalistic perspective. The first sections of this chapter aim at deepening our understanding of this concept, taking advantage of an evolutionary framework that is ideally linked to the considerations I have provided in chapter one, focused on the role of coalition enforcement in illustrating violence as a natural (animal and human) behavior.

Human beings usually make decisions and solve problems relying on incomplete information (Simon, 1955). Having incomplete information means that 1) our deliberations and decisions are never *the best* possible answer, but they are at least *satisfying*; 2) our conclusions are always *withdrawable* (i.e. questionable, or never final). That is, once we get more information about a certain situation we can always revise our previous decisions and think of alternative pathways that we could not "see" before; 3) a great part of our job is devoted to elaborating conjectures or hypotheses in order to obtain more adequate information. Making conjectures is essentially an act that in most cases consists in manipulating our problem, and the representation we have of it, so that we may eventually acquire/create more "valuable" knowledge. It is obvious that a great part of human conjectural activity is devoted to guessing moral hypotheses about situations and events able to help subsequent decisions and actions. Conjectures (and thus "moral" conjectures) can be either the fruit of an abductive selection in a set of pre-stored hypotheses or the creation of new ones, like in scientific discovery.[1] In order to make conjectures, human beings often need more evidence/data: in many cases this further cognitive action is the only way to simply make possible (or at least enhance) a way of reasoning that relies on "hypotheses" that are often hard to produce successfully.

[1] See chapters one and two of my book on abductive cognition (Magnani, 2009).

L. Magnani: Understanding Violence, SAPERE 1, pp. 115–169, 2011.
springerlink.com

Consider, for instance, diagnostic settings: often the information available does not allow a physician to make a precise diagnosis. Therefore, she has to perform additional tests, or even try some different treatments to uncover otherwise hidden symptoms. In doing so she simply aims at increasing her *chances* of making the appropriate decision. There are plenty of situations of that kind: for example, scientists are continuously engaged in a process of manipulating their research settings in order to get more valuable information, as I have illustrated in a previous book.[2] Most of this work is completely tacit and embodied in practice. The role of various laboratory artifacts is a clear example, but also in everyday life people face complex situations which require knowledge and manipulative expertise of various kinds – no matter who they are, whether teachers, policy makers, politicians, judges, workers, students, or simply wives, husbands, friends, sons, daughters, and so on. In this sense, human beings can be considered *chance seekers*, because they are continuously engaged in a process of building up and then extracting latent possibilities to uncover new valuable information and knowledge.

The idea I will try to analyze in depth in the course of the first sections of this chapter is the following: as chance seekers, humans are *ecological engineers*. Not only technologies and other artifacts are part of this ecology but also morality and, of course, violent modes of problem-solving. That is to say, humans (like other creatures) do not simply live *in* their environment, but they actively shape and change it while looking for suitable chances. In doing so, they construct *cognitive niches*[3] through which the offerings provided by the environment in terms of cognitive possibilities are appropriately selected and/or manufactured to enhance their fitness as chance seekers. Hence, this ecological approach aims at understanding cognitive systems in terms of their *environmental situatedness*.[4] Within this framework, "chances" are that kind of "information" which is not internally stored in memory or already available in an external resource, but that has to be "extracted" and then *picked up* upon occasion.

It is well-known that one of the main forces that shapes the process of adaptation is natural selection. That is, the evolution of organisms can be viewed as the result of a selective pressure that renders them well-suited to their environments. Adaptation is therefore considered as a sort of *top-down process* that goes from the environment to the living creature.[5] In contrast to that, a small fraction of evolutionary biologists have recently tried to provide an alternative theoretical framework by emphasizing the role of niche construction.[6]

According to this view, the environment is a sort of "global market" that provides living creatures with unlimited possibilities. Actually, not all the possibilities offered by the environment can be exploited by the human and non-human animals populating a peculiar environment. For instance, the environment provides organisms with water to swim in, air to fly in, flat surfaces to walk on, and so on. However, there are

[2] (Magnani, 2001).

[3] Cf. (Tooby and DeVore, 1987) and (Pinker, 1997, 2003).

[4] Cf. (Clancey, 1997) and (Magnani, 2005).

[5] (Godfrey-Smith, 1998).

[6] Cf. (Laland et al., 2000, 2001; Odling-Smee et al., 2003).

no creatures able to take full advantage of all of those possibilities.[7] Moreover, all organisms try to modify their surroundings in order to better exploit those elements that suit them and eliminate or mitigate the effect of the negative ones.

This process of *environmental selection* (Odling-Smee, 1988) allows living creatures to rebuild and shape "ecological niches". An ecological niche can be defined, following Gibson, as a "setting of environmental features that are suitable for an animal" (1979). It differs from the notion of habitat in the sense that the niche describes *how* an organism lives its environment, whereas the habitat simply describes *where* an organism lives.

In any ecological niche, the selective pressure of the *local* environment is drastically modified by organisms in order to lessen the negative impacts of all those elements toward which they are not suited. Indeed, this does not mean that natural selection is somehow halted, rather, this means that an adaptation cannot be considered only by referring to the agency of the environment, but also to that of the organism acting on it. In this sense, animals are ecological engineers, because they do not simply live their environment, but they actively shape and change it (Day et al., 2003).

4.1.2 Cognitive Niches as Carriers of Aggressiveness

It is important to clarify the concept of cognitive niche that is at the basis of the possibility to grasp human moral and axiological systems in a naturalistic way, and the intertwined violence, which in this perspective still appears in all of its "banality". The recent book by Odling-Smee, Laland and Feldman (2003) offers a full analysis of the concept of cognitive niche from a biological and evolutionary perspective. "Niche construction should be regarded, after natural selection, as a second major participant in evolution. [...] Niche construction is a potent evolutionary agent because it introduces feedback into the evolutionary dynamics" (Odling-Smee et al., 2003, p. 2).[8] By modifying their environment and by their affecting, and partly controlling, some of the energy and matter flows in their ecosystems, organisms (not only humans) are able to modify some of the natural selection pressure present in their local selective environments, as well as in the selective environments of other organisms. This happens particularly when the same environmental changes are sufficiently recurrent throughout generations and selective change: "Even though spiders' webs are transitory objects [...] the spiders' genes 'instruct' the spider to make a new one" (Odling-Smee et al., 2003, p. 9). The fact that spiders on a web are exposed to avian predators suggests that webs can be a source of selection that

[7] In a way, it can be argued that, thanks to material culture, human beings have managed somehow to take advantage of most environments on Earth (and outside of it), but it is a partial success which requires a continuous implementation of resources and knowledge in order to maintain those achievements as persistent.

[8] Attention is drawn for the first time to the idea of niche construction by important researchers like Schrödinger, Mayr, Lewontin, Dawkins, and Waddington. Firstly in the field of physics and subsequently in the field of the theory of evolution itself. Waddington particularly stressed the influence of organism development.

produces further phenotype changes in some species, such as the marking of their webs to enhance crypsis or the creation of dummy spiders probably to divert the attention of the birds that prey on them. Hence, also spiders adopt what humans call *cheating* and cognitively alter their cognitive niches to this aim. Cheating is part and parcel of aggressive predatory behavior.[9]

It is of course not appropriate and clearly anthropomorphic to call these kinds of non human animal behavior "violent", but it remains clear that both in human and non human – especially gregarious – animals the construction of cognitive niches is related to the importance of triggering cooperation and of attacking, more or less violently, other living beings. So the cognitive niches also play, constitutively, the role of carriers of aggressiveness, and in humans, who intentionally build them, they can be legitimately called "moral" and "violent". The whole treatment of coalition enforcement I have given in the first chapter showed how for human collectives it is important to build stable, learnable, and teachable artifactual cognitive systems (niches) related to morality, punishment, and violence.

In summary, general inheritance (natural selection among organisms influences which individuals will survive to pass their genes on to the next generation) is usually regarded as the only inheritance system to play a fundamental role in biological evolution; nevertheless, where niche construction plays a role in various generations, this introduces a second general inheritance system (also called *ecological inheritance* by Odling-Smee). In the life of organisms, the first system occurs as a one-time, unique endowment through the process of reproduction (sexual for example); on the contrary, the second system can in principle be performed by any organism towards any other organism ("ecological" but not necessarily "genetic" relatives), at any stage of their lifetime. Organisms adapt to their environments but also adapt to environments as reconstructed by themselves or other organisms.[10] From this perspective, acquired characteristics can play a role in the evolutionary process, even if in a non-Lamarckian way, through their influence on selective environments via cognitive niche construction. Phenotypes construct niches, which then can become new sources of natural selection, possibly responsible for modifying their own genes through ecological inheritance feedback (in this sense phenotypes are not merely the "vehicles" of their genes). Of course we have to remember that humans are not unique in their capacity to modify their environment, as we have already seen when referring to the case of the spiders that build "dummy spiders": other species are informed by a kind of proto-cultural and learning process that is very often intrinsically social, even if we have to say that animals seem to lack the ability to accumulate information as seen in the human cultural/technological case.

---

[9] On deceptive strategies see chapter three, section 3.3, this book.

[10] This perspective has generated some controversies, since the extent to which modifications count as niche-construction is not clear, thus entering the evolutionary scene. The main objection regards how far individual or even collective actions can really have ecological effects, whether they are integrated or merely aggregated changes. On this point, see (Sterelny, 2005) and the more critical view held by (Dawkins, 2004). For a reply to these objections, see (Laland et al., 2005).

It has to be noted that cultural niche construction alters selection not only at the genetic level, but also at the ontogenetic and cultural levels as well. For example the construction of various artifacts challenges the health of human beings:

> Humans may respond to this novel selection pressure either through cultural evolution, for instance, by constructing hospitals, medicine, and vaccines, or at the ontogenetic level, by developing antibodies that confer some immunity, or through biological evolution, with the selection of resistant genotypes. As cultural niche construction typically offers a more immediate solution to new challenges, we anticipate that cultural niche construction will usually favor further counteractive cultural niche construction, rather than genetic change (Odling-Smee et al., 2003, p. 261).

However, if some counteractive cultural reactions (which of course also include moral adopted options) fail to reduce natural selection pressures (usually because of costs, ignorance, or ethical limitations), genotypes that are better suited to the unchanged cultural modified environment could increase in frequency.

More powerful than sociobiology and evolutionary psychology, the theory of niche construction simultaneously explains the role of cultural (and so moral) aspects (transmitted ideas), behavior (and so moral behavior, which directly orients the construction of niche construction itself), and ecological inheritance (artifacts, to be intended also as moral/violent mediators, as I will better explain in the following subsection 4.4). Of course niche construction may also depend on learning. It is interesting to note that several species, many vertebrates for example, have evolved a capacity to learn from other individuals and to transmit this knowledge, thereby activating a kind of proto-cultural process which also affects niche construction skills: it seems that in hominids this kind of cultural transmission of acquired niche-constructing traits was ubiquitous, and this explains their success in building, maintaining, and transmitting the various cognitive niches in terms of moral systems of coalition enforcement. "This demonstrates how cultural processes are not just a product of human genetic evolution, but also a cause of human genetic evolution" (Odling-Smee et al., 2003, p. 27). From this viewpoint the notion of *docility* I have illustrated in the first chapter[11] of this book acquires an explanatory role in describing the way human beings manage ecological and social resources to make their own decisions.

In the previous chapter[12] I referred to the concept of viscosity to provide an explanation of the gap between the absolutism of morality and the empirical evidence that moral regulations are often infringed with no major consequences either for the whole moral system, or for the very individual who performs the infraction – alas, generating conflicts and violence. Viscosity is certainly constrained by docility, which favors the formation of "the state of being thick, sticky" but also of the state of being "semifluid in consistency, due to internal friction". I said that the fact that morality is viscous hints at its thickness and being glue-like, thus meaning its

[11] Cf. subsection 1.3.2.

[12] Cf. subsection 3.2.2.

capability to be deformed, stressed, pulled apart and reassembled without showing decisive harm to its own stability and reproducibility: this aspect also relates to docility. Viscosity and docility explain how

1. our objectified moral cognitive niches are stable, and at the same time
2. also vulnerable and modifiable, so that it is easy to see in a human individual
    a. the stability of moral convictions depending on his stable moral niches, "together with"
    b. the spontaneous attitude to "disengage" them – for example resorting to a "re-engagement" in other moral conducts which are not dominant in his present moral cognitive niche, but still present as vestigial traces of previous – no longer dominant – moral cognitive niches, as I will illustrate in chapter five.

Woods (2012) touches upon a similar problem, related to docility, when, analyzing fallacious reasoning, he stresses the fact that "Whether full or partial, belief states are not chosen. They befall us like measles", in other words, "say so" induces belief (doxastic irresistibility). Similarly moral cognitive niches too "befall us like measles". The problem is related to the effect of what Gabbay and Woods call *ad ignorantiam rule*: "Human agents tend to accept without challenge the utterances and arguments of others except where they know or think they know or suspect that something is amiss" (Gabbay and Woods, 2005, p. 27). The individual agent also economizes by unreflective acceptance of anything an interlocutor says or argues for, short of particular reasons to do otherwise, by applying the *ad verecundiam* fallacy. Accordingly, the reasoner accepts her sources' assurances because she is justified in thinking that the source has good reasons for them (the fallacy would be the failure to note that the source does not have good reasons for his assurances). Peirce contended, in a similar way, that it is not true that thoughts are in us because we are in them; "beings like us have a *drive* to accept the say so of others" (Woods, 2012).

It is noteworthy that all these information resources do not only come from other human beings. This would clearly be an oversimplification. Indeed, the information and resources that we continuously exploit are – so to speak – *human-readable*. Both information production and transfer are dependent on various *mediating structures*, which are the result of more or less powerful cognitive delegations, namely, niche construction activities. Of course, it is hard to develop and articulate a rich culture as humans did, and still do, without effective mediating systems (writing, artifacts, material culture, etc.). Hence, we can say that, first of all, docility is more generally concerned with the tendency to lean on various *ecological* resources, which are released through cognitive niches. Secondly, social/moral learning cannot be seriously considered without referring to the agency of those mediating structures, whose efficiency in storing and transmitting information far exceeds, from many perspectives, that – direct and non-mediated method – of human beings.

It is well-known that, from the point of view of physics, organisms are far-from-equilibrium systems relative to their physical or abiotic surroundings.[13] Apparently they violate the second law of thermodynamics because they stay alive, the law stating that net entropy always increases and that complex and concentrated stores of energy necessarily break down. It is said that they are open, dissipative systems (Prigogine and Stengers, 1984), which maintain their status far from equilibrium by constantly exchanging energy and matter with their local environments. Odling-Smee, Laland and Feldman quote Schrödinger, contending that an organism has to "feed upon negative entropy [...] continually sucking orderliness from its environment" (Schrödinger, 1992, p. 73). To create cognitive niches is a way that an organism (which is always smartly and plastically "active", looking for profitable resources, and aiming at enhancing fitness) has to stay alive without violating the second law: indeed it "cannot" violate it. In this sense cognitive niches can be considered *obligatory*: "To gain the resources they need and to dispose their detritus, organisms cannot just respond to their environments [...] to convert energy in dissipated energy" (p. 168).

Evolution is strictly intertwined with this process and so it has consequences not only for organisms but also for environments. Sometimes the thermodynamic costs are negligible (like in the heat loss caused by photosynthesis that is returned to the universe, "which is in effect infinite"– p. 169), sometimes they are not, in this case *abiota* of the environment have no capacity to contrast the niche-constructing activities of organisms (like for example, the atmosphere, which is in a new physical state of extreme disequilibrium in relation to exploitation of the Earth's limited resources). The only no-costs exception is when organisms die – and lose their far-from-equilibrium status). In this case the dead bodies are returned to the local environment in the form of dead organic matter (DOM), still a kind of niche construction, so to say, also called "ghost niche construction" (Odling-Smee et al., 2003, p. 170). Of course *biota* can resist any thermodynamic costs imposed on them by other niche-constructing organisms, often performing counteractive niche-constructing activities.

4.1.3 Cognitive Niches as Moral Niches

In the previous subsection I have tried to show that the concept of cognitive niche is an extremely appropriate intellectual instrument to grasp human moral and axiological systems, and their violent counterpart, in a naturalistic way. Indeed, before

[13] It is important to note recent research based on Schrödinger's focusing on energy, matter and thermodynamic imbalances provided by the environment, draws the attention to the fact that all organisms, including bacteria, are able to perform elementary *cognitive functions* because they "sense" the environment and process internal information for "thriving on latent information embedded in the complexity of their environment" (Ben Jacob, Shapira, and Tauber (2006, p. 496)). Indeed Schrödinger maintained that life requires the consumption of negative entropy, i.e. the use of thermodynamic imbalances in the environment. On bacteria and the second law of thermodynamics cf. chapter five of my book (Magnani, 2009).

we move on, it is important to present the moral and potentially violent dimension of cognitive niches, which will be the central topic of the next sections. I have said that the activity of niche construction may enter evolution insofar as it modifies the selective pressures humans and other animals have to cope with. From this we can draw two major consequences.

First of all, the activity of cognitive niche construction potentially affects all those who participate and live in the same local environment in terms of cognitive chances made available (or not). That is, eco-cognitive modifications – brought about collectively (like herd-like behaviors) or by certain groups – may affect our shared cognitive repertoire amplifying it but also constraining or even impoverishing it. On certain occasions, eco-cognitive modifications may be considered by some individual (or particular groups of individuals) as threatening, impoverishing, or detrimental for their possibility to solve problems. Basically, they can perceive their cognitive system as if it is externally *hacked* so that they have to partly re-engineer their relationship with the environment, for instance, by modifying their previous habits or simply forcing them to cope with habits perceived as maladaptive or threatening for them or their group.[14]

The second point deals with the role of coalition enforcement in cognitive niche maintenance. In fact, the construction of cognitive niches and the preservation and their maintenance through coalition enforcement has indeed a moral (and thus violent) dimension: that is, punishment, control and persecution of in-group free riders, and regulation of out-of-group conflicts.

From this perspective, the role played by morality (and, thus, violence) is manifold: any activity that involves and signals a commitment toward cognitive niche construction and maintenance is potentially perceivable as violent against concurrent niches. To develop and to maintain some eco-cognitive modifications typical of a certain community implies that those modifications are indeed worth being preserved because they are perceived as good and useful, which immediately clashes with other possible ways of organizing an homologue cognitive niche. If a cognitive niche displays a univocal relationship with the group who developed it and cares for its maintenance, participating in the niche also involves a more or less public endorsement of the group that supports it. Of course one can partake of several niches (and hence of several groups) as long as they do not compete (or are perceived as

[14] In the course of this chapter (cf. section 4.6) I will come back to this issue by referring to the notion of bullshit introduced by Frankfurt. As I will explain later on, the emergence of certain cognitive niches (the result of long-term eco-cognitive modifications) are perceived by some individuals or groups as potentially damaging for the collective they belong to. For instance, they may consider certain cognitive niches dependent on and favored by a certain method or attitude towards truth and/or knowledge that is conceived as tending to cause harm. The claims they make immediately have or acquire a "moral" tone that most people tend to obliterate or overlook. Accordingly, even the most genuine scientific disagreement implicitly or tacitly bears a moral one. I have already stressed this point in chapter one, section 1.3 and in chapter three, section 3.2: when somebody says that something is true he tacitly implies that everybody else ought to believe the same.

not competing) in the same area, since no matter how polite the context may be, any conflict is ultimately about violence.

Morality can be considered as part of the niche's distributed knowledge, and it precisely concerns violence insofar as it regulates (also violent) relationships between individuals in the niche and with those that are confronted with it without actually partaking of it. Such a regulating activity is permitted by the dimension of violence embodied in rules and regulations and related punishments but also tacitly conveyed by the cognitive as we just observed: the most patent case of such in-niche morals are deontological codes typical of highly specific cognitive niches, but to different degrees they are traceable in every cognitive niche. Of course, the explicit dimension of normativity is concerned in this characterization of the cognitive niche as moral knowledge expressed in the different registers of rules and regulations is one of the pillars of niche maintenance. Even if a niche is not primarily involved in prescribing certain behaviors to its members, a contextual decency is required in order to obtain a state of homeostasis in intersubjective relationships. Should a niche seem to be totally devoid of general normativity, it would thrive insofar as it was laid upon a wider cognitive niche that is in turn heavily concerned with morals and norms, namely, religions, political and legal institutions and so on.

The violent potential constitutively embedded in any cognitive niche actually displays the underlying dimension of that structural and symbolic violence illustrated in chapter one. Structural violence is seen as morally legitimate insofar as it plays a crucial role in the activities of niche maintenance. Immediately we have to note that when parents, policemen, teachers and other agents inflict physical or invisible violence for legal and/or moral reasons, those reasons do not cancel the violence perpetrated and violence does not have to be condoned in so far as it is not always perceived as such. On the other hand it must be analyzed how in the case of structural violence those perpetrating agents do not seem to act only on their own behalf but on that of larger institutions that can be political, industrial, economic or religious. Such institutions populate structural violence not with actors but rather with what we call "violent mediators" (or in the extreme case of human beings that have turned themselves into violence mediating socio-cultural "artifacts", as in the role of the policeman in the framework of structural violence).

The regulatory dimension of structural violence is often diluted in the pervasive form of narratives: the fairytales that are told to children from early youth, novels, plays, dramas and more recently motion pictures are all involved in the dissemination of some moral, economic or spiritual teaching. Human beings' innate talent and taste for storytelling may have benefited from the language decoupling feature in order to convey so-called "moral templates", that can be tacitly or consciously shared not only in defined activities such as in the writing of a novel or a screenplay, but also during the most unaware and casual gossip about absent third parties.

Structural violence may acquire its most subtle and omnipresent form as the symbolic violence perpetrated by language. As a device of social mediation language is necessarily a cognitive niche mediator (and hence distributor) of violence as well. The violent nature of language is a fact too easily admitted to allow serious reflection, as if every speaker were aware of this horrible truth and wanted to get rid of it as soon as possible, even by simply acknowledging it and leaving it at that. As I already pointed out in chapter two, a gentle cluster of speech forms innocently distributes harmful, abusive, destructive, and damaging roles, commitments, inclinations and habits. Language, which is the very moral medium of cooperation and non-violence, also involves unconditional violence even against the speaker herself, insofar as by language one acquires and imposes dominion not only over fellow human beings but also over one's conscious and less conscious self, framing thoughts and emotions in the rigid crystallization brought about by words. The importance of symbolic violence should not be disregarded for one very simple reason: the only requirement to become a perpetrator is easy to meet as it consists in a basic knowledge of the niche language, and the very fact of speaking a language makes the speaker both potentially and actually violent in the symbolic dimension. Culture, knowledge and more highly developed speech abilities do not really help but conversely turn one into an even more subtle perpetrator of violence.

All of this moral knowledge conveyed by the cognitive niche, regulating inward and outward violence, can determinate a grade of moral viscosity which regulates the trade-offs between the intersubjective level of the coalition dimension and the individual level.[15] The point at stake here regards the collective reasons according to which a behavior, potentially dangerous for the community individuated by the cognitive niche, has to be sanctioned or rather allowed to go unpunished for the benefit of the perpetrator. This dimension of morality, related to the particular cognitive niche an individual is embedded within, should not be confused with the general conception of morality that the same individual endorses. The two should not be contradictory but some aspects which are relevant for general morality could be indifferent to niche-related morality, and vice versa.

4.2 "A Sense of Purposefulness" in Evolution: Copying with Maladaptive Artifactual Niches

At an intermediate level between biota and abiota, *artifacts* – as parts of a cognitive niche – present negative entropy because they are highly organized but have no active ability to defend themselves and prevent their dissipation. Of course, and this is the case for various human artifacts, they can originate niches that seem maladaptive[16] rather than adaptive: "Contrary to common belief, environmentally induced novelties [induced by phenotypes] may have greater evolutionary potential than do mutationally induced ones. They can be immediately recurrent in a population; are more likely than are mutational novelties to correlate with particular environmental conditions and be subjected to consistent (directional selection); and, being relatively immune to selection, are more likely to persist even if initially disadvantageous" (West-Eberhard, 2003, p. 498). In the case of modern humans the problem

[15] I have already discussed the problem of moral viscosity in the previous chapter – subsection 3.2.2 – and at the end of the previous subsection, when dealing with the concept of docility.

[16] Cf. (Laland and Brown, 2006).

of managing these maladaptive artifactual niches immediately relates to the relationships between moral axiological systems and related moral conducts and the appropriate exploitation of knowledge in our technological world.[17]

A large part of the niche construction process is intrinsic to the Darwinian framework. The information that basically drives niche construction is of course at the level of semantic information encoded in DNA and provided by evolutionary processes as the result of natural selection. However, niche construction is also *active* and not reactive, *profitable* and not goalless, like natural selection and, moreover, it is always an informed selective process (governed by memory and learning). In this last sense niche construction is related to cognitive processes which are abductive in themselves, because it formulates hypotheses about the *chances* offered by the environment and the possible subsequent active *changes* performed through the niche.

It is interesting to note that recent research on population collapse and/or extinction, which takes advantage of an individual-based evolutionary model, concludes that, in environmental conditions dominated by low-frequency variation ("red noise"), extirpation may be an outcome of the evolution of cultural/moral capacity. In "red noise" environments, whose main character is that "variation is concentrated in relatively large, relatively rare excursions", individual learning – that is creative as opposed to conformist social learning – may be selected by the population. Indeed when social learning systems come "to lack sufficient individual learning or cognitively costly adaptive biases, behavior ceases tracking environmental variation. Then, when the environment does change, fitness declines and the population may collapse". It seems that risk of population collapse can be reduced by morally promoting individual learning and innovation (that is especially *creative abductive cognition*) over cultural conformity and conservatism.[18]

Finally, through inherited selected genes organisms are also informed about past natural selection, that is, their niche construction activities are informed *a priori* by past natural selection. This is not enough however, the semantic information being further retested and updated because of current natural selection pressures and then passed on in the genes of future generations: the best adaptive niche constructors have more chance of being selected.

Indeed, as I have already illustrated, it has to be recognized that first of all "[...] evolution depends on two selective processes, rather than one. A blind process based on the natural selection of diverse organisms in populations exposed to environmental selection pressures, and a second process based on the semantically informed selection of diverse actions, relative to diverse environmental factors, at diverse times and places, by individual niche-constructing organisms" (Odling-Smee et al., 2003, p. 185). Of course the second process was not described by Darwin: in this process selection selects for *purposive* organisms, that is, niche-constructing organisms. West-Eberhard fittingly observes: "Incorporation of environmental modifications

[17] I have fully analyzed this topic in the recent (Magnani, 2007b).

[18] (Whitehead and Richerson, 2009, p. 261).

into the genetic theory of natural selection greatly increases the power of Darwinian argument by showing that it does not depend entirely upon 'random mutation' but can capitalize on preexisting adaptive plasticity and re-organizational novelty in response to recurrent environmental induction" (West-Eberhard, 2003, p. 498). In this sense variation is *blind* but also *constructed* because "[...] which variants are inherited and what final form they assume depend on various 'filtering' and 'editing' processes that occur before and during the transmission" (Jablonka and Lamb, 2005, p. 319). This does not have any Lamarckian or teleological flavor.

An interesting example of active new selection pressure generated by a cognitive niche is that menopause and longevity (i.e. devoting more energy to raising existing children – also acting as grandmother – rather that producing new offspring) may be not a simple manifestation of normal mammalian aging but instead the result of a successful adaptation "via whatever mechanism" (Allen et al., 2005, p.). Changes in diet involving extracting and hunting food that accompanied the origin of *Homo* could have placed young children in a position of requiring more assistance from maternal adults of both generations. Both female longevity and brain size, which correlate significantly across mammal species, would have "co-evolutionarily" conferred increased fitness to organisms (offering greater chances to store intergenerational information about resources and threats) to cope with the variable environmental cognitive niches characterized by the presence of extended families deriving from "moral" established forms of cooperation. Moreover, the "cognitive reserve hypothesis" further states that, in a niche already characterized by intergenerational transfer of information about food, the development of technological and social intelligence – language is probably fundamental in this case – could have formed an increased cognitive reserve in aging. This is reflected in favoring the longevity of old, healthy human brains and their capacity to resist injury and diseases like dementia and Alzheimer (but also in favoring, as a secondary factor, the selection of brain size itself).

On the introduction of "a sense of purposefulness" – words resonating with the sphere of moral commitments, typical of recently "civil and civilized" humans – in evolution Turner (2004, pp. 348–349) says that purposefulness is embodied in the phenomenon of homeostasis: "Evolution then becomes less a province of one class of arbiters of future function – genes – and more the result of a nuanced interplay between the multifarious specifiers of future function". Kaplan and colleagues (2000) further develop the "grandmother hypothesis" showing and describing more clearly how intelligence and brain size would have coevolved with the dietary shift toward high-quality, calorie-dense, difficult-to-acquire food resources. The attainment of skills and abilities requires time and thus an extended learning phase, during which productivity is low, later compensated by higher productivity during the adult period (for example males further enhancing the hunting-extraction-feeding cognitive niche to support women's reproduction) and an intergenerational flow of food from old to young. Lowered mortality rates and greater longevity are thus selected. Of course the new feeding niche further promotes social/moral aspects (and partnership between men and women) such as food sharing, provision for juveniles, and a lowered predation risk, which in turn promote longevity and lengthening of the juvenile period. And, of course, the related ineluctable violent conducts. As we have seen in the section 1.3 of the first chapter, Bingham (1999, p. 140) proposes the higher level speculative "coalition enforcement hypothesis", which, taking into account moral and violent aspects (the latter sometimes being disregarded in evolutionary anthropology), aims at providing a unifying explanation of the various phenomena I have just illustrated, from lowered mortality to longevity, from brain size to dietary shifts.

Finally, as I have already said, organisms can gain from themselves *a posteriori* – at the individual level – more information through learning from their own experience (only possible if the organisms possess the required gene-informed subsystems which allow them to do this and develop the primordial forms of social instinct). On the role of learning, social learning, and culture as adaptation and maladaptation in evolution it is interesting to quote the work by Richerson and Boyd (2005). The authors provide an in-depth description of culture as an unusual system of phenotypic flexibility – through dramatically increased non-parental and parental cultural transmission – that can accumulate adaptive information more rapidly than selection could change gene frequencies. They usefully discuss the "big-mistake hypothesis" in explaining maladaptations, and contrast it to the explicit cultural evolutionary explanations, considered as empirically more adequate. Following the big-mistake hypothesis it seems that most of the information necessary to construct what we call culture is latent in genes shaped in Pleistocene environments, when decision-making/deontological systems evolved. However, in the post-Pleistocene epoch a sudden acceleration of cultural change modified the environments – thanks to cognitive niches – so that they are now far outside the ranges of evolved decision-making systems. The favored counter-hypothesis is that once cultural traditions create novel environments also through various moral commitments and conducts, these can affect the fitness of alternative genetically transmitted variants both in animals and humans, so that genes and culture are joined in a coevolutionary dance, as indicated by the cognitive niche theory. In this case the maladaptation generated by the culturally established tendency towards global cooperation might be explained in the following way: "Humans are quite adept at cooperating in large groups with strangers and near strangers, while the theory of selection on genes suggests that cooperation should be restricted to relatives and well-known nonrelatives" (Richerson and Boyd, 2005, p. 190).

Beyond the effect created by learning and culture in evolution, a kind of "supergenetic" transmission (a phenotype which affects genotypes at higher levels of organization – cells, organism, group) can be hypothesized. This transmission would be brought about through the range of possibilities granted by the genetic systems, making many phenotypes possible. Even if there is still a reluctance to recognize that it plays a role in evolution, some hypotheses now focus on the the so-called "epigenetic inheritance systems" (Jablonka and Lamb, 2005). Making use of a musical analogy, they say:

> We suggested that the transmission of information through the genetic system is analogous to the transmission of music through a written score, whereas transmitting information through non-genetic systems, which transmit phenotypes, is analogous to recording and broadcasting, through which particular interpretations of the score are reproduced. A piece of music can evolve through changes being introduced into the score, but also independently through the various interpretations that are transmitted through the recording and broadcasting system. [...] A recorded and broadcasted interpretation of a piece of music could affect the copying and future fate of the score in two different ways. First, a recorded interpretation could directly bias the copying errors that made. [...] A second, more indirect effect would occur if a new and popular interpretation affects which versions of a score are copied and used as the basis for a new generation of interpretations. [...] Epigenetic systems could have either or both types of effect on the genetic system: they could directly bias the generation of variation in DNA, or they could affect the selection of variants, or they could do both (2005, pp. 245–246).

At the level of the influence of learning and culture in human and animal evolution (which coincides with a fundamental part of niche construction), the authors distinguish between merely "behavioral" effects (for example thought, observation, imitation, and the role of lifestyles, not necessarily "social" or belonging to a shared morality), and "symbolic" ones, which are of course related to the effect – in the case of human beings – of language and other semiotic social communication systems and information exchange.

On the epigenetic inheritance system and the role of language as a cognitive niche embedded in a wider niche, considered as a semiotic network which includes symbolic and non-symbolic artifacts it is interesting to quote the analysis provided by Sinha (2006). It must be stressed that all human artifacts – both material and symbolic – are situated and can be re-situated in semiotic fields that have a cognitive value. In this framework epigenetic developmental processes are those "[...] in which the developmental trajectory and the final form of the developing behavior are a consequence as much of the environment information as of the genetically encoded information. [...] regulatory genes augmenting epigenetic openness can therefore be expected to have been phenogenotypically selected for in the human genome, permitting further adaptive selection for domain-specific learning in the semiotic biocultural complex, in particular for language". But it seems no innateness for language can be hypothesized in humans, as Clark also contends: "Note however, that in an epigenetic perspective, any developmental predisposition for learning language is unlikely either to involve direct coding of, or to be dedicated exclusively to linguistic structure" (Sinha, 2006, p. 113).[19]

We will further see in the following sections that moral cognition is fundamental in niche construction. Moral cognition contributes to coalition-maintenance but also to the building of various innovative habits which lead to the creation of new niches and to the modification of the previous ones. The problem is that at the same time axiological systems and associated deontological processes promote conflicts and related violence.

[19] On the so-called "extended phenotypes" and "extended organisms" cf. (Turner, 2004).

4.3 Distributing Morality and Violence through Cognitive Niche Construction

My contention is that the notion of niche construction is fruitfully applicable to human cognition. More precisely, I claim that cognitive niche construction can be considered as one of the most distinctive traits of human cognition. I have already said that, if we also recognize in animals, like many ethologists do, a kind of nonlinguistic thinking activity basically model-based (i.e. devoid of the cognitive functions provided by human language), their ecological niches can be called "cognitive": for example complex animal artifacts like landmarks of caches for food are fruit of "flexible" and learned thinking activities which indeed cannot be entirely connected with innate endowments.[20] The psychoanalyst Carl Gust Jung, who is aware that also animals make artifacts, nicely acknowledges their cognitive role proposing the speculative expression "natural culture": "When the beaver fells trees and dams up a river, this is a performance conditioned by its differentiation. Its differentiation is a product of what one might call 'natural culture', which functions as a transformer of energy" (1972a, p. 42).[21]

Cognitive niche construction emerges from a network of continuous interplay between individuals – eventually variously assorted in groups – and the environment. Individuals more or less tacitly manipulate what is occurring "outside" at the level of the various structures of the environment in a way that is suited to them. Accordingly, we may argue that the creation of cognitive niches is *the* way cognition evolves, and humans can be considered as ecological cognitive engineers.

Recent studies on *distributed cognition* and *extended mind* seem to support my claim.[22] According to this approach, cognitive activities like, for instance, problem solving or decision-making, cannot only be regarded as internal processes that occur within the isolated brain. Through the process of niche creation humans extend their minds into the material world, exploiting various external resources. For "external resources" I mean everything that is not inside the human brain, and that could be of some help in the process of deciding, thinking about, or using something. Therefore, external resources can be artifacts, tools, objects, and so on. Problem solving, like general decision-making activity, for example, is unthinkable without the process of connection between internal and external resources.

In other words, the exploitation of external resources is the process which allows the human individual cognitive system to be shaped by environmental (or, at any rate, contingent) elements. According to this statement, we may argue that external resources play a pivotal role in any cognitive process. Something important must still be added, and it deals with the notion of representation. The traditional notion of representation as a kind of abstract mental structure is old-fashioned and misleading. If some cognitive performances can be viewed as the result of a smart

---

[20] See (Magnani, 2007a).

[21] As far as psychoanalysis is concerned, I will illustrate Jung's ideas about cognitive/moral externalizations in artifacts below in section 4.8.

[22] Cf. (Zhang, 1997; Hutchins, 1995; Clark and Chalmers, 1998; Wilson, 2004; Magnani, 2006, 2007a, 2009).

interplay between humans and the environment, the representation of a problem is partly internal but it also depends on the smart interplay between the individual and the environment.

Finally, a note about the role of the Darwinian framework in the case of distributed cognition has to be added. As I have illustrated in the last section of chapter one, it is still highly controversial and hotly debated whether aspects of human culture and sociability should be framed within an evolutionary landscape. I have stressed that from an evolutionary perspective, social behaviors – and hence *moral* ones – do affect the fitness of both the individual that enacts such behaviors and of other individuals as well, but the available research which takes advantage of the Darwinian framework is often vague about what is biologically important, and what is not.

Shapiro (2010) raises thoughtful doubts about the possibility that we recognize patent Darwinian evolutionary patterns in the birth of a species that basically offloads its cognitive burden onto the environment. I agree with Shapiro, who warns us that "[...] one lesson to draw from the present discussion is that the relationship between a theory of mind and the theory of evolution is complicated. Whereas a theory of mind that is inconsistent with evolutionary theory may be easy to spot, a theory of mind that actually follows from the theory of evolution is much harder to spot" (p. 417). Here Shapiro casts some doubt on the probability that we will succeed in interpreting the idea of extended mind in the light of Darwinism, as the relationship between embedded or situated cognitive systems and natural selection is not as linear as some supporters of distributed cognition and cognitive niches contend. Of course, Shapiro's take does not contradict the basic claim postulated by the theory of cognitive niches. In fact, the creation of cognitive delegations onto the environment co-evolutionarily enters the Darwinian selection in its strict sense, that is by occasionally altering the selective pressures. For instance, as mentioned above in subsection 4.2, the development of mechanisms designed for the exploitation and manipulation of the environment might have been favored in a Darwinian fashion by the selection of regulatory genes augmenting epigenetic openness. As contended by Sinha (2006), the epigenetic openness, secured by evolution in human genome, granted the human being further adaptive selection for domain-specific learning in a richer biocultural complex.

4.4 Moral Mediators as Conflict Makers and Triggers for Structural Violence

Before more directly turning to this book's central theme of regarding morality and violence, it is useful to think about how ethical knowledge can affect an entity's moral status. Many kinds of entities – both living and nonliving, some already existing that were once considered much less valuable than they are today, others more recently introduced by human beings, have acquired new, different kinds of moral worth: an intrinsic value, or value as an end in itself. They are *moral mediators* – see below, subsection 4.4.1 – a concept I derived from that of the epistemic mediator, which I introduced in my previous research on abduction and creative and explanatory reasoning. For example women, some animal species, works of art, some flags and symbols, important databases, tools related to medical therapy, financial institutions, etc. An entity's intrinsic value, of course, arises not from a change in the thing itself but from changes in human thinking and knowledge; if various acts of cognition can imbue things with new moral value, I will submit in the second part of this section that certain undervalued human beings can reclaim the sort of moral esteem currently held by some "external things", like endangered species, artworks, databases, and even some overvalued political institutions. It is in this sense that they mediate the possibility to grant new ethical values to human beings.

From the philosophical viewpoint I am offering in this book, these entities, endowed with new strong intrinsic moral worth, exhibit

1. an additional chance for more human moral conflicts and possible related violence – they indeed extend the range of moralization we are faced with;[23]
2. a constitutive violence which is perceived by those humans that feel themselves as endowed with less moral worth than other "things".

4.4.1 Moral Mediators

First of all, these external entities are purposefully constructed to achieve particular ethical effects, but other aspects and cognitive roles are equally important. Moral mediators are also beings, entities, objects, and structures that objectively, even beyond the intention of human beings, carry ethical consequences, and, consequently, various kinds of conflicts, punishments, and structural violence. The concept of moral mediator also serves as a window on the idea of the distributed character of morality and as such it is central to moral reasoning and the problem of distributed punishment and violence. In many countries an electric chair is a paradigmatic example of a moral/legal and at the same time violent mediator, it morally – and of course legally – punishes and at the same time violently executes a human being. External entities of this kind function as components of a memory system that crosses the boundary between persons and environment. For instance, when a society moves an abused child into a foster home, it is seeking both to protect her and to reconfigure her social relationships; in this case, the new external setting functions as a moral mediator that changes how she relates to the world – it can supply her with new emotions that bring positive moral and psychological effects and help her gain new perspectives on her past abuse and on adults in general. But I will describe below how the new established moral setting can also trigger new conflicts and violence, just because it is a further ethical setting that clashes with the old ones. New ethical values always involve a new possibility of being jeopardized: when negotiation fails, violence can easily arise.

I was saying that we are surrounded by human-made, artificial entities, whether they are concrete objects like a hammer or a PC or abstractions like an institution or

[23] On the various problems generated by what I call "overmorality" cf. chapter six, section 6.5.

a society; all of these things have the potential to serve as moral or violent mediators. For this reason, I say that it is critically important for current ethics to address not only the relationships among human beings, but also those between human and non human entities. Moreover, by making a smart use of the concepts of "thinking through doing" and of "manipulative abduction" (Magnani, 2009), we can see that a considerable part of moral acting (and its possible violent counterpart) is performed in a tacit and structural way, so to say, "through doing". Part of this "doing" can be considered a manipulation of the external world in order to build various moral mediators that function as enormous new sources of ethical information and knowledge. I have called these schemes of action "templates of moral doing" (Magnani, 2007b, chapter six).

I use the term "model-based cognition" to indicate the construction and manipulation of certain representations, not strictly sentential and/or formal, but mental and/or related to external mediators: obvious examples of model-based inferences include building and using visual representations, conducting thought experiments, and engaging in analogical reasoning. In this light, an emotional feeling can also be interpreted as a kind of model-based cognition. Of course, abductive reasoning – the process of reasoning to hypotheses – can be performed in a model-based way, either internally or with the help of external mediators. Moreover, I can use manipulation to alter my bodily experience of pain; I can, for example, follow the behavior template "control of sense data", during which I might shift – often unconsciously – the position of my body.

Through manipulation I can also change my body's relationships with other humans and non humans experiencing distress – so using my body as a moral mediator – as for example Mother Theresa did, whose rich, personal moral feeling and consideration of pain were certainly shaped by her physical proximity to starving and miserable people and by her manipulation of their bodies. In many people, moral training is often related to the spontaneous (and sometimes fortuitous) manipulation of both sense data and their own bodies, for these actions can build morality, immediately and in a non-reflective way, "through doing". Moreover, part of this "doing" can be seen as a manipulation of the external world as a way to establish *new* "moral mediators" (which of course belong to more or less stable moral cognitive niches) in order to achieve certain ethical effects. Alas, we already know that moral mediators are also beings, entities, objects, structures that bring unintentional consequences – either ethical or violent.

Throughout history, women have traditionally been thought to place more value on personal relationships than men, and they are often regarded as more adept in situations requiring intimacy and caring. It would seem that women's basic moral orientation emphasizes taking care of both people and external things through personal, particular acts rather than relating to others through an abstract, general concern about humanity. The ethics of care does not consider the abstract "obligation" to be essential; moreover, it does not require that we impartially promote the interests of everyone alike. Rather, it focuses on small-scale relationships with people and external objects, so that, for example, it is important not to "think" of helping disadvantaged children all over the world (as men tend to aim at doing) but to "do" so when called to do so, everywhere. My philosophical and cognitive approach to moral model-based thinking and to morality "through doing" does not mean that this so-called female attitude, being more closely related to emotion, should be considered less deontological or less rational and therefore a lower form of moral expression. I contend that many of us can for example become more intuitive, loving parents and, in certain situations, learn to privilege "taking care" of our children by educating our feelings – maybe by heeding "Kantian" rules.

The route from reason to feeling (and, of course, from feeling to reason) is continuous in ethics. Many people are suspicious of moral emotional evaluations because emotions are vulnerable to personal and contextual factors. Nevertheless, there are moral circumstances that require at least partially emotional evaluations, which become particularly useful when combined with intellectual (Kantian) aspects of morality. Consequently, "taking care" is an important way to look at people and objects, and, as a form of morality accomplished "through doing", it achieves its status as a fundamental kind of moral inference and knowledge. Respecting people as things is a natural extension of the ethics of care; a person who treats "non human" household objects with solicitude, for example, is more likely to be seen as someone who will treat human beings in a similarly conscientious fashion. Consequently, using this cognitive concept, even a lowly kitchen utensil can be considered a moral mediator. When I clean the dust from my computer, I am caring for the machine because of its economic worth and its value as a tool for me-as-a-human-being. When, on the other hand, I use my computer as an epistemic or cognitive mediator for my research or in my didactic activities, I am considering its intellectual prosthetic worth.

Artifacts serve as violent mediators in many situations, as in the case of certain machines that can violently affect privacy. Implicitly, these artifacts channel and represent, from many perspectives, the morality of freedom of tolerant liberalism: thus, they play the related role of moral mediators for the same culture that produced them, but at the same time they can paradoxically jeopardize privacy, which is one of the main ethical values of that same culture, typical of constitutional democracies. For example, internet mediates human interaction in a much more profound way than do traditional forms of communication like paper, telephone, and mass media, even going so far as to record and keep tract of interactions in many situations. The problem is that because the internet mediates human identity, it has the power to affect human freedom, thus possibly jeopardizing various subjects' autonomy aspects. People who feel their identity within privacy negatively affected can also think they have become victims of violent attacks due to the intrinsic ethics that the internet mediates thanks to the massive profiling processes. In fact, thanks to the internet, our identities today largely consist of an externally stored quantity of data, information, images, and texts that concern us as individuals, and the result is a "cyborg" of both flesh and electronic data that identifies us. I contend that this new complex "information being" depicts new ontologies that in turn involve new moral problems and potentially more or less violent conflicts.

Moreover, we can no longer apply old moral rules and old-fashioned arguments to beings that are simultaneously biological and virtual, situated in a three-dimensional local space and yet "globally omnipresent" as information packets. Our cybernetic locations are no longer simple to define, and increasing telepresence technologies will exacerbate this effect, giving external, nonbiological resources even greater power to mediate ethical endowments such as those related to our sense of who and what we are and what we can do. These and other unwelcome effects of the internet, which conflict with the modern ethics of freedom – almost all of which were unanticipated – are powerful motivators of our duty to construct new knowledge, with the aim of efficaciously managing the insurgence of these conflicts and trying to avoid the violence that could arise. I believe that in the context of this abstract but ubiquitous technological presence, certain moral approaches that ethics has traditionally tended to disparage are worth a new look.

An issue in moral mediators is the concept of intrinsic value, or the worth an entity has for its own sake, and this idea is central to my approach. Analyzing how people have ascribed intrinsic value to various entities is ever more urgent given the important role of moral mediators in the lives of modern people. Much of the behavior we conduct through learned habits is devoted to building vast new sources of information and knowledge: external moral mediators. Let us return to our previous, edifying, example of those human beings who create, originally in a tacit and embodied way – without a clear project – foster homes, which play a moral role: they facilitate a foster child's recovery as they allow him or her to rebuild moral perceptions that were damaged by previous abuse.

These "new" creatively built moral mediators carry new ethical values that can be jeopardized by other values, and so become prone to be the source of further abuse and/or strong conflicts which can resort to violence. For example they can conflict with alternative preexistent and dominant ethical views centered on the value of that other stable moral mediator that is the *traditional Western family*, which is thought by many people as the only carrier of basic moral values, even if pervaded by abusive behaviors. Here, the artifact (the foster home) is an example of a moral mediator in the sense that it mediates – objectively, "out there", in an external structure – positive moral effects. But it is at the same time a potential violent mediator, for example when it is seen as a trigger of conflicts with other preexistent moral views and conducts. Many complicated external moral mediators can also redistribute moral effort and generate those collective axiologies – available "out there", crystalized in various "external" artifacts – that I will describe in the following section. Collective axiologies allow us to manipulate objects and information in a way that helps us to overcome the paucity of internal moral options (principles and prototypes, etc.) currently available to us.

I also think that moral mediators can help to explain the "macroscopic and growing phenomenon of global moral actions (and violent outcomes) and collective responsibilities resulting from the 'invisible hand' of systemic interactions among several agents at the local level" (Floridi and Sanders, 2004). Using moral mediators is more than just a way to move the world toward desirable goals: it is an action that can play a moral role and therefore warrants moral consideration. We have said that when people do not have adequate information or lack the capacity to act morally upon the world, they can restructure their worlds in order to simplify and solve moral tasks. Moral mediators are also used to reveal latent constraints in the human/environment system, and these discoveries grant us precious new ethical information.

Imagine, for instance, a wife whose work requires long hours away from her husband, and whose frequent absences cause conflict in their relationship. To improve their marriage, she restructures her life so that she can spend more quality time with her spouse, an action that can cause variables affected by "unexpected" and "positive" events in the relationship to covary with informative, sentimental, sexual, emotional, and, generally speaking, bodily variables. Before the couple adopted a reconfigured "social" order – that is, increased their time together – there was no discernible link between these hidden and overt variables; a new arrangement has the power to reveal important new "information", which, in our example, might come from a revitalized sex life, surprisingly similar emotional concerns, or a previously unrecognized intellectual like-mindedness. A realigned social relationship is just one example of an external moral mediator. Of course the new – morally mediated – reconfigured relationship can soon produce new unexpected conflicts and failures, and so other potential violent outcomes.

4.4.2 Moral Mediators and Structural Violence

I have said above that in the philosophical perspective I am offering in this book these entities, endowed with new strong intrinsic moral worth, exhibit

1. an additional chance for more human moral conflicts and possible violence, as they indeed extend the range of moralization we have to face;
2. a constitutive violence which is perceived by those humans that feel themselves as endowed with less moral worth than those things.

I still have to explain the second item above. Indeed, there is a second role of moral mediation that can be played by those entities, which humans have empowered with strong moral worth. Indeed, in my previous book on morality and technology[24] I have extensively contended that, given the fact that only human acts of cognition can add worth to or subtract value from an entity, revealing the similarities between people and things can help us to attribute to human beings the kind of worth that is now held by many highly valued, non-human things. This process suggests a new perspective on ethical thinking. Indeed, these objects and structures can mediate moral ideas and recalibrate the value of human beings by playing yet another role as, what I call, moral mediators.

What exactly is a moral mediator in this last sense? Moral mediators can become instruments of a cognitive process which helps in overcoming the violent effect they cause some people to perceive, people that, as human beings, are not endowed with the moral worth that is instead reserved to those external entities. First of all, moral

[24] (Magnani, 2007b).

mediators can be exploited to teach us how to extend value from already-prized things to human beings, as well as to other non human things and even to "non-things" like future people and animals. To overcome the human feeling of being violated by that excessive off-loading of moral value onto things, I believe that we can use these things as moral mediators that serve a sort of "copy and paste" function. Thus we can take the value of, say, a library book (which we look after very carefully and want to give back just as we received it) and transfer it to a person. Using moral mediators in this way, however, will require the construction of a vast new body of knowledge, a new way of looking at the world and it is in this perspective that I contend that "knowledge is duty". "External" things and commodities with desirable values – levels of importance that many people cannot claim – can become vehicles for transferring worth from non-human to human entities. When things serve this role they mediate moral ideas by revealing parallels between things that are more valued and people who are less valued, thereby granting us with precious ethical information and values.

Think for a moment of cities with extensive, technologically advanced library systems in which books are safely housed and carefully maintained. In these same cities, however, there are thousands of homeless people with neither shelter nor basic health care. Thinking about how we value the contents of our libraries can help us to reexamine how we treat the inhabitants of our cities, and in this way, the simple book can serve as a possible moral mediator. To make a case for respecting people as we respect computers, we can call attention to the values human beings have in common with these machines: (1) humans beings are "tools", albeit biological ones, with economic and instrumental value and, as such can be "used" to teach and inform others in much the same way we use hardware and software, so often human beings are instrumentally precious sources of information about skills of various kinds; and (2) like some important computers, people are skillful problem-solvers imbued with the moral and intrinsic worth of cognition. Moral mediators play an important role in reshaping the ethical worth of human beings and collectives and, at the same time, facilitating a continuous reconfiguration of social orders geared toward the rebuilding of new moral perspectives and, consequently, possibly weakening existing excessive chances for conflicts, punishment, and violence.

In sum, we are surrounded by man-made and artificial entities, not only concrete objects like screwdrivers and cell phones, but also human organizations, institutions, and social collectives. All of these things – insofar as they may trigger and favor moral conflicts – also have the potential to serve as powerful distributors of structural violence. For this reason, I strongly contend that it is crucial for current ethics to address the relationships not only among human beings, but also between human and non human entities. The following is a simple example: every crucifix in Italian classrooms certainly symbolizes (and so it "is" a moral mediator of) catholic morality. Firstly, this mediator tends to conflict with the values of the Italian constitutional democracy, which guarantees equal worth to every religion and its related morality. Secondly, it can also easily conflict – often very violently – with other people's religious and moral beliefs, especially in our globalized world.

4.5 Collective Axiologies as Moral Niches, and the Derived Violent Conflicts

A list of moral niches that at the same time constitute potential chances for violence is really easy to compile. A lot of sociological literature has stressed the role of conflicts in triggering violence. In this perspective, nurture, more than nature and something innate and instinctual, seems to be at the root of conflict. People fight for various entities that mediate the moralities they care about. Indeed resources like jobs, status (prestige), turf (geography, neighborhood), to which are attached values such as Christian family, civil morality, various religious principles, the American way of life, ideals of masculinity and of the right sex, contribute to the chief delineation of the moral *identity* of people belonging to a group: "Habits are the very stuff our identity are made of" (Žižek, 2009, p. 140). When in intergroup or intersubjective relationships these values are jeopardized, and negotiations tend to fail, violence arises: fear, anger, and hate easily trigger Manichaean mentality, scapegoating, harassment, killing, xenophobia, sexual abuse, homophobia, discrimination, humiliation,[25] etc. Beyond negotiation and consequent accord, it is violence that redefines social relationships and this often occurs in unpredictable ways. It has to be said that in several cases the perpetrated violence is later rendered acceptable through the exploitation of the typical still cognitive violent tools which are devoted to disguising violence: for example, trivialization, sanitization, relativization, blaming the victim, dehumanization.[26]

It is in this perspective that we see, at the beginning of the third millennium in the rich Western societies, the decline of Social Democracies, of the welfare state and the rise of New Right populist governments, characterized by an increase of more or less "new" violent processes, from demagogy (characterized by the constitutive practice of bullshitting in political argumentation) to torture and wars to export democracy.

Žižek observes:

> Abu Graib was not simply a case of American arrogance toward Third World people: in being submitted to humiliating tortures, the Iraqi prisoners were effectively initiated into American culture. They were given a taste of its obscene underside, which forms the necessary supplement to public value of personal dignity, democracy and freedom. [...] The clash between the Arab and American civilization is not a clash between barbarism and respect for human dignity, but a clash between anonymous brutal torture

[25] A collection of papers contained in (Guru, 1989) describes the wealth of violent ways to inflict humiliation, taking advantage of the analysis of the structures of humiliation in India. Humiliation is especially described as a method for 1) damaging self-respect 2) representing others as inferior (and making them feel as such), 3) creating people that lack the minimum moral capacity to protest. It is interesting to notice the number of moral and legal ways to validate and justify humiliation, which are illustrated in the book. Degradation, dehumanization, instrumentalization, humiliation, and nonrecognition, as violations of human dignity through rape, torture, poverty, exclusion, labor exploitation, and bonded labor, are illustrated in the recent (Kaufmann et al., 2011).

[26] Many cases and examples are described in (Oppenheimer, 2005).

> and torture as a media spectacle in which the victims' bodies serve as the anonymous background for the grinning "innocent American" faces of the tortures themselves. It seems, to paraphrase Walter Benjamin, that every clash of civilizations really is a clash of underlying barbarisms (2009, pp. 149–150).[27]

I agree with Žižek, the reality of globalization is both the affirmation that capitalism, and its liberalist ideology, are "effectively universal" and so is the need for the edification of new walls – paradoxically, those factors jeopardize liberalism itself, which is sad, but completely true "This is a clear sign of the limit of the multiculturalist 'tolerant' approach which preaches open borders and acceptance of others. If one were to open the borders, the first to rebel would be the local working classes" (Žižek, 2009, p. 88). In this difficult but eloquent philosophical lexicon the effective universality of capitalism is seen as "a neutral matrix of social relations. [...]. This is why the 'leave us our culture' argument fails. Within every particular culture, individuals *do* suffer, women *do* protest when forced to undergo clitoridectomy, and *these protests against the parochial constraints of one's culture are formulated from the standpoint of universality*. Actual universality is not the deep feeling that, above all differences, different civilizations share the basic values, etc; *actual universality appears (actualizes itself) as the experience of negativity, of the inadequacy-to-itself, of a particular identity*" (p. 133).

At this point we have to address more deeply the problem concerning the status of what I call moral niches. In my present perspective they have to be fundamentally seen in their capacity to delineate possible threats to others. Rothbart and Korostelina (2006a) provide an useful framework, which helps in making clear what I am referring to:

> During periods of crisis, the threatened group denigrates the Other as uncivilized, savage, subhuman, immoral, or demonic. Negative images are retrieved from mythic stories of the past. In turn such images lend validity to notions of ingroup nobility, skill, and virtue. Via these narratives, violent actions against a perceived "enemy" become sanctified, and the agents of such actions glorified. This dynamic is no less true for today's terrorist threats than for the Corinthians during the Peloponnesian war (p. 1).

This situation is also true for peaceful and cooperative communities, where group differences can be exaggerated. The narratives are full of reference to race, gender, politics, religion or nationality. These narratives reflect the existence of normative patterns, relatively stable in a collective and in individual minds, which resort to characteristic ways of doing and being, of feeling and acting, of speaking and entertaining social relationships, mutual obligations, complicated duties. The identity group itself is mediated by the adopted categories of right and wrong, good and

[27] On the relationships between torture and democracy cf. the extremely rich historical and political analysis given by (Rejali, 2007). Various aspects concerning the relationship between terrorism, non-violence, and democracy are illustrated in (Chenoweth and Lawrence, 2010), which point out changes in the balance of power between states and among states and non-state actors in generating uncertainty and threat, thereby creating an environment conducive to violence. The approach usefully deemphasizes the role of ethnic cleavages and nationalism in modern conflict.

bad, true and false, virtues and vices: "Of course, every protagonist group declares that its cause is just, and the sacrifices of other necessary. Over time, such declarations embed a group's collective axiology. For example, national identity often rests on claims of original habitation of the land. Such a claim is often used to legitimize control, rights, and privileges. National identity is inseparable from normative judgments about group virtues and outgroup vices. *Underpinning every social, religious, and national group is a polis, and every polis blends politics with value-commitments that confer a collective axiology on individuals*" (p. 4).

This collective axiology is a moral system of value-commitments that delineate which actions are prohibited, and which actions are necessary for specific tasks. Not only this, it also delineates a sense of life and world, provides perceptions of (good or bad) actions and events, and "provides the basis for evaluating group members [...] boundaries and relations among groups and established criteria of ingroup/outgroup membership". Often the collective axiology is based on a sacred past, myths or magic. Finally, the analysis of narratives that substantiate a collective axiology provides the criteria for individuating all the threats and their permissible responses, to punish group members that deviate, and to react against out-group challenges.[28]

The same problem is confronted by Žižek (2009), in a philosophical and Lacanian perspective, when he locates an important cause of violence in the narcissistic *fear of the Neighbor* – and his consequent potential *inhuman* dimension (he is often seen as jeopardizing our own values): immigrants, criminals, sexually depraved or "deviated", believers of other religions, etc. Even in tolerant democratic societies, the morality that we are all human and share the basic hopes, fears and pains, is *de facto* perversely suspended when it leads to the conviction that "the Other is just fine, but only insofar as his presence is not intrusive, insofar as this Other is not really other": a kind of "fetishist disavowal" (pp. 35 and 45). After all, every moral perspective is open to its pervertibility, even the most humanitarian, so to say: "The Christian motto 'all men are brothers', however, also means that those who do not accept brotherhood *are not men*" (p. 46). As I have illustrated in chapter two, the problem of conflicting axiologies is obviously linked to the problem of language (and of course of other more model-based[29] kinds of human cognitive activities, such as visualizations or emotions) and "its" unconditional violence, which is embedded in its capacity to carry moral signifiers as potential dividers: "it is because of language that we and our neighbours [can] 'live in different worlds' even if we live

[28] An obvious example of threat narratives which are converted into dogmatic ideologies of terror is provided by the case of violent totalitarian regimes. Subtle analysis of the role of identity salience and of cultural factors in the construction of the threat narratives are given in (Rothbart and Korostelina, 2006b; Korostelina, 2006b,a). The role of identity in political violence such as terrorism – and the role of values (moral) both in forming terrorist activity and in responding to its challenges – is analyzed, also taking advantage of various historical and up to date examples, in (Karawan et al., 2008). Criminological aspects of political violence are illustrated in (Ruggiero, 2006).

[29] Cf. footnote at page 28.

in the same street. [...] Reality in itself, in its stupid existence, is never intolerable: it is language, its symbolization, which makes it such" (p. 57).

I would integrate[30] the above vision of collective axiology usefully provided by Rothbart and Korostelina with a further note. The concept of collective axiology clearly describes the moral nature of group life and of intergroup relations – and it is further confirmed from the perspective of cognitive and anthropological research I will illustrate in the following chapter – but it also provides an insight into the private moral behaviors of individuals within the same group. We have said that collective axiology is conferred to individuals to concur to the formation of a complete identity of the group. We also know that in our societies individuals do not only belong to the whole group of the nation or ethnicity, but also to various subgroups (for example related to political parties, religions, professions, genders,[31] etc.), which in turn are characterized as permanent sub-axiologies. These sub-axiologies identify the subgroup itself and therefore that *part* of the individual which is identified with the subgroup in question. Moreover, these sub-axiologies delineate sub-threats and a myriad of more or less violent punishing activities against those individuals that *in-subgroup* actors perceive as dangerous for the subgroup identity. Obviously this situation is especially related to violence perpetrated by individuals of a subgroup against individuals of the same subgroup, for example in the case of political and professional conflicts. Another obvious example of sub-threat narratives which are converted into violent actions is the case of a father who overvalues the vulnerability of his patriarchal sub-axiology (and so of his authority), faced with a "disrespectful" son's behavior (who shares "another" alternative sub-axiology, for example related to the value of a non-repressive family).

Barnett and Littlejohn (1997) analyze incommensurate moral conflicts as vexing disputes that ordinary discourse and normal tactics of communication are not appropriate to resolve. They provide a pragmatic analysis of various ways of promoting understanding – *expressing* – of moral conflicts and their possible communicative mediation, beyond the mere violent modes of persuasion and repression, which instead aim at *suppressing* the conflict. For example, "changing the context" and "coordinating communication" can help negotiation, especially at the level of "public discourse". Other ways of fixing moral conflicts are described, which would be welcomed in our postmodern era: they resort to an active (and often creative) process of "transcending", where a rebuilding of the – supposed to be incommensurate – stories that substantiate the conflicting moral perspectives is at stake, to the aim of promoting mutual understanding and respect. New languages can be activated, beyond mere

[30] Of course established collective axiologies can also reflect the existence of political and economical orders. Recent studies stress the role of violence in maintaining and restoring order through the management of threatening conflicts (Kalyvas et al., 2008).

[31] During times of peace, denigration of women and related "moral models" of women directly lead to sexual violence, which is further deepened in case of armed conflict: "Rape, then is one of strategic tools of combat" (Cheldelin, 2006, p. 283) and sexual violence is almost inevitable in war; cf. also the horrifying analysis of sexual violence in World War II, Bosnia-Herzegovina, Sierra Leone, Israel/Palestine, Sri Lanka, El Salvador given by (Wood, 2008).

rhetorical eloquence of normal discourse, able to promote peacekeeping, compromise, and withdrawal. Tools that can be used are: silence and engagement, reframing the contradictions and encapsulating discussion about conflicts in small groups. In brief, they aim not only at a mere commitment to the rejection of absolutes and the celebration of differences, but also at a commitment to reintegration and new ways of knowing.[32]

4.5.1 Nonviolent Moral Axiologies, Pacifism, and Violence

I have just said that, with the aim of avoiding conflicts and preventing or extinguishing violence, new languages and new actions can be activated to promote peacekeeping, compromise and withdrawal. Silence and engagement, reframing the contradictions, but also courage, capacity to suffer, and indirect threat constitute part of the toolkit of so-called *nonviolence*. Nonviolent moral axiologies are very complicated – and historically multifaceted. Nonviolent morality – sometimes called "the politics of ordinary people" – certainly aims at implementing innovative and efficient ways of knowing the features of a conflicting situation to promote the reintegration of some agreement, avoiding usual violence. Dodd (2011, p. 152), noticing the – candid – immediate translucent moral halo which is usually attributed to nonviolence, maintains that "We all too often fall into the vagaries of opposing the cold instrumentality of violence with the human morality of nonviolence, and fall to recognize their common root". From the point of view of the treatment of violence I am illustrating in this book it is not difficult to grasp the intrinsic meaning of these words. Nonviolence, in so far as is substantiated by a precise emancipatory/moral axiological framework, is far from being exempt from relationships with violence, at least in two senses. First of all nonviolence is always "defined" by violence, second, it is always a trigger of possible violence, self-directed or indirect/reactive. Dodd usefully adds: "[...] we could perhaps argue that nonviolence is dependent on something that we have scarcely begun to understand about violence, at least theoretically: namely, a *constitutive dimension* of violence, that is quickly obscured if we conceive of violence in pure instrumental terms" (ibid.).

"Taking a stand for the best", nonviolent moral axiologies share with any other morality that structural core inherited from religion as I will clearly illustrate in the first sections of chapter six: on one side religions are a way of explaining the genesis of violence and a way of escaping it but, at the same time, insofar as religions are carriers of moral views (and take part in their "secular" construction) they constitute a great part of those collective axiologies I have illustrated above, which are possible triggers of punishment and violence. Hence, nonviolence patently reverberates that structural core of religion – so to say – at its highest limit, because more *explicitly* and *peculiarly* it establishes itself *against* violence by "directly" refusing it, thus fascinating everyone with this unintentional lexical "subterfuge", favoring naïve enthusiasm and moral force.

[32] A philosophical analytical study of moral conflicts and disagreements, which resorts to an opposition to moral realism, is given in (Tersman, 2006).

I am not engaged in ridiculously denying the potential of nonviolence in jeopardizing and diminishing the "actual" overall quantity of violence (which would be otherwise possible in absence of nonviolent actions) and in the creation of efficient practices of "unlearning violence". I am just stressing the intimate intertwining of both (moral) nonviolent and violent aspects. Nonviolence is always "shadowed by the potential for violence", as Dodd nicely adds (p. 146) and I definitely agree with him. As an action that carries compulsion, force, struggle and threat (persuasion, symbolic protest, moral indignation, noncooperation, boycotts, civil disobedience, non-resistance, moral and passive resistance, etc.), and which incorporates the "martial" virtue of courage and the capacity to violently suffer, nonviolence aporetically and astutely – but not hypocritically – works within the scope of violence.[33] Furthermore, nonviolent axiologies are always in conflict with the opposed violent ones, and for this reason they are often the actual triggers of immediate self-damage or self-sacrifice and/or of further reactive violence, which are sometimes atrocious, as the cases of Gandhi and Martin Luther King Jr. eloquently teach. Usually, nonviolence strategically or pragmatically (on the basis of variegated religious but also secular moral axiologies) depicts a new situation for the adversary where often both open strong violent reactions and withdrawal from them (for example causing the consequent loss of standing in public opinion) are highly costly. It is clear that in both cases the adversary – that nonviolence does not consider an enemy[34] – is efficiently non violently de facto "violated". In brief, much more than in the case of passive or inactive pacifism, potential nonviolent victims tend to constitutively become at the same time carriers of violence against the adversary.

Last but not least, nonviolence always promotes what I can call a "moral epistemology":[35] new "regimes" of truth related to the "*inessentiality* of something, or its nothingness, in the form of illuminating its fragility and pursuing the orchestration of its collapse", are proposed (Dodd, 2011, p. 150). For example Gandhi's doctrine clearly acknowledges the intrinsic aggressiveness or force of truth: "Gandhi used to argue that 'nonviolence' should be understood not simply in terms of not doing harm, as suggested by the word *ahisma*, but as a force, specifically a force for truth – and in this vein he suggested adopting the word *satyagraha*, 'truth force'", that is truth through love and righteous action (cit., p. 148). The new advanced and performed regime of truth is lived by the adversary as a "symbolic" aggressive threat, even if it is not performed by means of powerful sanguinary weapons.

Finally, it must be remembered that also the morality offered by nonviolence is often obviously lived by the bearers of other moralities as negative, even if they are not direct adversaries in a determinate situation. For example, in the light of

[33] On the need of recasting militancy and the democratic right to civil disobedience in current Western democracies, and on the central problem of realizing nonviolent collective protection against the development of "monopolizations of legitimate force" (whether military, economic, and ideological) cf. the rich (Balibar, 2005).

[34] Nonviolence also seeks to engage the empathy of the perpetrator, for example through love, compassion, respect, and reverence – to make claims on his/her morality, or at least the morality of those the perpetrator represents.

[35] On this problem see also below, "Why epistemology matters to violence", subsection 4.7.1.

an anarchistic perspective, nonviolence is depicted as dangerous, ineffectual, racist/aristocratic, statist, patriarchal and inferior to more militantly-exposed activism (cf. for example (Gelderloos, 2011)).

The case of pacifism is slightly different from that of nonviolence, but the philosophical analysis provided in this subsection still holds true for it. The pacifist, who endorses a commitment towards peace and opposes war and killing, claims to be – more or less (i.e. "maximal" and "minimal" pacifism) – nonviolent (and to withhold the use of force), and to be this way for moral/religious, as well as strategic and pragmatic reasons, but the action is less structured and more passive than in the case of so called proper nonviolence. For example, pacifism is often seen as related to more personal virtuous decisions mainly based on moral or religious grounds, and detached from the desire for change at the sociopolitical level, but we cannot forget that many political parties (for example the ecologically-conscious – green – ones) contend that they favor pacifism. Even many modern democratic constitutions affirm a mild pacifist character when they pronounce a basic repudiation of war.

Sometimes pacifism presents a violent perverse and malign face, when its meaning is clearly divorced from any correlated idea of justice: Richard Nixon called himself a pacifist, even while he continued to support the highly controversial – from the point of view of its legitimacy – Vietnam War. Linking war to the subsequent peace is obviously trivial and can easily pervert the label "pacifist". All the more so, as the peace acquired can be the violent "peace" of slavery. We should anticipate a critique of some forms of radical pacifism that will be stressed again in chapter six: a blindly benevolent, charitable and overly peaceful attitude towards the world could sometimes be difficult to distinguish from a kind of mildly violent universal indifference. The excess of peaceful, positive – but passive – empathy with other human beings, mixed with the hypertrophic admiration of nonviolence and peace, can paradoxically lead to passively refraining from active intervention when violence is detected and calls for intervention. The initial shining empathy imperceptibly degenerates into actual indifference.[36]

4.5.2 What Does the Micro-sociological Theory of Violence Teach Us?

In the perspective of this book interesting teaching derives from the so-called micro-sociological theory of physical violence. Indeed, this theory is reductionist, and tends to reduce the role of cultural, moral, and ideological factors in triggering violence, and thus stresses some basic common factors of violence. Collins (2008) lists many different kinds of violence, from combat infantry to police violence, from competent hitmen, ace pilots, and clandestine terrorists to domestic violence, common bullies, gang initiations, mosh pits, carousing and sports pseudo-violence. The research takes advantage of the information now made available by the video revolution and points attention to micro "violent situations", which badly shape the emotions and acts of individuals who enter them. This theory describes the shaping

[36] Cf. also page 276, chapter six of this book.

of an emotional field of tension, fear, anger, forward panic or weakness of the target, as the invariable trigger for physical violence in conflicting situations. It also sheds further light on the hypocritical distinction between good violence, that is not seen as violence at all, because perpetrated in the name of a moral social order, and bad violence, perpetrated by people that come from lower social conditions.

The theory also usefully points out how "real" violence is hard to perform, and even violent people are not good at it, on the contrary "symbolic" violence is patently easy. Furthermore, this theory stresses the interesting fact that we have to distinguish between "hot" violence, usually performed in a bad way by incompetent people (crowd violence, bullies, domestic violence etc., always audience-conditioned and audience-dependent),[37] and "cold" – highly successful – violence (military snipers, competent police officers, organized crime, expert robbers, professional killers etc., concerning actors who do not need an audience) which is the fruit of very competent techniques and regards a small minority of people acting in specialized violent niches – an elite characterized by extremely specific knowledge. This situation indicates how even violence has structural limits, like the failures of hot violence indicates – also given the fact that, following the micro-sociological theory of violence – its very nature is basically that of being a product of an emotional field.

Finally, the theory usefully warns us about the excessive explanation of violence in macro-cultural terms, and contrasts the tenets of some aspects of the evolutionary psychology theory, which sees violence as an evolutionary propensity for males to fight over reproductive dominance. In front of the empirical analysis of violence, the claim that there is a genetic component to violence is not so likely anymore, given the fact that violence is not confined to young men in their reproductive age. Human beings have evolved to be especially emotionally susceptible to the dynamics of interactional situations: "Humans are hard-wired for interactional entrainment and solidarity; and this is what makes violence so difficult. Confrontational tension and fear [...] is not merely an individual's selfish fear of bodily harm; it is a tension that directly contravenes the tendency for entrainment in each other's emotions when there is common focus of attention" (Collins, 2008, p. 27).

In this light, Collins is able to contrast Elias' contentions about violence as a "decivilizing" process. Elias (1969; 1982) argues that aggression was regarded as a pleasure in middle-age Europe, but with the rise of court societies in the sixteenth century the independent warriors were brought under the control of a centralizing state, giving birth to the process of modern "civilization". But this does not mean that violence should be considered as primordial, a kind of Freudian primal reservoir

[37] Baumeister (1997, chapter eleven) amply discusses the role of bystanders in favoring violence or altering its outcome. Also, victim groups can provide assistance to their oppressors, for example favoring internal divisions and lack of cooperation, or even collaboration with the aggressors, because of disaffection inside the group itself or personal advantage. The "potential" role of passive bystanders (also "institutional" ones, like the police, organizations and governments) in implicitly favoring the moral approval of violence and – directly – violence itself, for example in genocide or in family abuse, is analyzed, also taking advantage of (Staub, 1989): on the role of silence in violence perpetration cf. chapter six, subsection 6.5.1, this book.

of aggressiveness: conversely, it is basically socially-constructed, and "civilization" does not tame it, given the fact that we face an increase in social techniques related to performing violence (Collins illustrates to this purpose football hooligans' violence, as the creation of a new sophisticated technique of violence).[38]

However, a recent sociological article by Mucchielli (2010) about the development of violent behaviors in interpersonal relationships in France (and, generally, in Europe) since the 1970's, usefully stresses how the pacifying process, implied by the related civilizing process starting back at the end of the Middle Ages (postulated by Elias), is still at work and has not been reversed. Only verbal violence (insults, threats) has increased in police and judicial statistics: "if a pacifying process tends to reduce recourse to violence, it is because its primary consequence is the stigmatization and delegitimization of that violence" (p. 213). The general acknowledgment that violent behaviors are on the increase would actually be just an impression, against empirical data. This has to be interpreted not only as a consequence of central states acquiring the monopoly of legitimate violence (and taxation) but also as true social pacification and discipline determined by bureaucratic processes and the gradual internalization of religious morality, as already contended by Max Weber, but also by the diffusion of literacy and education and the development of manufacturing and the capitalist-Fordist organization of labor with its related socio-economic factors. In this perspective it seems that macro-socio-economic factors do not play a minor role in the promotion of violence:

> In fact, starting from the mid 1970s, following a phase of broad consensus as to shared progress and the effective converging of standards of living thanks to the post-war boom and the development of the welfare state, French society has gradually experienced economic-social-spatial segregation processes sparking stress, depression, anxiety, frustration, resentment, aggressiveness and anger in a portion of the population. Associated with the constant expansion of consumerism, this state of affairs represents the main factor contradicting the overall process of pacification that seems to have every reason to continue or even quicken the pace of the expansion started at the end of the Middle Ages and pursued through the beginning of the modern era. The issue raised, then, by the analysis of violence in our "post-modern" society is also, *in fine*, that of the democratic distribution of wealth, which shapes living conditions (Mucchielli, 2010, pp. 825–826).[39]

In summary, I think it is important to clearly acknowledge that humans are especially prone to violence at a micro-level, like Collins' theory contends, and that, for instance, the macro-problem of poverty is not so relevant in triggering violence, as politicians often deceptively and emphatically state – chiefly to divert public opinion from their own failures. Collins' conclusions teach us – in a kind of renewed Stoic way – that through culture and moral education we can offer a chance to

[38] I have to add that this critique of Elias does not affect his analysis of the decivilizing processes in Germany and the rise of the Nazis. His books still preserve all the explanatory power of an informed historical explanation of the political and social transformations that create the condition of possibility for huge and extended episodes of violence.

[39] The violent attacks against democracy in Western countries we face in our current times are described in the previous chapter, sections 3.5 and 3.6.

those human beings "who do not like violence" (at least as individuals) to *choose* to shape – at their *micro*-level – both their capacity to control tension, anger, fear, forward panic and so on, and their capacity to prevent themselves from falling into situations where it is possible to become a victim. Of course I have to stress the fact that this outcome of *micro*-sociological theory of violence, which suggests wisdom, paradoxically resorts to emphasizing the relevant role of *macro* cultural and ideological aspects, given the fact that only culture, morality, and ideologies (and the "collective axiologies" I have illustrated above in this section) can provide the cognitive conditions and contents that an individual can pick up – obviously if they are available – to shape his ways of reacting to, controlling or avoiding violence. Admitting that Stoic wisdom can offer the moral and emotional instruments to control and try to avoid violence, frankly I do not see that kind of "culture" as so widespread, taught and learnt, in our contemporary moral niches.

4.6 Building Violent Niches through Bullshit Mediators

4.6.1 Degrading Truth: Bullshit as a Violent Mediator

Disregarding truth is a source of conflicts and violence. The reason is simple: this attitude contrasts with a considerable amount of various moral commitments, which instead value truth highly and invest it with a morally beneficial role. Actually, the community of bullshitters is very large and bullshitting agents very often – even when they are conscious of the fact that they are bullshitting – candidly think of themselves as just adopting a kind of moral conduct. Of course the effect they usually reach is a more or less violent deception and fraud.

The notion of bullshit introduced by Frankfurt (2005) may help describe a fundamental feature of people who are not docile:[40] that is, their systematic carelessness about truth (Misak, 2008; Olsson, 2008). That attitude towards truth favors deception and fraud, so that bullshitting can be related to a potential violent dimension. In fact, its violent dimension emerges when we contrast it for example with the epistemological morality of scientific mentality, which is usually bound up with the concepts of reason, experience, inquiry, empirical reliability and standards or rational belief.

According to Frankfurt, there is an important distinction to draw between a bullshitter and a liar. What differs between the two is that the liar has a general concern about the truth. And this is just because, in order to tell a lie, he has to know what the truth is. Although the liar fails to be cooperative with respect to the content of a certain state of the world, he is indeed cooperative with respect to his attitude towards truth. The fabrication of a lie may require a great deal of knowledge and it is mindful: it presupposes the guidance of truth. More generally, a certain state of mind – namely, an intention to deceive – is required of the liar while making his statement. This attitude is what makes his statement potentially informative. For instance, consider the case of a person telling us that he has money in his pocket, when

[40] On Simon's concept of docility cf. chapter one, subsection 1.3.4.

he has not. His lie is informative as we can guess whether he lied or not. What is interesting about lying is that there is always a reason why a person may not tell the truth: lies and deceit can be detected. Lying is not clueless.[41] People, for instance, have at their disposal both verbal and non verbal cues enabling them to debunk potentially deceiving situations. A minor detail about the way a person dresses may suggest to a man that his wife is cheating on him, and vice-versa. Sometimes people fear the consequences of knowing the truth – the so-called "ostrich effect", therefore they prefer not to investigate. But this does not mean that they would not succeed in their investigation. Quite the opposite!

According to Frankfurt, the case of bullshit is different, as the bullshitter is supposed to lack any concern or commitment as to the possible truth-values of what he says. He does not just reason from "incomplete data", like in major cases of ordinary practical reasoning (for example abduction), he is rather especially expert and smart in reasoning from "indifferent data". What turns out to be extremely puzzling is not the content, but his attitude. For instance, a liar voluntarily gets a thing wrong. But in doing so he conveys a certain commitment to the truth-value of what he claims. A bullshitter does not care about it. As just mentioned, a liar has a deceptive intention that can be detected. Whereas, the case of the bullshitter is different. When a person believes *P*, she intends to believe *P*. And this intention becomes meaningful to other people. In the case of bullshitter, he believes without any real intention to believe what he believes. So, what really defines a bullshitter is his attitude towards truth: he fails to be committed to truth. He simply does not care whether what he says is true or false, accurate or inaccurate.

This analysis about bullshit allows us to argue that bullshitters are basically not docile: they are simply careless about the beliefs they hold and thus they do not allow what they know to be passed on to other people. Roughly speaking, what comes out of the bullshitter's mouth is *hot air* or *vapor*, meaning that the informative content transmitted is *nil*.

As Frankfurt brilliantly argued, a bullshitter fakes things. And his faking has important consequences. For instance, it may completely mislead whoever partners him in the conversation. This leads to the role played by second hand knowledge and its connection with docility. As argued by Simon, we do lean on what other people say: in chapter three[42] I pointed out that gossiping fallacies is cognitively successful, as they are based on our making use of others as information sources. Trust, for instance, is not informatively empty. One decides to trust another person because she has reasons to do so. As already noted, there are a number of clues we make use of in order to consider a particular source of information (a person, for instance) as trustworthy or not. What happens then to a bullshitter?

A bullshitter does not really intend what he says he believes in. She does not have any concern about the source of what she chooses to believe in. It just so

[41] Curiously, Arciuli and colleagues (2010) maintain that even a pre-linguistic instance like "um" could be a clue pointing out that a person is lying. In a series of experiments they reported that, when lying, people used instances of "um" less frequently than those when telling the truth.

[42] Cf. subsection 3.2.2.

happens that she believes. If so, then information transmission becomes highly *noisy*. Here we come up with another fundamental difference from lying. As already maintained, a lie is not informatively empty, because people have various mechanisms for detecting lies. Our lying detector is based on our ability to read others' minds. Basically, we can guess that a person might lie, because we know that we can lie. We read people's intentions. Would we say the same about bullshitters? Do we analogously have a sort of bullshit detector? However trivial this question might be, our answer is that we have nothing like that.

Following Frankfurt, I claim that a bullshitter is defined by the kind of attitude he has towards truth: he exhibits no commitment regarding what he came to believe in. Our take is that we can infer that he is bullshitting only because we are already familiar with (or expert on) what the bullshitter is talking about. The cues that are meaningful to us are only related to what he is talking about. But, as one can easily note, in the case of second-hand knowledge, this is precisely what is missing. This seems to be a kind of vicious circle, as we would need what we lack (knowledge) in order to detect bullshitting.

The impossibility to detect bullshitting also concerns another messy aspect of human cognition: the epistemic bubble. As illustrated in chapter three[43] an epistemic bubble is a cognitive state in which the difference between *knowing that P* and *believing to know that P* becomes phenomenologically unapparent. The radical thesis held by John Woods is that this epistemic embubblement cannot be avoided by *beings like-us*. We may refer to bullshitting as a special case of epistemic embubblement. By definition we are *embubbled* with respect to a certain piece of knowledge. The embubblement is what allows us to make a sort of leap of faith. We should not believe that we know that *P*, but we actually do. In the case of bullshitting the object of our embubblement is believing, meaning that bullshitters simply believe they believe, when they do not. Or, as Frankfurt put it, the bullshitter "[...] misrepresents what he is up to". However weird this formulation might seem to be, it captures the very essence of bullshitting, that is, the absence of care about the checking mechanism of the plausibility of what one believes in.[44]

4.6.2 Deception through Confabulating about Truth

There is common ground with confabulating that is worth mentioning to further investigate the cognitive dimension of bullshit. In four words, confabulations can be described as: false reports about memories (Hirstein, 2009). Quite recently, a number of studies have been conducted in order to shed light on the very nature of confabulation. One of the most interesting conclusions worth mentioning is that confabulation would be due to a deficit in the monitoring of reality (Schnider, 2001; Fotopoulou et al., 2007). This deficit is explained at a neurological level by the effects of focal lesions to the medial orbitofrontal cortex (Szatkowska et al., 2007).

[43] Cf. subsection 3.2.2.

[44] On some more "positive" aspects of bullshitting, for example in favoring creativity and discovery through metaphors and imagery, cf. (Perla and Carifio, 2007).

Basically, confabulating patients lack those mechanisms enabling them to inhibit information that is irrelevant or out of date. For instance, they are not able to distinguish between previous and currently relevant stimuli (Schnider, 2001). As a consequence of this deficit, they are simply unable to control and assess the plausibility of their beliefs. Some confabulation is extremely puzzling due to its weirdness, and has already been recognized as a symptom of particular syndromes.[45] For instance, patients affected by Anton's syndrome deny being blind when they are, those suffering from Capgras' syndrome think that their relatives have been replaced by impostors. Amazingly, patients that have been diagnosed with Cotard syndrome think that they are dead. More generally, confabulation emerges as human beings have a natural tendency towards "coherencing" and filling gaps.[46] Confabulating patients have to cope with memories or perceptions that are basically false, but that they have been accepting as true because of the reality monitoring deficit they are affected by. Then, the need for "coherencing" creates explanations that result as completely implausible and unacceptable to other human beings. In this sense, like in the case of bullshit, confabulating is not lying, but confabulators are affected by what has been called *pathological certainty* (Hirstein, 2005). Basically, confabulators do not doubt, when they should doubt.

In bullshitting, all these effects we have briefly surveyed are present, although they are not resulting from a deficit but brought about by the mindless attitude bullshitters have regarding truth-value. Drawing upon Hirstein's (2009) phenomenological definition of confabulation we define bullshitting as follows. A person is bullshitting when:

1. She believes that *P*.
2. Her thought that *P* is ill-grounded.
3. She does not know that *P* is ill-grounded.
4. She should know that *P* is ill-grounded.

To capture the difference between confabulation and bullshit, points 3 and 4 are very important. In confabulating patients a person holds an ill-grounded belief, because of her neurological deficit. As already mentioned, she has a reality monitoring deficit that impairs those mechanisms inhibiting irrelevant or out of date information. If she were a normal person, she would know that *P* is ill-grounded. The case of bullshitter is quite different. What overlaps between the two cases is that both of them should know that *p* is ill-grounded, when they do not. However, what does not overlap is the reason behind that. Bullshitters have no lesions preventing them to meet a basic "epistemic" standard of truth. *P* is ill-grounded because of their careless attitude. So, in the case of confabulators they simply get things wrong, because they cannot discern relevant and up to date information from that which is not. Conversely, bullshitters do not get things wrong, but, as Frankfurt put it, they are not even trying. Once we distinguish between these two states of the mind, then we can analogically claim that a bullshitter is, in a way, confabulating, inasmuch she holds

[45] An exhaustive list can be found in (Hirstein, 2005).

[46] On confabulation and religious cognition cf. chapter six, section 6.2, in this book.

a belief that she would withdraw or, at least, re-consider, if she tried to get things right.

Last but not least, there is another feature characterizing the bullshitter. Usually, when arguing a person is supposed to have some expectations. A person enters a discussion for some reason. For instance, he wants to gain a favor or he wants to display a certain identity. At some other time, he wants to express his dominance over another or he simply wants to entertain himself or others taking part in the conversation. Usually, all of these reasons lead the arguer to *frame* what he is doing in a particular way rather than another depending on what his purpose is (Goffman, 1986; Hample and Young, 2009). Now, the question is: what kind of *frame* – to use Goffman's terminology – does the bullshitter select? What does he hold in his mind? What purpose?

My take is that the bullshitter speaks without planning and in doing so he simply *blurts* something out. As Hample and colleagues put it, blurting "comes simply out of cognition without alteration" (2009, p. 23). Basically, he acts as if he responded *unreflectively* and *mindlessly* to an invitation to have his say – no matter whether real or imaginary. Instead of admitting that he has nothing to say (and hence holding his peace), the bullshitter enters a debate or discussion without questioning if he actually has something to say. I previously mentioned that confabulators suffer from a sort of pathological certainty, meaning that they do not come to doubt when they should. In the case of bullshitters, they suffer from what could be called *pathological opining*, meaning that they cannot abstain from opining about topics clearly beyond their knowledge.

I have one last remark on frames and bullshitting. As already mentioned, a person entering a discussion usually bears in mind a certain understanding of what is going on. That is, he (more or less tacitly) interprets why others are arguing, i.e. their frame. Indeed, it is important to understand what frame a partner in conversation is using as it avoids misunderstandings and, upon occasion, facilitates cooperation and information exchange. In contrast, when two or more people do not share the same frame, cooperation is less likely to come about. One can even project a false frame for his partner(s) and then hack the conversation to have a private gain. When this happens, Hample and colleagues (2010) argue that the victim "is said to be *contained*" in the swindler's frame. I contend that something similar happens with the bullshitter. I argued that the bullshitter has no interest or purpose in his mind when entering a conversation but he simply responds to a (sometimes imaginary) invitation like reacting to a stimulus. Hence, he actually does not project any frame but, insofar as he is engaged in a conversation, he gives the impression he is employing a frame. It is worth noting here that the deceptive (and violent) nature of a bullshitter does not rest on the projection of a false frame, but on the fact that his victims are naturally inclined to assume that he actually projects a frame when he does not. The bullshitter's conversational contribution is nil, but nevertheless he is taken seriously, as if he were really contributing.

The major consequence is that an imaginary frame is projected and the bullshitter's victims get into it and are thus *contained*, just like in the case of a fraudulent attack. Indeed, once they realize that the frame was merely imaginary, they may experience all that as fraudulent and aggressive. As a consequence, the bullshitter may even try to justify his misbehavior, for instance, saying that he did not really mean to trick his conversation partners, but that he just wanted to play. This sort of post-projection does not solve the problem. Conversely, it may even make things worse, as the newly-projected meaning sharpens the dissonance felt by the victims.

In order to make this point clear, let me introduce the notion of lamination in connection with the aggressiveness of playful arguments. According to Goffman, when a certain activity is re-framed, or *re-keyed*, it is said that a *lamination* (or layer) is added to that activity – the so-called *natural strip* that is the base appearance and character (1986, p. 82). As Hample et al. (2010) put it, laminations cover behaviors and/or activities "in the same way a good veneer can cover inexpensive wood". Basically a lamination is a sort of new appearance of the natural strip that completes the process of meaning transformation.

Indeed, the prerequisite for a lamination to be perceived is that the partners in conversation apply the appropriate frame. So, a lamination does not conflict or erase all the pre-existing meanings that a certain activity – for instance, arguing – may have. But it somehow becomes a new layer that metaphorically *laminates* it providing new interpretations, but, at the same time, it also favors the emergence of dissonance, in case the participants are not able to successfully re-key the activity.

To make an example of lamination, consider a valued object, say, a ring. It can be transformed into a gift, a promise of marital fidelity, a bribe, or an unwelcome, improper advance that can lead to bad consequences. All these are laminations that are added to the natural strip – the valued object *per se*.

The case of playful arguments is another example worth considering here especially in relation with bullshit and its violent dimension. As already mentioned, one can start arguing for a number of reasons: to exchange information, to promote oneself, or to pass time without having any particular purpose in mind. This is the case of playful arguments, that is, when a person starts arguing just for entertainment, reiterating arguments that are not really meant to reach a conclusion. This may be the case of a bullshitter, as I mentioned above. In response to possible complaints regarding his uncooperative attitude (his contribution is informatively empty), the bullshitter may shrug and kindly say that he just wanted to be playful.

Indeed, a playful argument can be beneficial in promoting cooperation. When both partners are able to apply the right frame, they can really *enjoy* arguing. And, insofar as they enjoy it, reciprocally rewarding feelings usually encourage people to strengthen their bonds and, of course, adopt a cooperative attitude. But this cannot be taken for granted. A playful argument – considered as a lamination – can be completely misunderstood if one fails to apply the right frame and use the correct key. So, as far as one misunderstands it, a playful argument may turn out to be a verbal source of aggressiveness signaling, for instance, competitiveness and dominance or, as in the case of bullshitting, carelessness about the truth and disregard for one's partner in a discussion (Hample et al., 2010).

This may appear quite paradoxical. We are usually told that arguing – and so playful arguments – are alternative to violence, as they are thought to be the result of civil attitude and cooperative temperament. Interestingly, Hample and colleagues

(2006; 2009) reported in a series of experiments a correlation between sensitivity to playful arguments and a variety of measures related to aggression and uncivil interpersonal relations. They also suggested (Hample et al., 2010) that arguing is closely related to *verbal combat* and *verbal force*. More precisely they claim – using Goffman's terminology – that verbal combat might be the natural strip for arguing – its base appearance. What follows is that more civilized approaches to arguing – i.e. those considering it as a means for finding a mutually accepted solution, for persuading, or for exchanging valuable information – are basically transformations of the natural strip. Consequently, they require people to develop quite sophisticated skills in order to be successfully employed. For instance, an argument might be cooperative, but the partners involved in conversation should put their respective interests on a par. This simply means that they should put aside any competitive or uncooperative dispositions that may threaten mutual exchange and cooperation in general.

In sum, contrarily to what one might think, a playful argument can be violent and aggressive as a lamination, only partly covering the natural strip of arguing. It is *opaque*. Accordingly, it would partly reveal the natural strip of arguing, which is basically a verbal way of putting somebody off in competition for a certain resource. This is in line and perfectly coherent with what I argued in connection with the violent nature of language and fallacious reasoning in chapter three above and in my previous book *Abductive Cognition* (Magnani, 2009, chapters seven and eight).

4.7 Violence and the Right and the Duty to Information and Knowledge

It is clear that bullshitters (intentional or unintentional) do not care about knowledge and truth. They are not inclined to true knowledge and do not possess any, so to say, "epistemological morality" (Frankfurt, 2005). The conflict with truth-believers is ineluctable. Now, my question is, are human beings ever *really* inclined to truth? Can we consider ourselves "epistemological/logical creatures"? An analysis of the problem of the right and duty to information and knowledge can help us deepening our understanding of the issue.

How is knowledge (and information in general, as bounded up with the concepts of reason, evidence, experience, inquiry, reliability, and standard or rational belief) typically regarded in current ethical debates? Do we already recognize the intrinsic value of knowledge? These are open questions requiring much discussion. Some ethicists contend that an interest in information could indeed have intrinsic moral significance, and they pose questions about a possible fundamental right to information *per se* that can serve as a foundation for information ethics (Himma, 2004). The conclusion seems to be that there is not, in fact, a general right to information, that we might have a right only to some kinds of information – religious, for example. In this scenario, the right would derive from an interest in specific content rather than from the content of other rights – for instance, the fundamental right to life and liberty. This would mean that the right to that information is fundamental but not general, so that "providing a comprehensive, plausible theoretical analysis of our information rights will be a complex and nuanced undertaking" (ibid.)

Moreover, there are two kinds of intrinsic value we must consider. The value of being regarded as an end in itself (intrinsic value in the descriptive sense) is different from the value ascribed because respect is seen as obligatory (intrinsic value in the normative sense): unfortunately, assigning intrinsic value to X is neither a necessary nor a sufficient condition for having a fundamental right to X. We might have a right to property, for instance, even if that property lacks intrinsic value. It seems there is no fundamental right to true information *per se* despite the fact that we intrinsically value information.

Anyway, having an informative or cognitive nature is not sufficient for having intrinsic value *per se*, and, consequently, intrinsic value ascribed in this way to knowledge *as duty* must be considered only circumstantial. While one day people may intrinsically value all true information and knowledge, we are not those people. On the basis of such considerations, it seems that there is no fundamental right to information *per se* despite the fact that we sometimes value information intrinsically.

Information also has an extraordinary instrumental value in promoting the ends of self-realization and autonomy.[47] But not all attitudes toward information promote these ends. The questions of whether and how people attribute instrumental value to information are empirical. Usually, it is that part of information we call "knowledge" that endows information with intrinsic value.

Unfortunately, there is no evidence that knowledge is universally regarded positively – some people, for instance, regard higher education as an elitist waste of time and money, other people are constitutive bullshitters – and the result is, at best, a nebulous idea of information's moral significance.

We can endorse a moral commitment toward information and knowledge at an intuitive level. An example related to our discussion of knowledge may be useful here: people ought to value a piece of knowledge either for its capacity to describe in an interesting new way some part of the world or for its ability to detect the unethical or violent outcomes of some technology or technological product that has been discovered, for example, to be a potential carrier of global damage. That is to say, there is no right to information that is both general and fundamental – there is no fundamental right to information *per se*. This does not imply, however, that we have no moral rights, general or otherwise, to information. From this ethical perspective, the moral meaning of my motto "knowledge as duty" becomes clear. Knowledge "has to be" a duty, and it relates to specific rights I aim at affirming.

I am engaged in assessing extended and circumstantial rights to information and knowledge. Why? Because knowledge – even if we know that not only bullshit and fallacies but also sound, true, rational, scientific, knowledge can promote violence – at least in certain circumstances typical of our technological era – can prevent violent results. Olsson usefully notes that knowledge is distinctively valuable in *all normal cases* and it is *normally* instrumentally valuable, both for individuals and societies (2008, p. 102). The right to attain knowledge in particular and information

[47] I have devoted my whole book (Magnani, 2007b) to stressing the importance of knowledge in our technological era: "knowledge as a duty".

in general is an important entitlement that I hope will become a generally accepted, objective value.

I am perfectly aware that a commitment to truth and knowledge might sound just like a dream dating back to the Enlightenment. Olsson also notes that "only a fraction of [society's] members are successful truth seekers and truth tellers, none of whom is capable of detecting signs of deception" and that, in spite of this fact, society can still converge on the truth, at least in the perspective of research results of social epistemology (2008, p. 106): the rest of human beings would be unreliable in the way they form beliefs, which also makes them unfit for telling the truth. In this sense it has to be clearly stressed that these people, being unreliable as informants, are bullshitters in a peculiar sense of that word, because they are not intentionally bullshitting – which makes them, in a way, more similar to confabulators. However, Olsson concludes, quoting the research on social epistemology by (Hegselmann and Krause, 2006):

> But I strongly suspect that the main philosophical point that can be extracted from Hegselmann and Krause's work is still valid: There may be informational exchange process which, combined with the social pressure to average among peers, compensates for the fact that a part of the population lacks direct contact with reality by ensuring that that part nevertheless enjoys an indirect access to the way things are via the beliefs and reports of reliable peers, whose views they are forced to take into account. None of this assumes any special ability on the part of the individual inquirers to detect signs of fraud. Pace Frankfurt (2006, p. 13), communal convergence on the truth does not require that "we can reasonably count on our own ability to discriminate reliably between instances in which people are misrepresenting things to use and instances in which they are dealing with us straight"

In this perspective both intentional and unintentional bullshitters would be – so to say, at the limit – still docile, and so their threat to docility which I illustrated in the previous section is counteracted by the global and indirect social effect of informational exchange, which would favor a convergence on the truth. After all, bullshitters also definitely need to have a contact with truth and knowledge, even if they are indisposed toward them to such an extent.

If we want knowledge to be considered a duty of the supranational society, then the goal should be to generate, distribute, and use knowledge to encourage economic and social development. Doing so, however, demands a far greater deal of research about ways of creating and constructing knowledge for public and private use and, subsequently, on how to apply this knowledge to decision-making activities. The very future of the human species is at issue: if we accept that it is our duty to construct new bodies of knowledge, both public and private, we will be better able to avoid compromising the conditions necessary to sustain humanity on Earth.

Unfortunately, at the beginning of the twenty-first century, we can see that in Western European and North American societies, at least, knowledge and culture do not appear to be a priority. Even though there are funds, institutions, and human resources devoted to reproducing cultural activities, the vitality of culture remains in question. For example, we face a real breakdown in many humanistic traditions, in spite of their potential in being of great help to many of the aforementioned issues, research in humanist fields – from philosophy to sociology to cognitive science – is voiceless before the problems of society. The intellectual exchange among these disciplines occurs mainly within universities and, as a result, creates a kind of academic short circuit. Intellectuals outside universities are similarly isolated, and their work is rarely woven into the fabric of society in general. It seems that people who do not partake of intellectual circles value research only as far as it relates to technology or to some branch of science related to technology, but such knowledge, sadly, is not sufficient to build a knowledge-based society. Technology, while important, is blind to many problems we will face in the future, and knowledge built only from *strictu sensu* scientific fields has limited value, as knowledge drawn only from humanities would have.

Indeed, both scientific and technological innovations may jeopardize our own safety. In this sense, a precautionary principle is usually called for. However, as the sociologist Frank Furedi (2002) has recently argued, an exaggerated presentation of the destructive side of science and technology may limit people's aspirations and dissipate their potential. Rather than exaggerating the negative side and celebrating safety, as mass media often do, we should acknowledge that it is knowledge and science that give us all the instruments to prevent possible destructive outcomes. As Furedi put it, "it is the extension of human control through social and scientific experimentation and change that has provided societies with greater security than before".[48] Knowledge is a duty, but not just any knowledge, it must be a well-rounded, varied body of information drawn from many disciplines, and even then it must be useful, available, and appropriately applied.

Indeed, we must remember that even in rich countries, knowledge gleaned only from science and technology can be difficult to obtain and exploit. Let us imagine a country with many atomic power plants that require securing. To achieve this goal, highly qualified employees must be produced and maintained, but doing so is contingent upon whether or not collectivities make adequate resources available to ensure the specific training and general education of suitable experts. Seen in the light of the Soviet Union's decline, the disaster at Chernobyl nuclear power plant is a striking real-world example: it is extremely probable that the nation's dissolution weakened agencies charged with training workers at nuclear sites and diminished the general public awareness of technological and scientific issues. In analogous cases, of course, problems have arisen not only from lapses in training but also from corruption and/or a lack of appropriate ethical knowledge. Moreover, even collectivities with highly qualified personnel and an advanced level of scientific-technological "know-how" can be vulnerable to accidents when that knowledge is used inappropriately, not just when there is a failure of ethical (and, of course, political and legal) controls and knowledge.

Another important issue is related to so-called unintentional power – that is, the power to violently harm others in ways that are difficult to predict. This problem, which is widespread in computing systems, can certainly result from a lack of knowledge about possible outcomes: "one difficulty with unintentional power

[48] On the so called culture of fear, see also (Glassner, 1999).

[...] is that the designers are often removed from situations in which their power (now carried by the software they have designed) has its effect. Software designed in Chicago might be used in Calcutta" (Huff, 1996, p. 102). The following are eloquent examples of unexpected behavior in computing systems that show how knowledge becomes a more critically important duty than ever before and must be used to inform new ethics and new behaviors in both public policy and private conduct. Huff eloquently describes some particularly bad cases:

> In the summer of 1991 a major telephone outage occurred in the United States because an error was introduced when three lines of code were changed in a multimillion-line signaling program. Because the three-line change was viewed as insignificant, it was not tested. This type of interruption to a software system is too common. Not merely are systems interrupted but sometimes lives are lost because of software problems: the violence perpetrated is hidden and invisible. A New Jersey inmate under computer-monitored house arrest removed his electronic anklet. "A computer detected the tampering. However, when it called a second computer to report the incident, the first computer received a busy signal and never called back" [...] While free, the escapee committed murder. In another case innocent victims were shot to death by the French police acting on an erroneous computer report [...]. In 1986 two cancer patients were killed because of a software error in a computer-controlled X-ray machine[49] (Huff, 1996, p. 102).

4.7.1 Why Epistemology Matters to Violence?

Furthermore, a recent paper addresses, in an amazing way, the issue "Why epistemology matters to intelligence", which we could thus paraphrase "Why epistemology matters to violence": Bruce (2008) examines how understanding epistemology can highlight how knowledge in "intelligence" is created, why some ways of producing it are more reliable than others, and how bad knowledge, from the epistemological point of view, can produce violence. Focusing on an important failed National Intelligence Estimate (NIE), he identifies epistemologically induced sources of error (and so of consequent harm and/or fraud) in the analysis, and explores possible correctives, also showing how self-corrective mechanisms typical of scientific method and mentality can improve reliability and should become a more integral part of the some important analytical processes.

A tragic case of fatal error is reported and analyzed: the ill-fated October 2002 Nation Intelligence Council (NIC) on Iraq's Weapon of Mass Destruction (WMD), which certainly was among the most important factors that triggered the subsequent war. The NIC was not really "intelligent", at least in the epistemological sense of the word: four "bad" ways of knowing each contributed to the NIC's flawed judgments, as well as the consequences of failing to incorporate error-detecting and reducing measures. They were – classically – various forms of fallacious reasoning: i) the appeal to authority (faulty), ii) the unquestioned referral to habits of thought (mixed with a lack of evidence and a barrier to alternative analysis), iii) flawed and merely

[49] (Gotterbarn, 2001).

persuasive reasoning (not seeing weapons seemed to provide evidence that Iraq had them), iv) minimal role of empirical evidence (further explained away as denial and deception or discounted because it did not support the habitual knowledge of a robust and active WMD capability). NIE was in an epistemological "perfect storm", the author concludes (p. 182), generated by a constitutive lack of any process of self-corrective cognition.

Similarly, a perfect "epistemological/methodological storm" accompanied (and was among the cardinal "ingredients" of) various business disasters due to the confirmation bias of key individuals who influenced, or outright determined, the path undertaken by companies in trouble. Let us quote this eloquent passage by Cohan (2007, p. 66):

> For example looking back doomed January 2001 merger of AOL and Time Warner at a price of 207 billion, I'm by no means the only one who asked why such a "perfect" merger went on to lose billions of dollars in net income. Almost seven years later, the company is reportedly worth 126 billion less than it was at the time of the deal. But the cause of this calamity is not subterfuge, it's confirmation bias. At the core of the problem was Gerald Levin, then Time Warner CEO, who acted alone – deliberately ignoring the advice of his subordinates who did not want to do the deal – because Levin was determined to leave Time Warner with a memorable legacy. [...] He kept all but a small number of Time Warner executives in the dark [...] about the pending deal until hours before it was announced. Moreover, even fewer Time Warner executives supported the proposal.

To summarize, in this case the decision makers presented a tendency to seek out only information that confirmed and reinforced their belief in the overwhelming need to conclude the transaction quickly, also putting aside information, when inconsistent with their plan, as a kind of irrelevant noise. Of course the result was a violent business disaster, which usually incorporates violent structural outcomes, for example job loss.

Finally, a subproblem regarding knowledge as duty concerns various processes associated with concrete activity in scientific research. For example, biomedical research seems so important that it warrants the moral obligation not only to pursue it but also to participate in it, at least when minimally invasive and relatively risk-free procedures are at stake. In this case, the duty concerns individuals who would be required to participate in serious scientific research. The issue is fraught with a variety of moral and legal controversies and violent chances: pragmatically balancing dangers, risks, and benefits; vigilance against, and the ability to identify, wrongdoing; fully informed consensual participation or possible (justifiable) enforceable obligation; the interplay between individual human rights and the interests of society; the possibility of industrial profits from moral commitment; the potential exploitation of poor people; the role of undue inducements; the special case of children, and so on (Gotterbarn, 2005). When one participates as a subject in a scientific experiment, the commitment to scientific knowledge is not necessarily derived from the idea that knowledge is a good thing in itself, but rather from the related duty of beneficence, our basic obligation to help people in need.

4.8 Constructing Morality through Psychic Energy Mediators

4.8.1 Moralizing Nature, Moralizing Minds

I have contended above that human beings continuously delegate and distribute cognitive functions to the environment to lessen their limits. They build models, representations, and other various mediating structures, that are thought to be good for thinking. In building various mediating structures and designing activities, such as models or representations, humans alter the environment and thus create those *cognitive niches* I have described at the beginning of this chapter. What is the role of this environment modification in the perspective of some psychoanalytic insights and in the process of moralization of a collective? In the following sections I will illustrate some fundamental aspects of the aforementioned interplay in the light of Jungian observations about the cognitive role of what we can call external *energy mediators*. I will outline some basic sources of moral/violent behavior, which the psyche can easily find and pick up in the environment, after a process implying its moralization. In the perspective of this book the following analysis further stresses, "also" taking advantage of a psychoanalytic framework, the naturalness of morality and its emergence in a distributed interplay between minds (even in an *unconscious* way) and suitably cognitively reconstructed environments, so favoring group identification and further chances for conflicts and violence.

Recent research on "identity" processes in individual human beings have stressed the role of the non-human environment, for example in the case of plants, animals, wind, and water. Let us consider the case of trees. Jung already sees the tree as participating in the processes of the formation of the psyche (and in moral aspects of it) in so far as it constitutes an archetype in the collective unconscious. Darwinians sometimes speak of an innate emotional affiliation of humans with other living organisms so that their preferences would have been shaped over millennia through interactions with features of the environment helpful to the survival of the species in its early development (for example they offer prospect in predation and refuge).

The so-called phenomenological approaches rely on various metaphors, which can be endowed with *moral* meaning: roots, trunks, and canopies mirror the infernal, earthy, and heavenly domains; flowers, fruits, and colors supply subsidiary arguments for human identity – in this respect trees certainly offer more than grass, the most universal and successful of plants. Gibsonian ecological psychologists emphasize that trees are the source for humans of multiple innate affordances that provide various action possibilities (climbing or playing hide and seek just to mention a couple) and the possibility of making artifacts (rope swing, tree forts) and of satisfying human needs (shelter, food, fuel, and medicine) (Sommer, 2003).[50] Finally, eco-psychologists maintain that, beyond the individual self, there is an ecological self that is "nurtured through the contact with the natural environment" (2003, p. 191):

[50] In this last case, in modern humans, the satisfaction of needs occurs through the mediation of sophisticated cultural schemes (such as in the exploitation of trees to counter pollution, to protect privacy and limit noise and to shape the person-home-neighborhood interplay on which survival of the individual depends).

from this perspective trees are important for city residents because they provide contact with natural rhythms, life forms, seasonal markers, and the gentle motion and sound of rustling leaves.

Moreover, through anthropomorphic interpretation (for example in children) an external object such as a tree or a squirrel is perceived as being similar to oneself and humanlike in certain respects, where the identity of both the object and the observer is reached (Gebhard et al., 2003). What is important for us is that this interpretation also includes moral aspects, such as freedom or the perception of nature as affectionate and caring, which are attributed in an isomorphic way to organisms or natural objects making it possible to grant greater independence and environmental moral value to them (Kahn, 2003). In turn, in the case of physiomorphism, human experience can be interpreted in terms of non-human nature or natural objects. Both anthropomorphism and physiomorphism can start a never ending, cyclic process of mirrors:

> Thus we may draw upon experiences with natural objects to understand ourselves (physiomorphism), but in turn our representations of these natural objects will have arisen by interpreting them in terms of ourselves and our personal experience (anthropomorphism).[51] [...] But attaching subjective meaning to an object, the object and the self become mentally intertwined and a unique relationship between the two is established. It is perhaps in this manner that external objects contribute to the formation of personal identity (pp. 104-105).

In this perspective perceiving an object as humanlike can be related to the problem of its possible *moral worth* (thus activating a kind of micro morality at the personal level) and, consequently, for example in the case of animals or trees, it can lead to the more extended awareness of the need for their protection and preservation (a "biocentric" perspective that expresses a kind of transpersonal macromorality): killing plants can be seen as analogous to killing humans (Rest et al., 1999). Human bodily and mental characters play, through anthropomorphism, the role of moral mediators, that can permit nature to be moralized, at least at the level of micromorality.

Following Jung, a "differentiated" psyche is a hybrid product of the interplay between internal and external sensory representations. Conversely, external representations are continuously re-built through the delegation of psychic contents. They play the role of moral mediators like, for example, in the case of the ancient religious images such as the Trinity or the mystery of the Virgin birth, which are clearly endowed with a moral worth. These representations are placed and produced "out there", hybridized with and within material supports from the external environment,

[51] It is interesting to note that also René Thom recognizes this fact in the framework of his mathematical catastrophe theory: "Men (as well as prehominids) were early incited by group living to build up some representation of the behavior of their kind, a representation, in particular, of the paths of their affective regulation. As a result, any external entity thought of as being individuated tends, by empathy, to be imagined after the manner of a living being" (1988, p. 16).

made autonomous with respect to their human origin but more or less available,[52] so they in turn can suitably be "picked up" by human beings and re-represented in their brains, forming moral attitudes and dispositions. Of course they are basically picked up to favor extension and "consolidation of consciousness" and "to keep back the dangers of the unconscious", the "peril of the souls" given the fact that "mankind always stands on the brink of actions it performs itself but does not control. The whole world wants peace and the whole world prepares for war" (Jung, 1968a, p. 23). They are also picked up through imitation and sometimes to contrast what Jung calls "individuation" processes. "[...] the more beautiful, the more sublime, the more comprehensive the image that has evolved and been handed down by tradition, the further removed it is from individual experience. We can just feel our way into it and sense something of it, but the original experience is lost, [the images] have stiffened into mere objects of belief" (p. 7). In these externalized images that became symbols of dogmatic archetypical ideas – "collective unconscious has been channelled [...] and flows along like a well-controlled stream in the symbolism of creed and ritual" (p. 12).

From this perspective, symbolisms (as for example the ones embedded in ritual sacrifices) implicitly constitute an objective collective cultural structure carrying axiological and moral-deontological significations, also beyond any awareness of the agent that is affected by them. They can unconsciously restitute the possibility to pick up moral attitudes and dispositions, like in the case of the scapegoating mechanism, which indeed appears to modern morality as a vestigial aspect of ancient moral behaviors.[53] Thus modern spontaneous forms of scapegoating, such as bullying and mobbing, which just present a violent face to the modern ethical mentality, are de facto linked to ancient forms of moral organizations of collectives in both gregarious non-human animals and in humans.

For Jung consciousness is basically made up of suitably internalized external "observations". In sum, psyche is a hybrid product of the interplay between internal and external representations. Conversely, once representations are internalized, we have them at our private disposal for rebuilding – possible creative and new – external representations through delegation (projections) of their psychic contents. First, this process presents the growth in complexity of the psyche where archetypes manifest themselves through their capacity to – unconsciously – organize images and ideas; later on archetypes, thanks to the above external/internal interplay, "by assimilating ideational material whose provenance in the phenomenal world is not to be contested, [...] become visible and *psychic*" (Jung, 1972b, p. 128).

From the perspective of the individual psyche Jung's remark is central and clearly stated. Acknowledging the importance of the above hybrid interplay, he says that the word "projection" is not really appropriate, and the role of the senses is fundamental "[...] for nothing has been cast out of the psyche; rather, the psyche has attained its present complexity by a series of acts of introjections. Its complexity has increased in proportion to the despiritualization of nature" (p. 25). Even more clearly, Jung

[52] Jung says that "[...] archetypal images are so packed with meaning in themselves that people never think of asking what they really do mean" (1968a, p. 13).

[53] Cf. chapter two, subsection 2.1.5.

points out the relevance of the hybridization processes: "The organization of these particles [of light] produces a picture of the phenomenal world which depends essentially upon the constitution of the apperceiving psyche on the one hand, and upon that of the light medium on the other" (p. 125).

I have already said in section 4.2 above that we can hypothesize a form of coevolution between the structural complexity of the psyche and that of externally delegated cognitive systems. Brains build external representations in the environment, learning new meanings from them through interpretation (both at the model-based and sentential level), after having manipulated them through motor actions.[54] When the internal fixation of a new meaning is reached through internalization – like in the example above, concerning the formation of a religious icon endowed with its moral/pedagogical value, that for example fixes the character of a suitable rite – the pattern of neural activation no longer needs a direct stimulus from the external representation in the environment for its construction. In this last case the new meaning (for example, a moral one) can be neurologically viewed as a fixed internal record of an external structure (a fixed belief in Peircean terms) that can also exist in the absence of such an external structure. In Wachandi's example that I will illustrate in the following sections it will become clear how, for instance, a representation (for example embedded in an artifact) can become creative by giving rise to new meanings and ideas in the hybrid interplay between psyche and suitable cognitively delegated environments.

4.8.2 Cognitive/Affective/Moral Delegations to Artifacts and Subsequent Reinternalization

We now have a good conceptual tool at our disposal which can shed light on Jung's considerations illustrated in "On psychic energy". The manipulation of external mediators – and so of external *moral* mediators – (and of the related external representations) is active at the level of moral cognition. However, it is actually a creative way of producing/extending every kind of cognition, in the widest sense of the term, involving rational, emotional, affective, attentional, doxastic, and other aspects of cognition. Jung acknowledges this fact: "The apperceiving consciousness has proved capable of a high degree of development, and constructs instruments with the help of which our range of seeing and hearing has been extended by many octaves. Consequently the postulated reality of the phenomenal world as well as the subjective world of consciousness has undergone an unparalleled expansion" (Jung, 1972b, p. 228).

Furthermore, the creative construction of artifacts (which embed various representations) as cognitive/affective/moral mediators can be seen in the framework of the "psychic energy" flow. The secret of cultural development is the *mobility and*

[54] Let me reiterate that representations and inferences can be sentential (based on natural language), model-based (that is sensory-related, formed for example by visualizations, analogies, thought experiments, etc.) or hybrid (a mixture of the two aspects above together with various manipulations of the world) (Magnani, 2001).

disposability of psychic energy (Jung, 1967), that appears – as a true "life-process" – in phenomena like "instincts, wishing, willing, affect, attention, capacity to work", sexuality, morality etc. It has to be noted that this mobility and disposability of energy is granted by the fact that Jung contends that man, more than other animals, "possesses a relative surplus of energy that is capable of application apart from the natural flow" (Jung, 1972a, p. 47). This "psychic energy flow" is at the core of the various possible processes of progression and regression as adaptations to the environment and to the inner world, as key tools able to satisfy the demand of individuation. In "civilized man" psychic energy is strongly canalized in that rationalism of consciousness, "otherwise so useful to him", but which in turn can become a possible obstacle to the "frictionless transformations of energy" (Jung, 1972a, p. 25). Psychic energy and its capacity to be extended onto the external world is appropriately compared by Jung to the fundamental primitive idea of *mana*, in its capacity to externalize and delegate meanings (or to perform "semiotic" delegations to the environment, in Peircean terms), and potentially consume everything.

In this perspective, "values" (and so moral values) can be considered quantitative estimates of energy that people can attribute to external things in various ways. For example, I have already noted above in section 4.4 that, in the case of ethics, the recent tradition of moral philosophy classifies things that display values attributed by human beings as endowed with "intrinsic values". These evaluations are of course attributed by consciousness, but also by the unconscious (Jung, 1972c, p. 10).

Once externalized and stabilized, values (and so moral values) are available to be picked up. From this perspective the constellations of psychic elements grouped around feeling toned contents (complexes) are related to both inner experience (intertwined with innate aspects of an individual's character and dispositions) and suitably sensory *external environment* representations picked up when available (p. 11). These constellations basically relate to both conscious and unconscious value quantities (affective intensities). Moral values (but also sexual, esthetic, rational etc.), externalized in artifacts, icons and so on, normally belong to an established collective framework (a cluster of collective axiologies, in the case of morality), shared by a relatively stable group of human beings: of course "repression" or "displacement of affect" can give rise to false estimates so that "subjective evaluation is therefore completely out of the question in estimating unconscious value intensities" (p. 10), in turn giving rise to various disequilibria.

Indeed, I have just said that the activation of moral motivations and attitudes is related to the constellations of psychic elements but "also" to both inner experience and suitably sensory external environment representations, picked up when available. I have also said that those constellations centrally relate to both conscious and unconscious value quantities (affective intensities). Consequently it is not difficult to see how, in one individual psyche, various moral disequilibria can happen. For example, it is very easy to find processes of disengagement of a conscious adopted moral framework, caused by the unconscious reengagement of another one, as I will extensively describe in the following chapter of this book. And the new reengaged morality often generates conducts that, in the light of the previous disengaged one, just appear unusually aggressive and violent to the human agent himself. In other words, it is in this interplay between conscious and unconscious that we can see what happens in the case of the so-called "disengagement of morality", when a conscious unbalanced moral commitment is deactivated to give room to another one, often without any clear awareness of the underlying psychic process, favoring various kinds of violent behaviors, either not necessarily perceived as such by the perpetrator himself, or if perceived, still considered by the subject as morally justified.

4.8.3 Artifacts as Memory, Moral, and Violent Mediators

An example of psychic energy delegation presented by Jung, that clearly expresses human culture as a "transformer of energy"[55] is the one related to an artifact built by the Wachandi of Australia:

> They dig a hole in the ground, oval in shape and set about with bushes so that it looks like a woman's genitalia. Then they dance round this hole, holding their spears in front of them in imitation of an erect penis. As they dance round, they thrust their spears into the hole, shouting "Pulli nira, pulli nira, wataka!" (non fossa, non fossa, sed cunnus!). During the ceremony none of the participants is allowed to look at a woman (p. 43).

It is a process of semiotic delegation of meanings to an external natural object – the ground, which applies energy for special purposes through the building of a "mimetic" artifact: an "analogue of the object of instinct", Jung says (1972a, p. 42). The artifact is an analogue of female genitals, an "object of natural instinct", that through the reiterated dance, in turn mimicking the sexual act, suggests that the hole is in reality a vulva. This artifact makes possible and promotes the related inferential cognitive processes of the rite. Once the representations at play are externalized (representations which are psychic values, from the Jungian psychoanalytic perspective), they can be picked up in a sensory way (and so learnt) by other individuals not previously involved in its construction. They can in turn manipulate and re-internalize the meanings semiotically embedded in the artifact:

> The mind then busies itself with the earth, and in turn is affected by it, so that there is the possibility and even a probability that man will give it his attention, which is the psychological prerequisite for cultivation. Agriculture did in fact arise, though not exclusively, from the formation of sexual analogies (Jung, 1972a, p. 43).

Artifacts (and the representations they incorporate) are produced by individuals and/or small groups and left *out there*, in the environment, perceivable, sharable, and more or less available to anyone. It is in this sense that we can classify an artifactual

---

[55] Animals also make artifacts, and Jung is aware of this fact when he acknowledges the role of what he calls "natural culture", see above page 129.

mediator of this psychoanalytic type as a *memory mediator*,[56] which mediates and makes available the *story* of its origin and the related *actions*, which can be learnt and/or re-activated when needed.[57]

Let us come back to the artifact above. Primitive minds are not a "natural home" for thinking about agriculture: together with the cognitive externalization and the artifact – and the subsequent recapitulations – certain actions can be triggered, actions that otherwise would have been impossible with only the help of the simple available "internal" resources. The whole process actualizes an example of that manipulative hypothetical reasoning I have described as a case of manipulative abduction. When created for the first time it is a creative social process, however, when meanings are subsequently picked up through the process involving the symbolic genital artifact and suitably reproduced, it is no longer creative, at least from the collective point of view, but it can still be creative from the perspective of individuals' "new" burdening cognitive achievements and learning. It is possible to infer (abduce), from the artifacts, the events and meanings that generated them, and thus the clear and reliable cognitive hypotheses which can in turn trigger some related motor responses. They yield information about the past, being equivalent to the story they have undergone. In terms of Gibsonian (Gibson, 1979) affordances we can say that artifacts as memory mediators – as reliable "external anchors" – afford the subject in terms of energy stimuli transduced by sensorial systems, so *maximizing abducibility*[58] and actively providing humans with new, often unexpected, opportunities for both "psychic" and "motor" actions.

Of course the meanings involved with artifacts such as the one I have just illustrated, are often intertwined with complex human behaviors which in turn involve rituals – very often endowed with religious and moral character. Girard eloquently describes a lot of these kinds of externalizations of symbolic artifacts as religious/moral ways of maintaining coalition enforcement in a collective.[59] He also

---

[56] As a kind of "memory store", in Leyton's sense. Leyton (1999; 2001) introduces this concept in a very interesting new geometry where forms are no longer memoryless as in classical approaches such as the Euclidean and the Kleinian ones, in terms of groups of transformations. From this mathematical perspective artifacts, in so far as they are expressed through icons, visual and other non linguistic configurations, are "memory stores" in themselves (Leyton, 2006). Of course in our case memory has to be intended in an extended Jungian sense, going beyond the explicit, linguistic or model-based aspects which are the main focus of the recent tradition of cognitive science: specific implicit structures are also at play.

[57] It is interesting to note that Turner (2005) identifies a range of "affordances" offered by a variety of mediating artifacts, including the life stories of recovering alcoholics in AA meetings (affording rehabilitation), patients' charts in a hospital setting (affording access to a patient's medical history), poker chips (affording gambling) and "sexy" clothing (affording gender stereotyping) (Cole, 1996). In this perspective mediating artifacts embody their own "developmental histories" which is a reflection of their use. I have illustrated in detail the relationship between abduction and affordances in (Magnani, 2009, chapter six).

[58] They maximize "recoverability" in Leyton's sense – cf. the following subsection.

[59] Cf. also chapter two, subsection 2.1.5, this book.

stresses the fact that at the same time – in appropriate social happenings and/or rituals – they work:

1. as activators of moral meanings, dispositions, and attitudes; and at the same time they also involve
2. sacrificial events in which human beings and animals are sometimes killed, where violence is disguised thanks to the active role played by the accompanying moral bubble in which the related collective is entrapped.

I will recall the problem of moral role of violent sacrifices in the last subsection of this chapter.

Progressively, the possible meaning that can be seen and learnt through the Wachandi artifact and the related rite, may become completely internalized and fixed so that referral to this externality – and learning from it – is no longer needed. Once internalized, the knowledge and the templates of action are already available at the brain level of (suitably trained) neural networks with their electrical and chemical pathways. When fixed and internalized they provide an immediate and ready "disposable energy": for example "We no longer need magical dances to make us 'strong' for whatever we want to do, at least not in ordinary cases" (Jung, 1972a, p. 45).

The semiotic process of externalization leads to the formation of a new meaning, which, as I have already said, Jung calls a symbol:[60] "The Wachandi's hole in the earth is not a sign for the genitals of a woman, but a symbol that stands for the idea of the earth woman who is to be made fruitful" (Jung, 1972a, p. 45). The artifact is a symbol, formed in the hybrid interplay between internal and external semiotic representations, and furnishes a "working potential in relation to the psyche" (p. 46). Another example is given by "[...] those South American rock-drawings which consist of furrows deeply engraved in the hard stone. They were made by the Indians playfully retracting the furrow again and again with stones, over hundreds of years. The content of the drawings is difficult to interpret, but the activity bound up with them is incomparably more significant" (ibid.).

4.8.4 Maximizing Abducibility of Morality through Symbols

Through artifacts and the representations they embed, the natural niche is transformed into a "cognitive niche"[61] by human consciousness and unconscious: symbols are provided as "libido analogues" that convert energy "ad infinitum" (Jung, 1972a, pp. 49–50), and thus cognition and culture. It is clear that symbols, in a Jungian sense, are artifacts that exhibit a *maximization of abducibility* regarding past

[60] The Jungian concept of a symbol is a little different from the one used in semiotics: it is of course richer in psychoanalytic value. However, it has to be noted that also according to Peirce's semiotics a symbol is just a kind of iconic sign, which is conventional and fixed, like the ones used in mathematics and logic, unlike generic signs, that are generally arbitrary and flexible.

[61] See the first sections of this chapter.

history, that is, of all the events that originate them. They are collective, stable, more or less available and sharable, related to both unconscious and conscious dimensions, strong *anchors* for thinking and for triggering action, which escape the fleeting nature of internal subjective representations.

These "symbols" go beyond the already considerable abducibility/recoverability force of various semiotic (for instance iconic) externalizations related to more basic survival needs – also present in many animals – (like caches of food, hunting landmarks, etc.) that are more constrained, if compared to symbols, in their capacity to trigger actions by promptly recovering various information and skills, such as the moral ones. These last externalizations, often plastically shaped by learning and not simply fruit of instinctual endowments, are more or less widespread in human and non-human animal collectives, but they are normally merely behaviorally oriented to the direct satisfaction of basic instincts, and thus they do not involve the broad cognitive role of symbols (in a Jungian sense). These artifacts basically play the role of strong cognitive remodeling of human and animal niches and, at the same time, present a regulative function of higher cognitive skills.

To make a simple example, the huge success of certain symbols of this kind, for instance the religious ones, can certainly be explained in terms of their capacity to maximize abducibility in a very extended domain of human "minds", even if endowed with various degrees of cognitive skills. They trigger hypotheses/thoughts and ways of morally behaving and of (violently) punishing or being punished that are supposed to be good, but the "user" of those symbols – available over there in the close environment – is in general very passive. The concept of "moral mediator" which I introduced previously in this chapter can be exploited to the aim of clarifying the use of some kinds of memory maximizing symbols, especially in social and collective situations where moral issues are at stake.

The Jungian focus on symbolic artifacts called *mandala*, made by patients during a psychoanalytic treatment, not merely based on the tradition of religious ones, but free creations determined by certain archetypical ideas unknown to their creators, is related to the problem of "seeing" for example stages of individuation, where step by step patients "give a mind to that part of the personality which has remained behind" (Jung, 1968b, p. 350). They are presented as "ideograms" of unconscious contents.

For Jung "the making of a religion" is a primary interest of the primitive mind and strongly relates to the production of symbols, which "enable man to set up a spiritual counterpole to its primitive instinctual nature" (Jung, 1972a, p. 59), that is, to implement a *moral collective framework*, as already clearly stated by Vico. For Vico, even the most "savage, wild, and monstrous men" did not lack a "notion of God," for a man of that sort, who has "fallen into despair of all the succors of nature, desires something superior to save him" (1968, 339, p. 100). This desire led those "monstrous men" to invent the idea of God as a protective and salvific agent outside themselves; this shift engendered the first rough concept of an external world, one with distinctions and choices and thus established conditions for the possibility of free will. In the mythical story, the idea of God supplies the first instance of "elbow room" for free will illustrated by (Dennett, 1984). Through God, men can "hold in check the motions impressed on the mind[62] by the body" and become "wise" and "civil."

According to Vico, it is God that gives men the "conatus" of consciousness and free will:

> [...] these first men, who later became the princes of the gentile nations, must have done their thinking under the strong impulsion of violent passions, as beasts do. We must therefore proceed from a vulgar metaphysics, such as we shall find the theology of the poets to have been, and seek by its aid that frightful thought of some divinity which imposed form and measure on the bestial passions of these lost men and thus transformed them into human passions. From this thought must have sprung the conatus proper to the human will, to hold in check the motions impressed on the mind by the body, so as either to quiet them altogether, as becomes the wise man, or at least to direct them to better use, as becomes the civil man. This control over the motion of their bodies is certainly an effect of the freedom of human choice, and thus of free will, which is the home and seat of all the virtues, and among the others of justice. When informed by justice, the will is the fount of all that is just and of all the laws dictated by justice. But to impute conatus to bodies is as much as to impute to them freedom to regulate their motions, whereas all bodies are by nature necessary agents (Vico, 1968, 340, p. 101).

Free will, then, leads to "family": "Moral virtue began, as it must, from conatus. For the giants, enchanted under the mountains by the frightful religion of the thunderbolts, learned to check their bestial habits of wandering wild through the great forests of the earth, and acquired the contrary custom of remaining hidden and settled in their fields. [...] And hence came Jove's title of stayer or establisher. With this conatus, the virtue of the spirit began likewise to show itself to them, restraining their bestial lust from finding satisfaction in the sight of heaven, of which they had a mortal terror" (504, p. 171.) "The new direction took the form of forcibly seizing their women, who were naturally shy and unruly, dragging them into their caves, and, in order to have intercourse with them, keeping them there as perpetual lifelong companions" (1098, p. 420).

We can add, maybe some women did not like to be segregated, out-group neighbors did not like those ways of segregating women and perhaps they also had a different kind of God, in-group free-riders did not like their religious rituals because they had a different religious "moral" concern and their own policies, and so on: the newly established moral framework is immediately a cause of conflict and of consequent possible violence.

---

[62] Indeed "That is, the human mind does not understand anything of which it has had no previous impression [...] from the senses" (Vico, 1968, 363, p. 110). "And human nature, so far as it is like that of animals, carries with it this property, that senses are its sole way of knowing things" (374, p. 116). Again, humans "in their robust ignorance" know things "by virtue of a wholly corporeal imagination" (376, p. 117). Aristotle had already contended that "nihil est in intellectu quod prius non fuerit in sensu."

4.8.5 Cultured Unconscious and Sacrifices

From the psychoanalytic perspective I have illustrated in the previous sections we can also understand how both i) externalized culture and ii) modern human beings comprehend within them "implicit" traces of each of the previous stages of cognitive evolution. The first case of externalized distributed culture is evident: remains, buildings, manuscripts, and so on, are fragments of ancient "cognitive niches" from which we can retrieve cultural knowledge.

In the second case it can be hypothesized that much of what Freud attributes to the unconscious is truly unconscious only in the cultural sense of the word, that is formed by "things that are not expressed or are repressed at the level of culture". It has to be acknowledged that, in recent cognitive science research, the unconscious is a solipsistic notion, not a cultural one and concerns a part of the human mind that is *a priori* outside the reach of consciousness, a golem, an *other side*, an "automaton world of instincts and zombies", like Donald eloquently says. An example is object vision: "It serves up all the richness of the three-dimensional visual world of awareness, gratis and fully formed. But we can never gain access to the mysterious region of mind that delivers such images. It lies on the other side of cognition, permanently outside the purview of consciousness" (2001, pp. 286-287).

In the case of psychoanalysis, the unconscious is constructed by drives, intuitions and representations that are shaped by the brain/culture symbiosis and interplay and so are not *a priori* inaccessible to awareness. It is well-known that Jung has also hypothesized the existence of a collective unconscious, i.e. that part of individual unconscious we share with others human beings, shaped by the evolution of the above interplay, which has ancient archetypes "wired" in it, that still seem to act in our present behavior. I have already said that an example can be the moral/religious "scapegoat" mechanism, typical of ancient groups and societies, where a paroxysm of violence would tend to focus on an arbitrary victim, and a unanimous antipathy generated by "mimetic desire" (and the related envy) would grow against the victim herself.[63] Let me reiterate what I have already stressed in chapter two of this book. These kinds of behavior are very common in human beings and often unconsciously performed. It is easy to hypothesize that they can be implicitly "learned" during infancy and then implicitly "wired" by the individual in a kind of – obviously collective – *cultured unconscious*. Consequently, they are there, available in our minds/brains, to be picked up and executed – paradoxically, given the fact we often mean to be civilized, modern human beings – as archaic forms of "social" behavior.

A similar conclusion is reached in the perspective of evolutionary ethics. Lahti (2003, p. 649) observes that – insofar as morality did not evolve to curb selfishness and that instead the character of morality as a "universal law" (which is recent in evolutionary terms) may function to update behavioral strategies which were adaptive in the paleolithic environment of our ancestors – it can be hypothesized that the

[63] On this archaic mechanism and its effect in the violence that characterizes ancient and modern societies cf. (Girard, 1977, 1986). Again, cf. also subsection 2.1.5, chapter two of this book.

emergence of archaic moral behaviors (for example a patriarchal template) is always possible, and thus moral conflicts are explainable from this point of view:

> If many of our psychological tendencies were shaped in a past situation that was different from that of today in some relevant ways, as the evidence suggests is the case, this raises the possibility that the internal conflict that accompanies the making of some of our moral decisions may result from a discordance between modern moral codes and the set of behavioral strategies that would have been adaptive during most of our evolutionary history. If this is true, one would predict a correlation between the degree of internal conflict we experience with regard to a particular moral guideline, and its variance from the adaptive strategies of hunter-gatherers living in small kin groups. Moreover, the moral law, in Ruse's sense of a set of universal, prescriptive, and non-subjective guidelines, is likely a recent phenomenon, postdating the hunter-gatherer period. This concept of a moral law may function to update our behavior to the present social environment from that of our paleolithic ancestors. A much older predisposition to obey parents and other leaders real and imagined may have been co-opted, with existing sources of moral authority being replaced by a universal God or a value-laden universe.

Chapter 5
Multiple Individual Moralities May Trigger Violence
Engaging and Disengaging Morality

5.1 "Multiple Individual" Moralities: Is "Moral Disengagement" in the Perpetration of Inhumanities a Reengagement of Another Morality?

An important research in the area of psychology that can enrich our perspective on the relationship between morality and violence concerns the so-called *moral disengagement* illustrated by Bandura (1999). The neglect of moral conduct is widespread in moral agents: moral standards, even if previously adopted as guidelines for self-sanctioning and to avoid self-condemnation or self-devaluation, are often contravened. The moral behavior, Bandura says, is both inhibitive – refraining from certain behaviour – and proactive – behaving according to that particular idea of humanity, which is embedded in the adopted moral standard. Unfortunately, the activation of the moral standard can be deactivated – that is what moral disengagement is about – so that the alternative behavior is no longer viewed as immoral, the possible consequent harm is minimized, expected positive consequences are overemphasized and victims are devaluated in their very nature as human beings. What is important to note from our perspective is that to "engage moral disengagement" people often construct *moral justifications* of the new actions so that the conduct is made personally and socially acceptable "by portraying it as serving a socially worthy or moral purpose" (Bandura, 1999, p. 195). The redefinition of killing is an amazing example of disengagement: shifts in destructive people's behavior is seen in military conduct, where a new conduct "is achieved not by altering their personality structures, aggressive drives or moral standards" (ibid.) Usually this justification also consists in a legitimation of violence!

In sum, becoming violent by freeing ourselves from self-censure is much easier that expected or imagined. In war settings, decent and ordinary people can become "horrible violent people" and see themselves:

> [...] as fighting ruthless oppressors, protecting their cherished values, preserving world peace, saving humanity from subjugation or honoring their country's commitments. Just war tenets were devised to specify when the use of violent force is morally justified. However, given people's dexterous facility for justifying violent means all kinds

L. Magnani: Understanding Violence, SAPERE 1, pp. 171–233, 2011.
springerlink.com

> of inhumanities get clothed in moral wrappings. Voltaire put it well when he said, "Those who can make you believe absurdities can make you commit atrocities". [...] When viewed from divergent perspectives the same violent acts are different things to different people. It is often proclaimed in conflicts of power that one group's terroristic activity is another group's liberation movement fought by heroic fighters. This is why moral appeals against violence usually fall on deaf ears. Adversaries sanctify their own militant actions, but condemn those of their antagonists as barbarity masquerading under a mask of outrageous moral reasoning. Each side feels morally superior to the other (ibid.).

Ideologies, religious convictions, nationalist commitments, ethnic stereotypes, even trivial personal needs – considered justified and ineluctable – are the main tools for moral disengagement and guilt-avoidance.

Bandura satisfactorily illustrates other psychological "tools" for violence, often performed with the help of amazing self-deceptive mechanisms: *euphemistic labeling* (we come back to the idea of language as a "knife" that I illustrated in chapter two), for example assaulting actions are verbally "sanitized" and concealed through camouflage and flagrant hypocrisy, bombing is "servicing the target"; "capital punishment is our society's recognition of the sanctity of human life", like a US senator one time said, a reactor accident is a "normal aberration", "we took care of somebody"[1] to see many other – almost, sad to say – funny examples.

Other tools are: *advantageous comparison*, when behavior is colored by what it is compared with, so that reprehensible acts can be seen as righteous, terrorists affront martyrdom thinking of the cruelties inflicted on the people they represent and defend; *displacement of responsibility*, that is the minimization or obscuring of the agentive role in the harm one causes, when for example a man kills simply as he carries out orders, like in the case of Nazi followers and their commandants; *diffusion of responsibility* (for example in group decisions or collective action – scapegoat mechanism, no one feels responsible for the perpetrated violence, groups tend to suppress private doubts); *disregard or distortion of responsibility*, thanks to the minimization of harm or through discrediting of the evidence of harmful consequences; *dehumanization* of the recipients of detrimental acts, i.e. when we see the harmed creatures as "less" human than ourselves and our peers (i.e. slaves, savages, strangers, "degenerates", etc.) it is easier to avoid suffering personal distress and self-condemnation; *blaming* one's adversary is also another means that can assist self-exonerative purposes (for example when we see ourselves as faultless victims and so the adoption of harmful behaviors is seen as justified because compelled by the circumstances).

Also bureaucratization,[2] automation, and urbanization play serious roles in this effect of dehumanization. Of course the psychological devices listed by Bandura are often combined – in appropriate contexts – to the aim of strengthening the moral disengagement of ordinary people. Moreover, as everyone clearly understands, moral

[1] Cf. (Bandura, 1999) and (Baumeister, 1997, chapter ten).

[2] Empirical results concerning interesting relationships between moral disengagement, work characteristics, satisfaction, and workplace harassment are illustrated in (Claybourn, 2011).

disengagement is gradualistic, so that the level of ruthlessness can increase and a high level of violence can become radical and routinized. Terrorists provide a clear example:

> The process of radicalization involves a gradual disengagement of moral sanctions from violent conduct. It begins with prosocial efforts to change particular social policies and opposition to officials, who are intent on keeping things as they are. Embittering failures to accomplish social change and hostile confrontations with authorities and police lead to growing disillusionment and alienation from the whole system. Escalative battles culminate in terrorists' efforts to destroy the system and its dehumanized rulers (Bandura, 1999, p. 204).

This gradualization also operates in *networks*, like in the case of the death industry – tobacco, guns, weapons – where, for example, we have to cope with the merchandizing of terrorism brought about "by unsavory individuals": "It requires a worldwide network of reputable, high-level members of society, who contribute to the deathly enterprise by insulating fractionation of the operations and displacement and diffusion of responsibility. [...] By fragmenting and dispersing sub-functions of the enterprise, the various contributors see themselves as decent, legitimate practitioners of their trade rather than as parties to deathly operations. [...] Such mechanisms operate in every day situations in which decent people routinely perform activities that bring them profits and other benefits at injurious costs of others" (p. 205).

Moral disengagement often occurs in a reciprocal interplay of personal, social, and institutional influences and pressures, for example it is favored by peer modeling and exposure to bad examples. Moral mediators of various types, available "out there" in human and artificial environments, can serve moral disengagement, providing new moral justifications and excuses. Also alcohol and various drugs are among the most universal and familiar mediators of escaping guilt and other inhibitions (Baumeister, 1997, pp. 540–542), that is, they are a useful tool for performing evil.[3]

Jensen (2010) makes an interesting analysis of six "demoralizing" processes in the context of the "adiaphoric company". These processes create a realm of "being-with", in which outcomes of human interaction are evaluated on rational grounds, and whether or not a particular action should be undertaken in accordance with stipulated ethical rules. Unfortunately the realm of "being-for", in which individuals are supported to take increased responsibility, is marginalized. The author concludes that not only does the process of disengaging morality systematically produce moral distance between humans, which weakens individual spontaneous outbursts of sympathy to take increased moral responsibility, but it also promises to release individuals from their moral ambivalence by declaring organized action to be morally indifferent. Organizational action reengages another morality (*de facto* the new, so to say, amazing "Nietzschean" company ethics) which appears evil in the light of

[3] The prototypical kind of moral disengagement by indifference to harmful realities such as ecological problems and population growth is also clearly described by Bandura in (2007).

many common moralities because – pathetically – it depicts itself as something "adiaphoric", beyond good and evil.

I would like to address a mild criticism to Bandura's thesis. The reader should agree that his analysis is extremely useful to understand moral human fragility and the variability of human reactions, in cases where the agent is still perfectly aware of the abstract moral rules he previously endorsed and then no longer applies. As I have illustrated above, Bandura lists "reasons" which explain the disengagement: are we sure that those reasons are just "extra-moral" or "immoral" triggers for disengagement, at least from the perspective of the agent itself, if not of course from our own or Bandura's perspective of observers and judges? I hypothesize that moral disengagement is basically a reengagement of another morality. I will address this problem in detail very soon, in the following section.

Baumeister (1997, pp. 276–277) usefully notes that various *moral subcultures* of the *irresistible impulse* constitute a way of favoring and justifying many violent acts. It is well-known that these subcultures are more or less widespread and implicitly disseminated by mass media and pseudo-moralistic gossip. It is a subcultural construction that amazingly extends the range of irresistible impulses beyond the area of biological necessities. Of course the urge to urinate can become irresistible, but unfortunately a kind of perverse social pedagogy teaches ordinary people they "can allow themselves to lose control of angry, violent impulses", even if biological needs are not at stake, thus encouraging violence:

> [...] people know just how much they can allow themselves to lose control. Yet that point is not firmly determined by natural law; rather, it is influenced by cultural beliefs. This is where the theory of a subculture of violence needs to be revised. It is not that cultures place a positive value on violence but that culture dictates when and where (and how much) it is appropriate to lose control. [...] A culture of violence does not have to place a positive value on violence. It can encourage violence merely by making appropriate to let oneself go in response to a broad range of provocations (p. 276).

It seems we are dealing with a kind of sub-morality which represents the exact *inversion* of Stoic morality, massively devoted to teaching the control of impulses and the importance of moral "indifference" when faced with puzzling or dangerous events and situations. Stoically, we should be indifferent to avoid being provoked by those situations into acting inconsiderately. Ordinary decent people learn to morally treat many impulses *as if* they were irresistible, as if they were biological needs, even when they are not. And these impulses quickly legitimate a disengagement of morality, or, better – as I will explain soon – a reengagement of another morality, in this case the *sub-morality* of the "irresistible impulse".

Further insight on the so called "Lucifer effect" is provided by Zimbardo (2009). The notable book is a treatise about the malleability of human nature and the rapidity of its shifts – so to say – from civility to malevolence, that is from a standard morality (Bandura is expressly quoted in the book) to what I think is *another* moral framework rather than a disengaged one. The book also takes advantage both of the empirical results of the Stanford Prison Experiment – paralleling this experiment to the case of the Abu Ghraib atrocities – which illustrates the overriding "power of the situation". The analysis permits the author to contrast the account of violence in terms of dispositions, which would transmute decent people into evildoers: an account that is basically ascribed to the Bible of the Inquisition, the *Malleus Maleficarum*.

Recent research into the relationship between war and morality adopt a new option, which can be usefully seen in the perspective of moral disengagement and of reengagement of morality. McMahan (2009) contends that common sense beliefs about the morality of killing in war are deeply mistaken. The predominant view is that in a state of war, the act of killing is ruled by different moral principles from those that rule acts of killing in other contexts, that is for self-defense: "This presupposes that it can make a difference to the moral permissibility of killing another person whether one's political leaders have declared a state of war with that person's country. According to the prevailing view, therefore, political leaders can sometimes cause other people's moral rights to disappear simply by commanding their armies to attack them. When stated in this way, the received view seems obviously absurd" (p. vii). Hence no disengagement of morality is legitimate, and there is now a reengaged "war morality". This contention clearly and provocatively shows how war disentangles common morality and pretends to implement a new one. Many war events can be usefully reinterpreted. The author certainly considers American participation in World War II justifiable, but also remarks:

> It is revealing about our attitudes in general that we sometimes do take combatants who have committed war crimes to be fully excused, or even justified, and not just in cases involving extreme duress, invincible ignorance, or insanity. Perhaps the most notorious case of this sort is that of General Paul Tibbets, who was the commander and pilot of the Enola Gay, the plane, named for his mother, from which the atomic bomb was dropped on the Japanese city of Hiroshima in August of 1945. According to the US Department of Energy, approximately 70,000 people were killed immediately, while more than 30,000 more died over the next few months from injuries and exposure to radiation. [...] This single act by Tibbets, with contributions by the other members of his small crew, had as an immediate physical effect the killing of more people, the vast majority of whom were civilians, than any other single act ever done. The law of war prohibits – and prohibited at the time of Tibbets' action – the intentional killing of civilians for the purpose of coercing their government to surrender. And all plausible moral theories, including even the most radical forms of consequentialism, prohibit the intentional killing of *that many* innocent people in virtually all practically possible circumstances. Tibbets' act is therefore the most egregious war crime, and the most destructive single terrorist act, ever committed, even though it was committed in the course of a just war. Yet he was congratulated for it by President Truman, who had given the order that he do it, and was awarded various medals and promoted from colonel to brigadier general. When Tibbets died in 2007 at the age of 92, the obituary in the *New York Times* carried a caption in bold type that read "A war hero who never wavered in defending his mission", and ten days later the same newspaper printed a further celebratory op-ed piece with this caption: "Paul Tibbets, the hero we wanted to forget" (pp. 128-129).

5.1.1 Reengagement of Another Morality

I strongly contend that very often the agent's disengagement of adopted and well-known moral rules is simply due to a shift to another moral framework, that prevails over the first one. Let me make an example, imagine a mentally healthy husband endowed with a moral catholic education who knows perfectly well that, on that moral basis, he must respect his wife and consequently he also acknowledges the consequence that he must not beat her. Unfortunately, in certain circumstances, he disengages his own moral principles and beats the wife and, at the same time, he also thinks he has done the right thing. I think that in this case the disengagement if due to a sudden moral change of perspective, for example we can hypothesize he "also" shares a patriarchal morality as a more or less conscious sub-morality, where the violent action is considered the right punishment when some of its prohibitions are infringed.[4]

I have already mentioned the "scapegoat" case of mobbing behaviors, that we can witness in widespread patriarchal behavior all over the world: it is a moral performance that is often partially unconsciously and spontaneously activated. In this case it is also easy to understand how it can be implicitly "learned" in infancy and still implicitly "pre-wired" in an individual's ideological *cultural unconscious* that we collectively share with other human beings. In sum, disengagement of morality is often a reengagement of another morality. As a matter of fact, people who observe the subject's newly adopted morality usually consider the new behavior as immoral and certainly more violent than the previous one, but this is not the interpretation given to the newly triggered violent actions by the agent himself. He still sees the new behavior as justified and right: the possible violence involved does not matter, it is perceived as ineluctable and very often not even perceived as such – the violence is only perceived by the victims and by some observers. It is likely that some observers – who share that same sub-morality – can easily consider that punishment appropriate, disregarding at the same time the subsequent violence. Again, this case sheds further light on what I called the basic equivalence between engagement of morality and engagement of violence, amazingly almost always hidden from the awareness of those very individual agents who perpetrate the violence.

The "altruistic lying" (studied in the case of romantic relationships by Kaplan and Gordon (2004)) is really interesting in the perspective of moral disengagement and of the interplay between morality and violence. Liars can construe their behavior as altruistically motivated and this case of course "applies more broadly to other types of transgressions in many different contexts and in varying degrees of severity". Indeed, the same underlying processes that operate during common interpersonal transgressions (like in the case of many lies) may help explain how more complicated transgressions can be interpreted as morally correct by the people who commit them. Often these kinds of liars take advantage of cross-perspective recall, insight, and empathy to avoid the masquerade of altruism: "[...] one's perspective is likely to be associated with a host of powerful retrieval cues that lead people to

[4] The same case here illustrated can be seen in the perspective of a psychoanalytical analysis concerning the so-called "structure of evil", as I will illustrate below in subsection 5.3.5.

recall previous instances in which they were in the same, rather than the opposite role. For example, lie receivers' feelings of anger, upset, and mistreatment may be enhanced because their current situation activates previous instances in which they were lied to" (p. 503). Another important constraint, that favors the moral justification process of altruistic liars, is constituted by the cognitive and affective demands of the situation at stake. This urgency may *deplete* the resources necessary for a consideration of the situation from another perspective.

McKinlay and McVittle (2008, pp. 163–171) illustrate how many violent offenders (for example in the case of rudeness or bullying) depict themselves as morally decent, for example contending that the victim was responsible for the aggressor's action: the aggressor's personal responsibility is abandoned and of course an immediate conflict arises, because (usually) the witnesses' account is considered in order to attribute responsibility to the aggressor. The aggressor often disguises violence simply denying that aggression took place, as in many cases of sexual abuse and homophobic bullying, in which violence is *morally* understood by the subjects as a good and just part of an exaggerated masculine identity (often male victims are suitably feminized to favor a justification for harming them).[5]

Another recent research by Kimhi and Sagy (2008) examines moral justification as a mediating mechanism for stress, used by Israeli conscripts who had served at army roadblocks in the West Bank. The study uses Bandura's model of moral disengagement and establishes that the greater the justification needed by the soldier to make roadblocks acceptable (be it a cognitive one, such as right-wing political attitudes and religious orientations, affective or behavioral, i.e. long service), the more he would feel adjusted to army demands, and he would see war goals with more clarity and acceptability. The results partially support the hypothesis concerning association between moral justification and feelings of adjustment at the end of army service.[6]

Furthermore, Goldman (2010) has stressed the fragility of moral deliberation taking advantage of the concept of "partial amorality", when for example we subtract ourselves to a moral requirement we had agreed with, for example not to eat meat for a certain reason. He also points out the role of weakness of will (*akrasia*) in moral deliberation and the importance of addressing ethics from a practical and not merely theoretical perspective, reaching skeptical conclusions which are certainly in tune with the philosophical perspective I am trying to promote in this book (even if I am even more skeptical than he is):

> But all this simply assumes moral motivation on the part of ourselves and others with whom we engage in moral dialogue. It is because we assume such motivation that we deliberate and argue about moral matters. And the noncynical view is that this assumption is justified and correct most of the time. But it remains an assumption. When it is correct, when people are morally concerned, they may well be rationally required to translate such concern into action or forbearance. But the fact that we assume a

[5] For further details on this issue cf. also (Lines, 2008).

[6] Recent research provides explanation on how in-group membership would offer absolution and justification for acts of "good violence", which are perceived by soldiers to clearly contrast with the enemy's "bad violence" (Alpher and Rothbart, 2006).

> capacity to respond to moral reasons when we argue with, advise, or hold others accountable does not show that such motivation is rationally required. In fact, we do need to train, socially condition, or educate people to be morally motivated, and the training is not primarily through argument. If it were simply a matter of replacing ignorance with knowledge, the social problem of immorality would be far more tractable, similar to that of illiteracy. And if it were only a matter of showing people how to avoid defeating the aims they really desire, it would again be a more tractable problem. But our deepest social problems result from the fact that moral indoctrination is only partly successful. The defect that must be remedied is not a defect in intellect or rationality, or a matter of ignorance. Sociopaths need not be fools or even philosophically unsophisticated. Nor do they have to struggle to overcome their rationally required moral motivations (p. 16).

I have to add that also in the case of illiteracy we faced a kind of disillusionment: the wide offer of education in Western countries, favored by the welfare state of the last decades, has demonstrated how difficult it is even to secure stable degrees of simple alphabetization as a definitive premise for changes in society, if only from the mere perspective of a simple cognitive efficiency of its members.

Moreover, the case of moral disengagement also illustrates the role played in it by the "moral bubble" effect which I have described in chapter three. The "embubblement" renders invisible and condones both the disengaged (or "reengaged") morality and the related violence. The *disengagers* are entrapped in a *moral bubble*, which systematically disguises their violence to themselves: as I have already said, this concept is also of help in analyzing and explaining why so many kinds of violent behavior in the world today are treated by the observers *as if* they were something else, for example as pure – unexplainable, unmotivated – evil. Lack of awareness of our disengagement of morality – or reengagement of another moral framework – is very often accompanied by lack of awareness of the deceptive/aggressive character of consequent verbal interactions (and behaviors), because the new moral assumptions which motivate violence and punishment are rarely seen and acknowledged as actually "violent".

Furthermore, moral disengagement/reengagement explains the reason why people justify violence and cruelty by derogation and dehumanization. An interesting appendix to moral disengagement/reengagement worth citing is provided by the theory of cognitive dissonance first introduced by Leon Festinger (1957). Basically, cognitive dissonance is a state of tension occurring when a person has two cognitions – that can be beliefs, opinions, moral judgments, and so on – that are felt as incoherent or inconsistent (Tavris and Aronson, 2008). Dissonance creates an irritation that needs to be settled down and then solved somehow by establishing a new belief or judgment able to accommodate the inconsistency between the two cognitions. The state of tension is therefore solved by resorting to self-justification, which contributes to restoring coherency in one's beliefs system. Cognitive dissonance may favor self-deception, and other cognitive biases that are ego-busting like, for instance, distorting memories of past events, softening our responsibilities, and so on.

As just mentioned, this engine of self-justification may also have extremely negative consequences on the perception of our own violent acts and thoughts, thus creating an escalating spiral of violence and hatred. As cognitive dissonance theory predicts, when engaging in a violent act, one may feel the urge to justify it. If cognitive dissonance is solved in favor of softening one's responsibility, then the self-justifying explanation will work like evidence confirming and validating that the violent action was due and that the victim deserved it. Therefore, the cognitive dissonance is reduced, and the person that has committed the violent act paradoxically becomes more certain in his conviction that he was right. In the case of violence generated by the moral shift due to cognitive dissonance, the aggressor may for example even persuade himself into thinking that his victim should be further punished generating a dangerous loop of positive feedback, in which the new adopted moral perspective, that triggered the original violence, begs for further justification that in turn begs for more violence.[7] We can add that this situation of dissonance reduction and violence activation reminds us of the scapegoat mechanism illustrated above at page 54, following Girard's perspective: we can guess that the final target of the aggressor is the brutal elimination of the victim to the aim of reducing the appetite for violence that had possessed her just a moment before. Violence just reduces the appetite, but does not extinguish it, violence has to be repeated just to feel it has actually reduced.

A further note on the ambiguous status of moral engagement and disengagement relates to the problem of the *epistemic* status of moral responsibility of wrongful acts. Following Sher (2000) it also seems necessary to contemplate the role of the epistemic conditions of these acts, and not only of the mere awareness of the agent. Indeed Sher criticizes the inclination – I have already illustrated – to center the epistemic condition of morally wrongful acts exclusively on conscious awareness. The book challenges this view, called the"searchlight view", by contending that the agent is responsible when, and because, his failure to respond to his reasons for believing that he is acting wrongly or foolishly has its origins in the same constitutive psychology that generally does render him reason-responsive. In this wide perspective, for example, "ignorant" wrongdoing – and thus moral responsibility – can nevertheless be attributed to the

[7] Interestingly, Wong and colleagues (2008) argued that a stronger escalating tendency might be triggered by a serious commitment to a rational thinking style. Rational thinking style is characterized by a number of elements including, for instance, a conscious and analytical attitude towards problem-solving in which emotional interference is usually minimized (Pacini and Epstein, 1999). The idea developed by Wong and colleagues is that those people who rely on rational thinking style might be escalating their commitment towards an idea or course of action, as they usually display stronger commitment towards their beliefs. Their thinking style, leaning on analysis instead of free reasoning, makes them more confident about their abilities and their ideas and beliefs, and thus less eager to abandon them. What is implicit here is that they seem to perceive, and subsequently exploit, the violent and aggressive element that more reasoned dispositions still possess. Such a violent and aggressive dimension tacitly embedded in a rational thinking style serves the purpose of reducing cognitive dissonance. That is, cognitive dissonance is reduced by means of transferring violence delivered by a rational thinking style directly to one's beliefs in order to suppress dissonance. Quite literally, that would happen like grasping a knife.

agent as long as it is accounted for by the agent's own "constitutive attitudes, dispositions, and traits" (p. 87). Lost of self-control, conflicting obligations (where for example the agent is in front of the choice of a possible wrong or of a definite wrong), the urgency of acting here and now (which triggers confused mental states), emotional distress, are all extremely knowledge-sensitive situations. In these cases the lack of knowledge – conceptual, emotional, manipulatory – or the incapacity to recover and exploit the suitable knowledge is a way of facilitating crime and violence.

5.1.2 Kant and the "Inverted Stoicism"

I think Immanuel Kant provides some observations that are of precious help in further understanding the concepts of disengagement and reengagement of morality, also related to the notion of moral bubble I have introduced in chapter three, section 3.2.2. Kant thoughtfully says:

> The human being (even the worst) does not repudiate the moral law, whatever his maxims, in rebellious attitude (by revoking obedience to it). [...] He is, however, also dependent on the incentives of his sensuous nature because of his equally innocent natural predisposition, and he incorporates them too into his maxim (according to the subjective principle of self-love). If he took them into his maxim *as of themselves sufficient* for the determination of his power of choice, without minding the moral law (which he nonetheless has within himself), he would then become morally evil. But, now since he naturally incorporates both into the same maxim, whereas he would find each, taken alone, of itself sufficient to determine the will, so, if the difference between the maxims depended simply on the difference between incentives (the material of the maxims), namely, on whether the law or the sense impulse provides the incentive, he would be morally good and evil at the same time – and this is a contradiction (as we saw in the introduction). Hence the difference, whether the human being is good or evil, must not lie in the difference between the incentives that he incorporates into his maxim (not in the material of the maxim) but in their subordination (in the form of the maxim): *which of the two he makes the condition of the other.* If follows that the human being (even the best) is evil only because he reverses the moral order of his incentives in incorporating them into the maxims (1998b, pp. 58–59).

So to say, there is a kind of slippery-slope process of continuous disengagement of morality (the maxim), thanks to its subordination to human "sensuous nature". This subordination actually governs man's "power of choice" and, I can add, quickly appears to the decisional agent as a new reengaged framework which is still and sincerely felt as really "moral", as if it would be due to the maxim.

Kant is even clearer in describing the corruption of morality (in our term its continuous "disengagement/reengagement"):

> He indeed incorporates the moral law into those maxims, together with the law of self-love; since, however, he realizes that the two cannot stand on an equal footing, but one must be subordinated to the other as its supreme condition, he makes the incentives

> of self-love and their inclinations the condition of compliance with the moral law – whereas it is this latter that, as *the supreme condition* of the satisfaction of the former, should have been incorporated into the universal maxim of the power of choice as the sole incentive (1998b, p. 59).

The propensity to this "inversion" – like Kant says – of the moral order is a natural propensity to evil, and it is even in itself "morally evil", radical: not a mere "malice" but a "perversity of the heart". The consequences for the perception of "guilt" on the part of the human agent reproduce both i) the problem of "moral bubbles", I have introduced in chapter three, and ii) the disengagement of morality I discussed above.

The moral agent falls into a perfidious and self-deceitful reengagement in a new decisional framework where evil is simply supposed to be good, and so morally justified (that is, it is exactly a reengagement of another morality, to use the term I introduced above, or in a sub-morality, to use the term I introduced when illustrating the logic of the "irresistible impulse". I also noted that in this case we face with a kind of "inverted Stoicism" and Kant too speaks of a similar "inversion" of moral order):

> The *innate* guilt (*reatus*) [. . .] can be judged in its first two stages (those of frailty and impurity) to be unintentional guilt (*culpa*): in the third, however, as deliberate guilt (*dolus*), and is characterized by a certain perfidy on the part of the human heart (*dolus malus*) in deceiving itself as regards its own good or evil disposition and, provided that its actions do not result in evil (which they could well do because of their maxims), in not troubling itself on account of its disposition but rather considering itself justified before the law (ibid.)

Furthermore, the moral agent falls into a perfidious moral bubble, where the evil is not perceived:

> This is how so many human beings (conscientious in their own estimation) derive their peace of mind when, in course of actions in which the law was not consulted or at least did not count the most, they just luckily slipped by the evil consequences; and [how they derive] even the fancy that they deserve not to feel guilty of such transgressions as they see other burdened with, without however inquiring whether the credit goes perhaps to good luck, or whether, on the attitude of mind they could well discover within themselves, they would not have practiced similar vices themselves, had they not been kept away from them by impotence, temperament, upbringing, and tempting circumstances of time and place (things which, one and all, cannot be imputed to us) (p. 60).

In sum, Kant eloquently concludes "This dishonesty, by which we throw dust in our own eyes and which hinders the establishment in us of a genuine moral disposition, then extends itself also externally, to falsity or deception of others. And this dishonesty is not to be called malice, it nonetheless deserves at least the name of unworthiness" (pp. 60–61).[8]

[8] The recent meta-ethical intricate discussion about the positive or negative role of self-interest in morality is illustrated in (Bloomfield, 2008).

5.2 Pure Evil?

I have just illustrated that Kant hypothesizes that an "immoral" human being practically does not exist. Let us repeat the quotation: "The human being (even the worst) does not repudiate the moral law, whatever his maxims, in rebellious attitude (by revoking obedience to it). The law rather imposes itself on him irresistibly, because of his moral predispositions, and if no other incentive were at work against it, he would also incorporate it into his supreme maxim as sufficient determination of his power of choice, i.e. he would be morally good" (1998b, p. 58): it is in this sense that it seems hardly acceptable that human beings are involved in what many authors (for example Baumeister (1997)) call "pure evil". The prototypes of human evil involve actions that intentionally harm other people thanks to a transgression of a moral rule perfectly present and approved in the agent's mind. On the contrary, various processes of moral disengagement and moral reengagement are usually at play, a kind of dynamic which is implicitly addressed by the Kantian words I quoted in the previous section and I cite again here: "This is how so many human beings (conscientious in their own estimation) derive their peace of mind when, in course of actions in which the law was not consulted or at least did not count the most, they just luckily slipped by the evil consequences; and [how they derive] even the fancy that they deserve not to feel guilty of such transgressions as they see other burdened with, without however inquiring whether the credit well discover within themselves, had they not been kept away from them by impotence, temperament, upbringing, and tempting circumstances of time and place (things which, one and all, cannot be imputed to us)" (Kant, 1998b, p. 60).

I think the postulation of the existence of an immoral human being, a true "pure evil", is a strong intellectual and moralistic idealization, which abstractly sees evil perpetrators as cunning, wicked, malicious, sadistic people who inflict senseless harm on innocent and weak victims. Actual violence perpetrators, ordinary – rarely mentally or physically ill – human beings, often see themselves, embedded in their "moral bubbles" as totally or abundantly justified, for example when they respond to the real or imaginary offensive attack from their victims. In this case the violence/evil the victims will suffer exists primarily only in the experience of the victim.

Contrarily to this perspective a classical and useful study on the problem of evil people is provided by Haybron (2002), where the main concern is to argue in favor of a robust bad/evil distinction, to avoid the conflation of evil persons with evildoers, and that we usually explain and understand the evil action in terms of its relation to evil character. He proposes an affective-motivational account of *evil character*, as a significant moral category, which marks just one end of a moral *continuum* that has, at the opposite pole, the *saint*. He maintains that "frequent evildoing" accounts confuse this moral space (labeled as "aretaic") with the one defined by the moral hero and the moral criminal ("mirror thesis"): "[...] the evil person is beyond ordinary moral criticism and dialogue: he has no better nature to which we can appeal. Morality has no significant foothold in him. He is arguably beyond redemption through rational deliberation; nothing short of a conversion or reprogramming, it seems, could rehabilitate him. He understands morality, and may be perfectly capable of moral decency, but he rarely if ever exercises this capacity. Because of this, the evil person is also beyond society: a moral exile. [...] The evil person is something of an alien, lying somewhere between the human and the demonic. We call her, not coincidentally, a *monster*" (p. 279). The aim of such a thesis must not be misunderstood: Haybron does not mean that evildoers are moral monsters, conversely, he means to stress how the utterly, intrinsically evil individual is as rare and hard to apprehend as a *monstrum*, just as it is hard to find a person who is a complete *saint*. No matter how trivial this may sound, it is important to bear in mind that from a moral point of view, much of our everyday "badness" falls within these two limits, whose theoretical pregnancy is necessary but does not imply actual possibility (apart from mythologies and theological accounts).

A list of the major aspects of the *myth* of pure evil are the following:[9] 1) evil involves the intentional infliction of harm to people; 2) evil is driven primarily by the wish to inflict harm merely for the pleasure of doing so; 3) the victim is innocent and good; 4) evil is the other, the outsider, the out-group; 5) evil has been that way since time immemorial; 6) evil represents the antithesis of order, peace, and stability; 7) evil characters are often marked by egotism; 8) evil figures have difficulty maintaining control over their feelings, especially rage and anger; 9) evil violent people tend to have highly favorable moral opinions of themselves (unfortunately, there is evidence that low self-esteem is the major cause of violence: threatened egotism – especially when directed to insecure people – is at the roots of a lot of violence and of violent revenge).[10]

Finally, also the *tragic* character of human action can accentuate the limits of our everyday conception of evil and, at the same time, of our moral and legal conceptions of evil responsibility, which focus on individual agency and guilt, seen as absolute. There is a gap between the structure of human action and the way we usually ascribe moral responsibility: the problem of *structural* violence which I illustrated in chapter one furnishes an eloquent example of the difficulty of ascribing complete moral responsibility to individuals. Following Kierkegaard's understanding of tragic action, and taking advantage of contemporary discourse on moral luck, poetic justice, and relational responsibility, Coeckelbergh (2010) argues in favor of an epistemologically richer reform of our legal practices based on a less "harsh" (Kierkegaard) idea of moral and legal responsibility and directed more towards an empathic understanding based on the emotional and imaginative evaluation of personal narratives. Rejecting the fatalistic conception of tragedy he ascribes to

[9] (Baumeister, 1997, pp. 72–74).

[10] A further skeptical note on the "myth of pure evil" is provided by Russell (2010, p. 45), who stresses, drawing on the book by Cole *The Myth of Pure Evil* (2006), that even if the concept of evil has the requisite form to be explanatorily useful, it will be of no explanatory use in the real world. Indeed Russell contends that the concept of evil implies an unrealistically dualistic worldview, with purely evil people on one side and ordinary people on the other, but he also admits that: "[...] even if we accept Cole's claim that no actual person is thoroughly or innately bad, it still seems very likely that some actual persons are evil, and hence that evil can be an explanatorily useful concept."

Nietzsche and Heidegger, and recovering Kierkegaard's interpretation, he recognizes that the tragic does not necessarily imply that one has to accept *fate*, but rather the tragic aspects of human actions that can be characterized as follows:

> [...] there is luck in the way things turn out, luck in the circumstances in which one finds oneself, luck in the traits one has, influenced by social-environmental factors, and luck in the way one is determined by antecedent circumstances (Nagel, 1979). There is no reason why we should suppose that the actions of anyone judged in court should be regarded in a different way than the actions of other people. Many "crimes" might have turned out differently, circumstances might have been different, one might have had a different upbringing, and so on. These factors may not "excuse", that is, relieve one of all responsibility. But these factors should be taken into account not only when determining punishment, but also when determining moral responsibility or legal guilt. And of course, the degree to which one is constrained by these factors might differ, for instance in the case of people who are said to be mentally ill. But this supports viewing their and our responsibility as a matter of degree; it does not justify ascribing absolute guilt or declaring them absolutely morally incompetent (pp. 236–237).

In sum, human actions (and human wrongdoing) are always "in between" the two poles of complete passivity and absolute activity: "just as the action in Greek tragedy is intermediate between activity and passivity (action and suffering), so is also the hero's guilt, and therein lies the tragic collision. [...] The tragedy lies between these two extremes. If the individual is considered as entirely without guilt, then the tragic interest is nullified [...]; if, on the other hand, he is considered as absolutely guilty, he can no longer interest us tragically" (Kierkegaard, 1944, p. 117). The suggestion is that this intellectual awareness may help our societies and communities to better cope with unacceptable deeds by individuals who are neither criminals nor patients, to make room for praise as well as blame and punishment, and to set up practices and institutions that do not completely rely on a plain conception of responsibility that is hard to bear for human beings.

5.2.1 Criminal Psychopaths' Morality and Ethicocentrism

I have contended in this book that human beings live with moralities of various kinds, and possess and adopt various moral frameworks (e.g. religious, civil, personal, emotional, etc., not to mention their intersections and intertwining) which they engage and disengage both intentionally and unintentionally, in a strict interplay between morality and violence. There are also private moralities and habits – perceived as fully moral by the agents themselves, which we can call *pseudo-moralities* if we compare them to the translucency of the modern moral frameworks, as they are usually described in books about moral philosophy (Kantian, utilitarian, religious, ethics of virtues, feminist ethics, and so on). These personal moralities can be very easily observed not only as the fruit of the emergence of archaic moral templates of behavior in mentally healthy human beings – that is, templates of possible moral behavior trapped in a kind of hidden moral collective unconscious – but also in the case of violent psychopaths, who suffer from a personality disorder involving a profound lack of empathy and remorse, shallow affect and poor behavioral controls:[11] psychiatrists and criminologists usually describe how extremely personal – often disguised, fragmented, and depraved – concerns and convictions, which are envisaged as "moral" in the subjective estimation of criminal psychopaths, are capable of triggering atrocious violence.

Kent Kiehl, a psychologist who focuses his research on the clinical neuroscience of major mental illnesses (with special attention to criminal psychopathy, substance abuse, and psychotic disorders such as schizophrenia), usefully observes that psychopathy immediately affects morality:

> Psychopathy is a personality disorder characterized by a profound lack of empathy and guilt or remorse, shallow affect, irresponsibility, and poor behavioral controls. The psychopaths' behavioral repertoire has long led clinicians to suggest that they are "without conscience" (Hare, 1993). Indeed, Pinel (1801), who is credited with first identifying the condition, used the expression "madness without delirium" to denote the lack of morality and behavioral control in these individuals, which occurred despite the absence of any psychotic symptoms or defects in intellectual function. Thus, the psychopath presents clinically as a "walking oxymoron". On the one hand, the psychopath is capable of articulating socially constructive, even morally appropriate, responses to real-life situations. It is as if the moment they leave the clinician's office, their moral compass goes awry and they fail seriously in most life situations (2008, p. 119).

I must immediately stress that when Kiehl says that criminal psychopaths present a "lack of morality", I prefer to suggest that they display a lack of *our* morality: the *ethicocentric* morality[12] of a civil, cultivated observer. It seems that the criminal psychopaths' acts result inconsistent with their verbal reports, like in the following case, still illustrated by Kiehl:

> I was working with a psychopath who had been convicted of killing his long-term girlfriend. During his narrative of the crime he indicated that the trigger that set him off was that she called him "fat, bald, and broke". After her insult registered, he went into the bathroom where she was drawing a bath and pushed her hard into the tile wall. She fell dazed into the half-full bathtub. He then held her under the water until she stopped moving. He wrapped her up in a blanket, put her in the car, drove to a deserted bridge, and threw her off. Her body was recovered under the bridge several days later by some railroad workers. When asked if what he had done was wrong, he said that he knew it was a bad idea to throw her off the bridge. When I probed further, he said that he realized that it was bad to actually kill her. This inmate was subsequently released from prison and then convicted of killing his next girlfriend. When I met up with him in the prison some years later, he indicated that his second girlfriend had "found new buttons to push". He was able to admit that he knew it was wrong to kill them.

---

[11] The publication of the Diagnostic and Statistical Manual of Mental Disorders third edition, DSM-III, changed the name of this mental disorder to *Antisocial Personality Disorder*.

[12] Analogously to ethnocentrism, ethicocentrism is the tendency to believe that one's ethical framework is centrally important, and the correct meter to measure all other moralities.

In the case I just reported, it seems that a "morality" of killing is activated: the victim is sacrificed because she *deserved* that punishment in the light of the psychopath's rigid morality. Being questioned, a morality of decency is advanced and verbally reported before the moral imperative not to take another person's life (the *wrong* deed consists in throwing the body in the river) and finally, the morality of not-killing is verbally proposed (the *wrong* deed consists in the killing itself).

In my perspective of disengagement and reengagement of morality, the first moral fragment (killing to punish) does not only trigger but also justifies violence, and plays a dominant role. However, it coexists with other moral fragments, that are reengaged and that sometimes disengage the dominant one. Many criminal psychopaths share multiple moralities with mentally "sane" human beings, moralities which play the role of more or less freely chosen "reasons", and they are involved in processes of disengagement and reengagement; these various shifts seem anomalous insofar as they display a strange sudden intermittence of changes or long delays, a lack of stability within the various stages or an excess of stability, and in some cases the – so to say – special "individuality" of the adopted structured morality is heavily at play.

Kiehl contends that many other psychiatric conditions (also some underlying criminal behaviors) are related to the aforementioned impairments in understanding moral behavior – still, some, paradoxically, are seen as "unencumbered by moral imperatives", as in the case of a schizophrenic who killed someone he thought had implanted a monitoring device in his head. The usual interpretation of this supposed lack of morality is the following: in the case above, through our twenty-first-century academic or forensic ethicocentric screen, the criminal schizophrenic could not be convinced that sacrificing his victim was a *bad* thing to do because he was unable to articulate that it was wrong to kill this person. I rather think that cases like this are better illustrated as characterized by the stability of a central and unique totally "subjective" moral framework, not sharable in a collective dimension, but still lived as "moral" by the human agent (i.e., if the schizophrenic could not be persuaded into acknowledging that his deeds were wrong, he probably kept thinking they were right, which is a moral stance). We are dealing with a kind of *personal morality*, as I have noted above, envisaged as a fully acceptable dominant morality in a subjective estimation, concurring with an anomalous absence of those multiple moralities which in my opinion characterize mentally healthy human beings.[13]

---

[13] A wealth of research is available on the psychometric aspects of various psychopathological behaviors, intertwined with glibness, superficial charm, low empathy, lack of guilt or remorse, and shallow emotions. Other studies concern the neural counterparts of psychopathological symptomatology, which are for example localized in malfunctions of orbital frontal cortex, the anterior insula, the anterior cingulate of the frontal lobe and the amygdala, and adjacent regions of the anterior temporal lobe. Other neural counterparts are being studied, especially regarding reduced activity in psychopaths during language processing in the right anterior temporal gyrus, the amygdala, and the anterior and posterior cingulate, in the case of attention, orienting, and affective processes, and the relevant role in psychopathy of the paralimbic system. Details are furnished in (Kiehl, 2008).

Relatively well-known research[14] about criminal psychopaths stresses the fact that they do not discriminate between moral and conventional rules (for example, mere etiquette and various social rules, such as which side of the road to drive on, or how to move the pieces in a game of chess), contrarily to non-psychopathic criminals and "normal" individuals. So to say, the criminal psychopaths rate the wrongness and seriousness of the respective violations in a similar way and as authority-independent. Moreover, in a second experimental result criminal psychopaths tended to treat all rules as "inviolable" in an effort to convince the experimenter that they were mentally healthy. The interpretation resorts to state a deficit of moral motivation together with a deficit of moral competence, as a direct result of the emotional deficit.

I consider this interpretation of results to be puzzling.[15] I do not agree with it. First of all, almost always *conventional* rules also carry the *moral* values of a group (for example, etiquette is not simply a morally-neutral rule), and so the experiment is biased by this aprioristic assumption of the experimental psychologist; second, the antisocial violent outcome is not necessarily due to impaired violence inhibition and to a general lack of emotional concern for others. In the perspective I have outlined above, the data obtained can also be interpreted in terms of a rigidity in the adoption of a given moral perspective and in the perseverance in applying the related violent (criminal) punishment, in lieu of a more open mechanism of moral disengagement and reengagement in other moralities, possibly less inclined to perform violent punishment. On the contrary, the supposed lack of moral emotion (Prinz, 2007, p. 45) seems to me intertwined – at first sight paradoxically – with the production of a lack of moral flexibility. In this sense criminal psychopaths do *not* have problems with morality because they are practically "amoral" and they lack moral (emotional) commitment, but instead because they are engaged in a kind of rigid *hyper-morality*, which is not open to quick and appropriate revisions.[16] One should wonder whether the emotion, in front of inflicted harm, is lacking because subjects are engaged in a rigid morality whose punishments are seen as just and deserved, or if it is the lack of emotion that promotes rigidity in the adopted moral perspective. It is not that criminal psychopaths do not master moral emotions and show reduced activation of areas involved in attention and emotional processing, but it seems instead they just master their moral emotions that way: in sum, they are just emotionally *retarded* in the light of our moral judgment of "normal" individuals or non-psychopathic criminals!

It is a real pity that psychiatric and psychoanalytic traditions, which are still obsessed by an excess of positivistic commitment, mostly refuse any interest in the moral aspects of mental illnesses. In this perspective psychiatrists often correctly complain about the tenacious persistence of a "moralistic" perspective in cases of childhood sexual victimization, in fact there has been a tendency in psychiatric

[14] (Blair et al., 1997).

[15] (Kelly et al., 2007) illustrate a growing body of evidence which justifies substantial skepticism about all the major conclusions that have been drawn from studies using the moral/conventional distinction.

[16] A further interesting point on psychopaths is provided by the psychoanalytical analysis of the so-called "structure of evil", as I will illustrate below in subsection 5.3.5.

professionals to *vilify*[17] those very patients who display abnormal sexual behaviors as a result of various kinds of sexual trauma (Van Slyke, 2006). I say, to respect the purported objectivity and freedom from moral bias in scientific evaluation, diagnosis, and therapy on the part of the psychiatrist is one thing, but the lack of consideration towards the moral life of criminal psychopaths and their victims is a totally different thing. After all, morality is no longer the "other" of scientific rationality, like it has almost always been considered over the last two centuries (science deals with what is the case, whereas ethics deals with what ought to be), but a legitimate object of rational analysis.[18] Prinz too seems perplexed: "These deviations suggest that they do not possess moral concepts; or at least that their moral concepts are fundamentally different from ours" (Prinz, 2007, p. 43).[19]

Here we may draw an interesting parallel with confabulating. As mentioned in chapter four,[20] confabulation results from the inability to discard beliefs or ideas that are patently false. This is due to the fact that confabulators seem to lack the mechanisms enabling them to inhibit information that is irrelevant or out of date. The main effect is that the process of belief monitoring and revision cannot take place, and the confabulator is simply trapped within his bubble. I argue that something similar may happen to criminal psychopaths. That is, they are trapped in a sort of moral confabulation resulting from the inability to discard a certain morality as unacceptable. In turn, such an inability would block the normal moral flexibility and thus the process of moral reengagement.

To sum up, usually perpetrators of evil do not regard themselves, like Kant had already stressed, as wrongdoers, neither in case of sound people nor of course, all the more so, in the case of mentally ill ones. Paradoxically, they often see themselves as victims, for example treated unjustly or aggressively, so that they think – perversely – they should deserve sympathy, support, and tolerance (if not praise).

5.2.2 *Mental Incapacity and the Fear of Decriminalization*

It has to be said that present-day legal judgments of psychopathological criminals strongly avoid the exploitation of "moral" considerations and also tend to disregard the possible "moral" aspects of criminal conduct. Currently, the attribution of

[17] It could be speculatively suggested that such a perspective is the remnant of a vestigial "honor" culture, which despises, insulates and punishes those guiltless troubled individuals just as a family would punish a girl who lost her virginity because of rape. The issue of honor cultures has been addressed in subsection 1.2.2 of chapter one.

[18] On Searle's contention that to say something is true is already to say you ought to believe it, that is *other things being equal*, you ought not to deny it cf. above at page 22, chapter one. This means that normativity is more widespread than expected

[19] On the moral content of what is called "personal construct theory taxonomy" of the acts of killers cf. (Winter, 2006). Winter quotes the serial killer Alan Brady (2001): "Serial killers, like it or not, can possess just as many admirable facets of character as anyone else, and sometimes more than average".

[20] Cf. section 4.6.

responsibility to criminals often takes advantage of the concept of *mental incapacity* so that, in these cases, the *moral judgment about moral conducts* of "psychopathological" criminals is potentially extinguished insofar as they are merely seen as affected by an overall mental incapacity, exclusive object of the psychiatric and legal technicalities. One must note that the attribution of responsibility changes over time, as Lacey (2010, p. 116) observes: "[...] patterns of responsibility-attribution relate to the roles and needs of a criminal justice system: to a political need for legitimation, and to a practical need to specify and co-ordinate the sorts of knowledge which can be brought into a court room". In the late nineteenth century, these patterns were finally affected by the diffusion of democratic sentiments and entitlements, thus transcending the traditional notions of "acting maliciously" or exhibiting a "bad moral character". Nowadays, the state's responsibility in proving not only conduct but also individual responsibility (i.e., psychological and internal, capacity-based, requirements of "mens rea", the *guilty mind* presupposed by criminal liability) is crucial for the legitimation of criminal law, not as a system of brutal, retaliating force but as a system of actual justice.

What is at stake is that "the treatment of what we would today call mental incapacity defences, in which what would become the psychiatric profession was emerging as an authoritative witness to the 'facts of the mental matter'" is related to the fact that "[...] in principle, the field of mental incapacity should reflect the most fully developed aspect of the 'inner' or 'psychological' model of criminal responsibility" (p. 119). In brief, it is evident that, in this perspective, the jury's commonsense *moral* assumptions about madness, which characterized the evaluative/character based practice of the past, decline. Currently, incapacity defenses which lead to judgments of non-responsibility focus on *cognitive* incapacities (for example "lesions of the will",[21] found in the factual conditions of mental, inner or neural states of individuals, where knowledge and consciousness are central), as opposed to *volitional* incapacities, that were considered as forms of *moral* insanity. Nevertheless, it is worth mentioning that Lacey concludes by acknowledging a kind of *resurgence* of character-based patterns of attribution of criminal responsibility:

> Emerging from their subterranean (though clearly important) position in the exercise of discretion at prosecution and sentencing stages, character-based principles are enjoying a revival not only in "three strikes and you're out" sentencing laws and paedophile registers but also in the substantive law, particularly that dealing with terrorism, and in the operation of evidential presumptions, detention rules and the renewed admissibility of evidence of bad character. Why, we might ask, has character suddenly become an acceptable explicit principle of criminalisation once again? And does this imply that its decline was more formal than real? (2010, p. 129).

The reason for this resurgence seems to be clear: further attention for capacity-based practices of responsibility-attribution better relates to the habit of considering individuals and their engaged capacities *per se* rather than their social status or appearance, that is to say, an attitude towards the whole practice of justice which

[21] Cf. below, subsection 5.2.4.

derives from certain standards of legitimation following the democratic acknowledgement of individual freedoms. Unfortunately, such a disposition would prosper in a world endowed with

> [...] some confidence in its institutional capacity to deliver such individualised judgments while maintaining adequate levels of social control. Such a world has arguably never existed. But that individualising impulse has most certainly had a significant impact on the form of (some parts of) criminal law over the course of its "modernisation". We might speculate that, at times when the sentiments underpinning norms towards equal liberties are fragile, perhaps because of fears about crime, or terrorism, or order more generally, explicitly character-based patterns of attribution tend to enjoy a revival (ibid.)

In sum, the revival of "moral character" in criminal law seems related to the renewed emergence of a *culture of control* in our anxious, fragile, and insecure world, which results in a potential greater criminalization, recognizable "overcriminalization", as I will illustrate in the following chapter.[22]

From a wider point of view, one could notice how both the hypertrophic diffusion of psycho-pathological insights in the appraisal of criminal responsibility and the revival of character-based criminalization are easy ways of escaping a more burdensome, yet richer, practice of criminal justice: as I just stated, stressing the criminals' moral character leads to the inescapable excess of *overcriminalization*, but similarly, a reductionist psychopathology of criminals yields the perverted fruit of utter *decriminalization* (at least from the point of view of social ideologies and everyday people's mentality), resulting in the impossibility of any guilt ever being attested.

So far, I have contended that many of the pernicious outcomes of the anomalous engagement and disengagement of moralities are caused by the "anomalous" engagement of more or less rigid personal, individual moralities – that only the agent himself recognizes as such – and by their abnormal consecutive replacement. The reader could ask, how can a morality that is private still be a morality? She should note that morality can be fragmented and private – in the sense that it is not shared with some specific group – because it is a vestigial remaining of more ancient moral concerns and axiological frameworks, as I have illustrated when dealing with the psychoanalytic concept of collective unconscious:[23] for example, mobbing and bullying behaviors are surely not explicitly labeled as "moral" in our civil Western countries, but still mentally "work" in people and are perceived as good motivations for supposed-to-be "moral" behaviors, exactly as they worked fairly well in ancient times, for example when the scapegoat mechanism was a perfectly approved, efficient, and justified conduct. Of course these behaviors were not necessarily labeled "moral" in the respective human groups, with the same meaning we now sophisticatedly and intellectually attribute to it, but certainly they *de facto* played a decisive role in that cooperative sense which works in the case of coalition enforcement.[24]

[22] Cf. section 6.5.

[23] Cf. chapter four, subsection 4.8.5.

[24] Cf. this book, chapter one.

5.2.3 *Gene/Cognitive Niche Co-Evolution and Moral Decriminalization*

The process of moral decriminalization of so-called criminal *psychos* I have just indicated certainly presents some puzzling problems. To shed more light on the problem of decriminalization let us illustrate the example of violent physical attacks: thanks to evolutionary biology we know that conspecific aggression is usually related to the establishment of sexual and social dominance and mostly exhibited by males. Physical aggression may have provided societal fitness in the stone age (to gain a good hierarchical position and sexual dominance), but physical aggression is morally and legally inhibited in many modern societies, which value cultural dominance over physical force, taking advantage of that process of civilization so clearly illustrated by Elias (1969). (Of course civilization is a recent cultural acquisition in cognitive niches of Western societies, and there are still countries where physical violence and aggression are largely admitted as a means of self-establishment).

Physical aggression is nevertheless still widespread in the "civilized world". Why? I have already noted that the speculative Jungian hypothesis about the existence of a collective unconscious (and thus of hidden archaic moral templates of behavior concerning right and wrong, for example based on morality of honor and of revenge, scapegoating, magic thinking, various religious beliefs, etc.) can be interpreted as the presence of pervasive and permanent cultural niches, whose information enters the brain automatically in early life. There are very few cases of *wise babies*, to use a term proposed by Ferenczi:[25] very young children that are able to protect themselves from the "barbaric" – archaic – and usually violent fluxes of cognition and affectivity which often impact newborn cognitive niches (families, for example). Consequently, once silently stabilized in neural networks, various archaic moral and non moral aspects of the collective unconscious can be reactivated and made explicit and effective, together with their violent outcomes. It is also in this perspective that the reader can better grasp the sense of that "multiplicity of individual moralities" I have inserted in the title of this chapter. One can hardly be both a primitive bloody avenger and a civil citizen of a constitutional democracy in the translucency of his consciousness, indeed the two aspects are in this case in explicit competition, but he can more easily be both if we take into account the hidden presence of the collective unconscious.

In our highly composed cognitive niches, a lot of physical violence tends to be made intelligible as a manifestation of mental retardation, susceptible genes, stress, low intelligence, brain damage, or in terms of other physical conditions that reduce the effectiveness of the frontal lobes concerned with conscious control of behavior. The management of these supposed-to-be pathological violent actions is often

[25] "The Dream of the Wise Baby" is a one-page text that Ferenczi wrote in 1923. It is a description of an usual adult dream. It describes a very young child, a neonate, a baby with glasses, who is teaching adults. It relates to the idea of young children, who had often been traumatized, and that had accelerated those developmental features that led them to acquire highly acute sensitivities and intuitions, or wisdom beyond their years.

reserved to the technicalities of forensic psychiatry, which in turn is sometimes affected by uninformed attitudes, confusions, and excess of business-oriented medicalization. In our terms, we can say that psychiatry maintains that in those individuals that are more susceptible to be permeable to various modes of archaic or subcultural moral behaviors (related to strong methods of punishment), a particular permissive state for physical violence can emerge. After all, the child of Western societies is often immersed in very (morally) violent cognitive niches: thanks to sensorial exposition to the moral/violent information absorbed through parents and caregivers (as a matter of routine practice), and later on, through peers and present-day technological media favored by globalized communication, which give high potential to cognitive niches made of written language and iconic features. Thus, various moral templates of behavior are absorbed (together with other "less moral" templates, related to other kinds of knowledge, belief, art, music, etc.) and implemented in individual brains, immediately available to consciousness or not, "stored in the neurons in binary fashion as bits of information, and made available at occasion" (Leigh, 2010, p. 122).

I have already highlighted how the "technicalization" of violent behavior leads to some degrees of decriminalization both in trials (excuse, mitigation, etc.)[26] through insanity defenses, and in people's mentality, and how this is crucial for the legitimation of criminal law, not as a system of brutal, retaliating force but as a system of justice. A lot of – more or less scientific – knowledge is available, for instance from psychiatry and neuroscience, and it came to be considered as an authoritative source for describing who is to be held responsible and who is not in the case of violent aggressive behavior. Unfortunately, even if in various cases the lack of responsibility is patent, in various other cases it is not. The gene/cognitive niche co-evolution can help us to understand why.

Let us better introduce the problem of gene/cognitive niche evolution, by first of all coming back to a point I made in connection with the "multiplicity of individual moralities". As argued above, different moralities may co-exist in a single person even though they are not all active at the same moment. A continuous process of disengagement and reengagement fairly describes the kind of plasticity characterizing our moral life. It is a complex theoretical task to treat the issue of moral plasticity from an evolutionary perspective. I already made quite a bold claim about it referring to the highly speculative yet intriguing idea of the collective unconscious: now I will try to further refine my proposal with respect to the gene/cognitive niche co-evolution.

My proposal is that the various archaic (both moral and non moral) aspects of the collective unconscious are more likely to emerge in connection with the impoverishment of the cognitive niches one lives in: that is, some moral templates relying on archaic modes of moral behavior are somehow re-activated or *re-enacted* as the result of a "moral sensory deprivation" caused by the pauperization of a cognitive niche.

[26] On the problems of psychiatric aspects of justification, excuse and mitigation in Anglo-American criminal law cf. (Buchanan, 2000).

As already maintained,[27] cognitive niches are also moral niches insofar as they implicitly specify the most suitable behaviors to activate in order to exploit the various cognitive chances provided by the environment. The specification of such behaviors is governed by moral affordances functioning like anchors that, in turn, facilitate or even suggest the activation of the most profitable behavioral "scripts" in a given cognitive niche. For example, consider the case of science: science is articulated in a cognitive niche that is constructed so as to deliver and ponder evidence and counter evidence around a certain hypothesis or idea. In order to fully exploit the potential of such a cognitive niche, scientists should comply – at least – with a number of rules/norms that are essentially moral (in many cases, also endowed with legal counterparts). For instance, they are not supposed to lie about the results of their experiment by hacking them so as to support their hypothesis. That would be considered as cheating.

More generally, I claim that the presence of a cognitive niche implicitly requires compliance – partly described by docility[28] – with certain moral behaviors that, if successfully activated, allow a person or a group to fully exploit the chances furnished by the local cognitive niche. Now, what happens when a morally valued cognitive niche goes through a process of impoverishment? The impoverishment of a cognitive niche can be described as the permanent loss of certain cognitive chances due to a pauperizing structural re-organization of the environment affecting the way external resources are accessed and moral sensory stimulation is nested. For instance, one of the first appreciable consequences is the increase of the social and moral costs associated with sustaining a certain standard of performance. As such costs go up, a person or, more likely, a group, is pressed into adopting supererogatory acts. Or, as a more economical alternative, the same person (or group of persons) may be prompted to deactivate a set of moral templates in use and re-activate some other templates typical of different contexts. The success of these strategies, aimed at the restructuring of the niche, is extremely contextual and cannot be guaranteed: consequently, if one fails in applying one of these two options the result may become extremely awkward from a moral point of view. A perverted morality of paroxysmal mutual help might drain the individual's psycho-cognitive energy out of excessive supererogations, while the reactivation of a hazardous or anachronistic set of templates might only quicken the crumbling of a niche by pulling the plug on the moral conventions underpinning the life of its members so far: the niche's morality then turns out to be "maladaptive" (at least, in a local – not Darwinian – sense).

Consider for instance the case of democracy. Democracy can be considered as a carrier of powerful cognitive moral niches permitting the flourishing and maintenance of fundamental pillars for our living, for instance, capitalism, science, modern civilization, *Rechtsstaat*. Its recent impoverishment in rich Western countries – that can be partly described as *ochlocratization*[29] – has favored the emergence of what I will describe further in this chapter as the "Fascist state of the mind". From the perspective

[27] See chapter four, subsection 4.1.3.

[28] See chapter one, subsection 1.3.2.

[29] See above, chapter three, sections 3.5 and 3.6.

of a sincere supporter of democracy, the Fascist state of the mind is a set of violent and archaic templates of moral behavior that drastically dumb down the moral character of those who adopt it, facilitating various episodes of strong violent punishment, such as aggressive propaganda, demonization of the opposition, mobbing, bullying, intellectual genocide,[30] political assassination, and so on. The recent financial crisis and its connection with globalization is a case in point that can support my claim. It is well-known that, during the last decade, the democratic niche – which also profitably assisted capitalism – has been weakened by the emergence of *deregulation* and other phenomena, chiefly the removal of government rules and regulations constraining the free market. Such a policy did not produce any benefit either for the market or the population. Quite the contrary, it literally deconstructed part of the democratic niche. It has inexorably pauperized society as a whole, favoring the re-activation of forces that are hostile to capitalism and democracy itself, and their related old-fashioned moral templates of behavior and axiologies.[31]

In order to clarify this point from an evolutionary perspective, thus dealing briefly with the hotly debated issue related to the relationship between culture and nature, I summarize here again the main points related to gene/cognitive niche co-evolution. In chapter four[32] I have introduced the co-evolution between genes and cognitive niches. I have described that general inheritance (natural selection among organisms influences which individuals will survive to pass their genes onto the next generation) is accompanied by another inheritance system which plays a fundamental role in biological evolution, where niche construction counts. It is the general inheritance system (also called *ecological inheritance* by Odling-Smee, Laland and Feldman (2003)). In the life of organisms, the first system occurs as a one-time, unique endowment through the process of reproduction (sexual for example); on the contrary, the second system can in principle be performed by any organism towards any other organism ("ecological" but not necessarily "genetic" neighbors), at any moment of their lifetime. Organisms adapt to their environments but also adapt to environments as modified by themselves or other organisms. From this perspective, acquired characteristics can play a role in the evolutionary process, even if in a non-Lamarckian way, through their influence on selective environments via cognitive niche construction. Phenotypes construct niches, which then become new sources of natural selection, possibly responsible for modifying their own genes through ecological inheritance feedback (in this sense phenotypes are not merely the "vehicles" of their genes).

In sum, from the point of view of niche-construction, evolution depends on two selective processes, a blind process based on the natural selection of diverse organisms in populations exposed to environmental selection pressures and a second process based on the semantically informed selection of diverse actions, relative to diverse environmental factors, at diverse times and places, by individual niche-constructing organisms. The second process was not described by Darwin. In this

[30] Cf. below subsection 5.3.3.

[31] As regards the finance-dominated accumulation regime, income distribution, and the current crisis, cf. the very clear account in (Stockhammer, 2009).

[32] Subsection 4.1.2.

process selection selects for *purposive* organisms, that is, niche-constructing organisms. Consequently, the process of transmission and selection of the extragenetic information that is embedded in cognitive niche transformations has to be considered *loosely* Darwinian for three main reasons (Odling-Smee et al., 2003, pp. 256-257.):

1. extragenetically informed behavior patterns are broadly adaptive and maladaptive;
2. variants occurring during genetic evolution are random, whereas those of extragenetic information are not. They are smart variants, because the response to both internal and environmental cues is targeted appropriately to behavioral repertoires. As I have said above, variation is in this case *blind* but also *constructed* because "[...] which variants are inherited and what final form they assume depends on various 'filtering' and 'editing' processes that occur before and during the transmission" (Jablonka and Lamb, 2005, p. 319). In this sense extragenetic information produces variants, which originate suitable cognitive stabilities in artifactual niches, which in turn "afford" humans and other organisms in many ways.
3. extragenetic information
   a. when neurally stored – both consciously and unconsciously – it allows a strong interaction of its elements, which can also be seen in selective terms, as suggested by neural Darwinism theory, which sees neurons as diverse populations submitted to loosely Darwinian effects at the level of both neural development and moment-to-moment functioning which interfaces with experience;[33]
   b. when stored in material devices it no longer presents the characters of a more or less Darwinian evolving creative population, like in the case of neural cells: instead it has an evolutionary impact insofar as it causes *persistent* modifications upon the environment.

Edelman's theory – which is also supported by the evidence provided by some artificial intelligence devices called "Darwin automata", expressly built to test it – is controversial, as Rose clearly explains, because of the way the theory considers how extragenetic information is transformed through the dynamics of synaptic modifications:

> During development there is thus a superabundance of synaptic productions, a veritable efflorescence – but if synapses cannot make their appropriate functional connections with the dendrites of the neurons they approach, they become pruned away and disappear. This overproduction of neurons and synapses might seem wasteful. It has led to the argument that just during evolution "natural selection" still eliminates less-fit organisms, so some similar process of selection occurs within the developing brain – a process that the immunologist and theorist of human consciousness Gerald Edelman has called "neural Darwinism". However, this transference of the "survival of the fittest" metaphor from organisms to cells is only partially correct. It seems probable that the whole process of cellular migration over large distances, the creation of

[33] Cf. (Edelman, 1989, 1993; Seth and Baars, 2005).

> long-range order, requires the working out of some internal programmes of both individual cells and the collectivity of cells acting in concert. [...] Overproduction and subsequent pruning of neurons and synapses may at one level of magnification look like competition and selection; viewed on the larger scale, they appear as co-operative processes (1993, p. 76).

Anyway, according to Edelman the brain would be a Darwinian "selection system that operates within an individual lifetime" (Edelman, 2006, p. 27) so as some synapses are strengthened and some are weakened through the experiential selection. What biases the brain system to yield adaptive responses is a process called reentry "[...] a continual signaling from one brain region (or map) to another and back again across massively parallel fibers (axons) that are known to be omnipresent in higher brains. Reentrant signal paths constantly change with the speed of thought. [...] consciousness is entailed by reentrant activity among cortical areas and the thalamus and by the cortex interacting with itself and with subcortical structures" (pp. 28 and 36). It is from this perspective that – through core reentrant neural integrative processes – many sensory and motor signals are linked together, thus providing various perceptual categorizations (also connected to memory) which originate a scene "in the remembered present of primary consciousness, a scene with which an animal could lay plans", and of course motor outputs (p. 36). Selectionist brains are certainly the effect of historical contingency, irreversibility, and the operation of non linear processes.

In the light of the previous considerations, given the fact there is

1. a co-evolution between genes and cognitive niches[34] during human evolution and, especially,
2. because of their specific coupling which occurs during the life of any individual,

the methods that are currently used in Western societies to discharge moral and legal responsibility seem to me unclear in their epistemic structure and so partially unreliable. Indeed, it is a fact that the brain is configured in a certain way, and there is evidence according to which the presence of genetic or anatomic disfunction (such as epilepsy, delirium, dementia, thyroid dysfunction, cerebrovascular disease, encephalitis, diabetes, etc.) promotes aggressiveness: the problem is that all this is often vaguely linked to a consequent lack or impairment of free will capacities, and nothing more. Indeed, philosophers of free will frequently refer to mental and brain disorders as conditions that compromise free will and reduce moral responsibility, and so does forensic psychiatry.[35] For example, what if some neural clusters were shaped during the personal history of an individual immersed in the aggressive morality of an honor

[34] Cognitive niches are made of fashion, makeup, culinary arts, painting, architecture, music, poetry, fiction, nonfiction, ideologies, morality, science, technologies, medicine, psychology and so on. Neurons in brains are constantly in interaction with other neurons but also with the exogenous information embedded in cognitive niches, acquired through senses (Leigh, 2010).

[35] I have extensively dealt with the problem of free will and of the ownership of our own destinies in the current technological world in (Magnani, 2007b, chapter three).

culture, so that he presents anomalous distribution of excitations in areas related to aggressiveness (even detectable through fMRI scans) with respect to "normal" agents? Does this authorize us to state that the person who embodies those neural networks is not responsible for his violent, illegal outbursts? Is the presence of certain genes, susceptible to the exposure to unlucky cognitive niches (for instance an abusive family), a reason which authorizes the philosopher or the forensic psychiatrist to subsequently hypothesize a lack or an impairment of free will in a criminal offender? Furthermore, on another level, does the fact that his brain did not have the chance to be exposed to the cognitive niche of civil morality, as embedded in law itself, make the criminal offender morally and/or legally condoned?

5.2.4 Can We Freely Decide to Kill Our Free Will? What Is Crime Commodification?

I agree with Meynen (2010) who contends that both in philosophy and forensic psychiatry "it remains unclear in what way free will is compromised by mental disorders". A connection between mental disorder and freedom is present – and proposed as scientifically granted – in the introduction of the fourth edition of the *Diagnostic and Statistical Manual of Mental Disorders* (DSM-IV).[36] It reads, "In DSM-IV, each of the mental disorders is conceptualized as a clinically significant behavioral or psychological syndrome or pattern that occurs in an individual and that is associated with present distress (e.g., a painful symptom) [...] or an important loss of freedom".[37] It is important to note that various notions of free will can be advanced – and are not devoid in themselves of ambiguities: 1) one must be able to act otherwise; one must have alternative possibilities; 2) one must be able to act or choose for a reason; 3) one has to be the originator (the causal source) of the action. Obviously free will is always related to moral responsibility. Various constraints, both standard and psychiatric, are supposed to create problems for free will, for example, diminished capacity, intoxication, unconscious drives, infancy, entrapment, duress or coercion, kleptomaniac impulses, obsessional neuroses, desires that are experienced as alien, post-hypnotic commands, threats, instances of *force majeure*, various psychopathological states, physical and genetic impairments. Such excuses typically find application in cases involving the ignorant, the misled, the coerced, the mentally insane, the intoxicated, the biologically abnormal. In these cases the *actus reus* tends to be conceded but *mens rea* is denied.

In the presence of these constraints, moral condemnation and/or legal punishment appear to be inappropriate. Meynen concludes that philosophers of free will have paid scarce attention "to identifying the precise reasons why (certain) mental disorders would diminish free will; a detailed analysis of what it is that mental disorders do that has such an effect on free will is lacking": this happens in the

[36] In Europe and other parts of the world, the ICD-10 Classification of Mental and Behavioral Disorders has been the predominant diagnostic system (World Health Organisation, 1992).
[37] (American Psychiatric Association, 1994, p. xxi, quoted in (Meynen, 2010)).

case of defining criminal responsibility in real subjects (for example related to psychosis), which leads to the choice of non-moral medical treatment instead of the fully moral/legal punishment which would normally follow a misbehavior. For example it is not clear when free will is *partially* compromised, and then when and to what extent responsibility can be actually discarded. The empirical fact that legal or psychiatric forensic technicalities can *de facto* solve ambiguities does not mean they are always based on serious scientific reasons.[38]

The man who killed his girlfriend that I mentioned above in subsection 5.2.1 acted for *reasons* as strong as moral imperatives, so in this respect his free will is preserved: his mental disorder does not affect the related meaning of free will. Similarly, can the capacity to choose *alternative possibilities* be jeopardized by mental disorder? There is no final answer to this question, yet. Finally, what about the *source/cause* of criminal violent action, which depicts the third sense of free will I have indicated above? Is it the guilt, the "proper person", his mental disorder, or his "biology"?

Professional psychologists and the so-called behavioral scientists argue for a broader range of ways in which psychology might be applied to criminal justice and, thereby, to law (Carson et al., 2007). They always contend that further "scientific" light can be shed not only on the problem of criminal responsibility, but also on eyewitness identification, investigative interviewing, credibility assessments and lie detection, fact finding, evidence, decision making and its discontents. They stress, for instance, that legal judgements, in particular, are influenced by short-cut, heuristical reasoning processes which have to be studied and clarified. I would like to note that psychology and other behavioral sciences do not have a *privileged* disciplinary status, for instance over philosophy or logic, that criminal justice "must" take advantage of. It is well known that too many psychologists just aim at promoting and diffusing their discipline as necessary everywhere, all the more in legal settings, even if the contribution often result scarce or counterproductive. Just to make an example, it is very sad that the study of abduction, so important in criminal

[38] A related problem concerns the fact that psychiatry confronts not only the sic Other but also the intercultural sic Other: how can psychiatry take these multiple otherness(es) into account without sacrificing its own rationalistic universal points of view? My friend Giovanni Corsico, a militant psychiatrist, usefully observes that psychiatry, to attenuate the eventually violent effect of theoretical universalization, has to give some precedence to systematic "observation", an observation that should occur when taking advantage of an "empathetic distance" with respect to the sic Other. Indeed, it has to be acknowledged that therapeutic treatment tends towards normalization and to the extinction of every manifestation of that same otherness and of its eco-cognitive existential uniqueness. In turn, my friend continues by warning us that the psychiatric judgement of insanity, which comes from the psychiatric (eventually forensic) technicalities – related to the embarrassingly powerful monumental/universal nosography – "should not disregard that sense of 'inbeing' (the existence – inherence – in something else)", which concerns the psychiatrist himself, together with a firm awareness of the "sense of shared culpability". Something similar is acknowledged by the attention given to the role of "listening and answering" proposed by the phenomenological approach to otherness, elaborated for example by Waldenfels (2007b; 2007a). On this "responsive phenomenology" cf. also the annotation given above at page 68.

investigation and in legal trials, is paradoxically disregarded by the psychologists themselves, even if studied in depth for example by philosophers, logicians, and AI scientists (Magnani, 2009). Some psychologists even acknowledge that "[. . .] unfortunately whilst work on abduction and defeasible arguments is exciting the interest of computational scientists interested in artificial intelligence it has provoked less interest amongst psychologists" (Carson, 2007).

Let us present a further interesting speculation. What about a person that, in presence of dysfunctional cognitive niches (poverty, abuse, and other various kinds of direct or structural violence), has in the beginning *freely* chosen and later on freely educated himself (and his brain's neural networks) to perform violent physical aggressiveness, fearlessly and repeatedly. Indeed, after years, he might have developed a criminal psychopathic personality and he can be described as such by a psychiatrist. In such cases the everyday language clearly expresses the same conclusion of the psychiatrist: "he is dominated by his impulses", so it is not him that performed the crime but his mental illness: get medical treatment!

A question arises: who (or what) transformed him into a person who lacks free will or has it impaired? He himself, his environment, his brain, his genes? I think we need more knowledge about puzzling situations like this.[39] Could it help our analysis to consider a person – for instance responsible for violent actions – who is supposed to be affected by a psychopathological lack (or impairment) of free will, yet who may also have freely brought himself to that condition? Maybe he freely chose a specific reaction in his coupling with cognitive niches, a reaction that later on conducted "him" to weaken or annihilate his own free will.

From this perspective we can see that people can be considered as responsible for dismissing the ownership of their own destiny. But, what about the responsibility for violent actions committed after that initial moral "choice", in the presence of the consequent impaired intentionality and free will? A similar problem is illustrated by Meynen himself (2010). I gladly submit his words to the reader:

> For instance, with respect to the person being the "genuine source of the action", I mentioned that the mental disorder-rather than the "person proper" – could be considered the cause of a crime. Yet, this raises the question, what is the person proper and how can one distinguish the person proper from a mental disorder? This line of

[39] I have to say that my philosophical discussion is not aimed at violently attacking the practice of providing legal excuses for those whose criminal acts were due to severe mental illness. I am aware that an often rudimentary attack on legal excuse and mitigation is already acting in society and legal settings, even in the case of serious psychotic disorders: I do not endorse such an attack at all. On this problem, in the US, cf. (Felthous, 2010). On the violent structural aspects of forensic psychiatric settings, the creation of a risk society by so-called corporate psychiatry, in which risks are controlled through chemical straight jackets and individual human rights disappear into the vacuum of individual immorality and corporate greed, and the so called "medicalization of evil" cf. (Mason, 2006). On the recent use of neurological evidence in trials and the cases of neuroimaging of aggression (that is the involvement of certain brain regions in the etiology of aggression and violence useful for the court system), data on competency, disposition, insanity, dangerousness, substance-abuse, neurotoxicology, also in minority groups cf. (MacNeill Horton and Hartlage, 2003).

questioning will, sooner or later, bring up the question, what exactly is a mental disorder? – a central topic in the philosophy of psychiatry. And if we focus on the "cause" of an event, then we must decide how to assess, among the manifold phenomena that contribute to the occurrence of a particular event (e.g., actions), which of these contributory phenomena count as an authentic "cause". For instance, did an addict's original decision to use heroin cause the heroin addiction and thus also cause the actions that subsequently resulted from the heroin addiction? In brief, a central issue will be, how do the person proper and the disorder relate and how can they be distinguished when it comes to the initiation of actions?

Is this attitude still reminiscent of the old-fashioned judgment based on *moral character*, that (it seems) we had abandoned in the nineteenth century, or is it an actual problem we need to address when evaluating crimes? In which cases should we condone a criminal and the violence he perpetrated? In case that we condone his crime, but the criminal had performed the violent action in a state of free will, are we not in the presence of a kind of perverse disguised forgiveness, a dressed up excuse, which does further wrong to all the other criminals who could not benefit from the same awkward forgiveness?[40]

Sometimes people adopt a sincere morality of love and compassion, which in our cultural niches always "pretends" to be harmless and familiar and instead turns out to be simply tragic. Indeed, some people, at the same time, implicitly adopt a culture of violence, like Žižek (2003, p. 30) clearly stresses, referring to the prototypical example of Che Guevara's diary.

> Let me say, with the risk of appearing ridiculous, that the true revolutionary is guided by strong feelings of love. It is impossible to think of an authentic revolutionary without this quality. This is perhaps one of the greatest dramas of a leader; he must combine an impassioned spirit with a cold mind and make painful decisions without flinching one muscle. Our vanguard revolutionaries [...] cannot descend, with small doses of daily affection, to the places where ordinary men put their love into practice.[41]

We face two possible explanations of *violent love*: direct and indirect killing.

1. It is the fact of having chosen – responsibly – a violent morality of love that *directly* generated your violent criminal actions ("I killed my girlfriend because I loved her", is a typical thought/explanation of too many sound human males).
2. It is that "originally" *free* choice of "absolute" love that later on generated in your neural brain an anomalous constitutive (and hardly reversible) inclination to give up any inhibition to aggressiveness. So, even after having killed your girlfriend, *you* say "I killed my girlfriend because I loved her": you are the one who says you loved her, but you are still the same one who killed her, and now you are glad that something *other* can be blamed for it, for instance your anomalous *brain* circuits full of "irresistible impulses" and "automatisms". And these were gradually implemented by none other than *yourself*, thanks to your "past" *free* coupling with cognitive niches. Against all odds, you are the one

[40] On the apories and possible violent aspects of forgiveness cf. chapter six, section 6.6.

[41] Quoted from (Anderson, 1997, pp. 636–637).

who, no matter how indirectly, killed his own girlfriend, thanks to a perverse and not-so-involuntarily dismissal of your free will.[42]

The case of the human male killing his female illustrates the typical story of violent outcomes of "fatal love" (Buchli, 2006): the killer consciously and responsibly chooses to be possibly (yet automatically) violent, and, embedded in that *moral bubble*,[43] he always – also after the crime – thinks of himself as a loving person, notwithstanding all the violence he perpetrated. Not only, when he provides an explanation of his "irresistible impulses", which of course he could not control (because they are out of the reach of his free will), he still consciously affirms he killed his female because he loved her, as he has not given up yet on applying the *love-morality* (but, in the background, the killer was another agent, his "anomalous brain'). "Ok guys, I killed her, but it is not me who killed her, it was my brain..."

Even if we do not have to fear the psychiatric *legal* decriminalization, which is in any case justified by the need for a "civilization" of the criminal law – beyond a brutal and monochrome force of blind punishment – it is worth stressing that psychiatric, psychological, and neurological knowledge is often rudimentary, obviously continually changing during the standard research processes of the involved academics, and often applied in settings where incompetence, excessive economic drives, greed and other variables endowed with possible violent outcomes are at play. What is really unfortunate, in my opinion, is that media and consequently public opinion became absolutely comfortable with insanity pleas, in spite of being conspicuously ignorant as far as the knowledge of forensic psychiatry is concerned. Still, the bits of information people acquired taught them the capacity to roughly classify almost any violent or bloody actions as the fruit of criminal psychopathologic individuals. In such a way – curiously – they are inclined to decriminalize them, nearly *a priori*.

In sum, for common people the violent individual is no longer responsible because he was – so to say – the real victim of a kind of mental infection due to a "parasitic" moral niche (i.e. poverty, a revengeful honor culture...), or because the real killer was "his biology" (an anomalous brain, for example). On one side the responsible party is the objective moral niche, on the other an unlucky biology.[44]

---

[42] On the perverse culture of irresistible impulse cf. above at page 174.

[43] I have described the concept of *moral bubble* in chapter three (subsection 3.2.2).

[44] Interestingly, Miller (2010, p. 731) addresses this problem of responsibility complaining about the "mistreatment of psychology in the decades of brain". The observed outside interest-group pressure on National Institutes of Health (NIH) in US would be "unquestionably well intentioned but misguided, and it is often motivated in part by the assumption that biological construals of mental illness reduce stigma. Such a prediction would make sense if people tend to be held less responsible for their biology than for their psychology. Why that should be the case is not apparent. Although such a notion of differential responsibility is commonplace, it is not obvious that we have less control over our biology than over our psychology. The genes one has are not one's responsibility, but we have considerable control over their interaction with the environment, and as argued above, that interaction is where the action is in psychopathology". The possibility of dangerously "inflating" the psychiatric concept of disease through neuro-genetic approaches is treated in (Schleim, 2009).

Responsibility for violent behavior is externalized and *everyone* is happy to think atrocious violence does not normally come from the core of an individual's free will. This is a way to sterilize violence and disregard it as something exogenous to our decisions. Indeed, as I have often contended in this book, it is obviously extremely painful to look violence in its own vivid eyes. Disregarding violence and, consequently, the punishment of violent behaviors is easier and also economically of benefit to present-day states.[45] Nowadays, every culture (and subculture) represents and defines how a person behaves or feels according to a suitable classification of anomalous mental conditions, and provides the related therapies. Our culture has allowed for various psychiatric effects of decriminalization, comprising some supposed-to-be civil legal excuses and more or less efficacious medical treatments, but also generates an ideological celebration of *fatal* irresponsibility, which affects common people's mentality and invests media.[46]

Lastly, I think I can draw a small final reflection: in our Western countries, *moral decriminalization*, *overmoralization* (cf. the following chapter, section 6.5) and an overt *commodification* of any domain of public life can indeed be said to thrive off each other. States have always used pecuniary sanctions as efficacious means of punishment, insofar as they both hurt the wrongdoer and benefit the nation's coffers, but this would not happen at the expense of a proportioned (often even excessive) moral-legal punishment. Nowadays, we witness a progressive moral decriminalization in most violent cases presumably related to criminal psychopathy and/or lack of conscious deliberation (mostly resorting to arguments about the psychopathological dimension of crime, as explained in the previous sections) being paired with an increasing *commodified overmoralization*. By this, I mean to stress how only the prosecution of misbehaviors that can be economically punished is increasingly pursued, while an embarrassing "dogoodist" moral decriminalization is applied to other felonies: for instance, those concerning offense and harm against individuals, ideals, and fundamental democratic institutions, from whose prosecution no particular economic public gain can be drawn. As a matter of fact, it seems much harder for a

[45] This last consideration also explains why we frequently face the sad feeling – also supported by some empirical evidence – that, too often, certain criminal patients cannot be included inside "corporate psychiatry". Also because of the degradation of welfare states almost everywhere in rich countries, such criminals are neither medically treated nor punished, and often left alone to harm again and again, faced with the indifference of forensic psychiatrists, of public opinion (until they become the victims of the moment and then play the role of scandalized citizens), and of all those who are more or less responsible for their condition. A kind of disruption of both the modern rational ambition towards the cognitive recovery of psychopathological criminals, and of the *Rechtsstaat*'s duty to administrate punishment: this in turn becomes extremely violent and unjust against those who are not lucky enough as to fall within some psychiatric *excuse*, and whose moral punishment becomes consequently harsher, turning them into awkward scapegoats.

[46] A study on lay recognition of psychopathy and beliefs about those behavioral manifestations, etiology, and treatments of psychopathy is illustrated in (Furnham et al., 2009). An expected strong confusion in distinguishing between cases of psychopathy, depression, and schizophrenia is shown: the authors conclude that educational programs are required to improve mental health literacy in relation to psychopathy among the general public.

fraudulent banker to plead mentally ill – and get away with it – than it is for a serial murderer. . . !

5.3 Fascist Morality and Happy Violence: "The Fascist State of the Mind"

Chapter nine of Bollas' book *Being a Character* (1993), entitled "The Fascist State of Mind" begins with a famous sentence attributed to Benito Mussolini (1935): "Our program is simple, they ask us for programs. But there are already too many. It is not programs that are wanting for the salvation of Italy but men and willpower": it is clear that the complexity of words and ideas of "programs" are merely considered stupidities, worthy of weak people. Bollas immediately says that "[. . .] fascism extolled the virtue of the state, an organic creation driven by the militant will of the masses [. . .]. 'Fascist' is now a metaphor in our world for a particular kind of person, and I wish to reserve this ironic scapegoating of the Fascist from the convenient movement of its personification of evil, as, like Wilhem Reich and Hannah Arendt, I shall argue that there is a Fascist in each of us and that there is indeed a highly identifiable psychic profile for this personal state" (pp. 193 and 196).

The Fascist state of the mind is seen by Bollas as "ordinary": how does one become fascist? Following Hanna Arendt, Bollas contends that, looking at the camps of Buchenwald and Dachau, terror becomes total "when it becomes independent of all opposition" (Arendt, 1976), some kind of gangs dominated by "a hierarchy of Hitler clones" watched each other commit atrocities in order to ensure that no one in the gang stepped outside the *ethos of terror*. Something similar is occurring at the level of the individual unconscious psyche where a kind of highly structured destructive narcissism, which provides a sense of happy superiority and self admiration, kills the loving dependent self and other capacities of it, such as empathy, forgiveness, and reparation. People that are not deterred by empathy can harm others more easily because they experience less or no personal – empathic – suffering at all as they witness the pain they caused; furthermore, thanks to reiteration of harm and "desensitization" (through which inhibitions that surround harmful acts are removed), they can more easily develop sadistic attitudes and even experience fun in harming (Baumeister, 1997, pp. 232–238, 285–291).[47]

Bollas contends that this process is occurring under the pressure of some particular intense drives (such as greed), forces (such as envy) or anxieties (for example fear of mutilation). A kind of doubling of the self is achieved, two functioning wholes are issued, so that part of the self acts as an entire self. Nazi doctors were killers and remained ordinary family men: "Nazi doctors escaped the sense of guilt arising from their evil actions by transferring the guilt from the ordinary to the 'Auschwitz self'" (Bollas, 1993, p. 199): a doubling that it is also occurring at the ordinary level,

[47] It has to be said that sometimes empathy can be used to the aim of cruelty, when the violent human has to know what the victim is feeling in order to maximize his suffering.

for example when a surgeon needs to shift from ordinary human self to the so to say "aggressive" doctor self, in order to perform operations.[48]

It is important to bear clearly in mind that the aforementioned *doubling of the self* does not coincide with Sartre's conceptualization of "bad faith" which, put simply, is a kind of falsehood that involves lying to oneself.[49] In the human condition of bad faith, people treat *themselves* as means; they tend to ignore or jettison the concept of choice in some respect because it is somehow vexing or burdensome, and in doing so they relinquish freedom and externalize responsibility (Magnani, 2007b, p. 128).

Usually, one sinks into bad faith for very specific reasons: for instance, instead of facing his responsibilities one might fall into bad faith by denying his own interests if confronted with economic distress, or pretending to be a free and magnanimous spirit when dealing with an unfaithful partner. This is not the place for a full analysis of bad faith: let it be enough to remark that bad faith, a condition in which many people live – alas – all their lives, originates from the *unwillingness* to adopt a certain course of action (because of ignorance, fear or other constraints) that would ensure in the long term a greater happiness, freedom and possession of one's future. In short, that person deceives *herself* by constructing a limited reality that does not take into account the full range of choices available to her. Of course, with our decisions we greatly affect each other's lives, but as far as bad faith is concerned, she is herself the first and main potential victim: it is from herself that she is hiding the truth; the deceiver and the deceived coalesce into a single consciousness in a way that must be distinguished from true mental illness or malfunction of consciousness. It has to be said that bad faith, weakness of will, and *akrasia* can all be involved in the process of moral disengagement described in the previous section, in various interesting ways which more or less unconsciously affect intentions, decisions/resolutions, (moral) evaluations, action executions, some of them still awaiting illustration. On some moral aspects involved in these psychic/mental processes cf. (Holton, 2009; Mele, 2010).

Conversely, as will be showed further on, the Fascist state of mind arises from a similar but opposite kind of *malaise*: a moral super-anxiety brought about by the lack of actual anxieties. Once Nazi rule was established, the reach of ideology was not restrained by any limitation and having defused the very idea of an actual opposition (to be engaged in a dialectical confrontation), the hypertrophic narcissist ego could label anything as an opposition (and therefore a threat and cause of anxiety) and start yet another annihilating war against it.

Many of those involved in trials for the crimes perpetrated by Nazi Germany (see the Nuremberg Trials, for instance) defended themselves arguing that they were *just obeying orders*: that could be read as the claim that Germany was a whole nation committed to bad faith, as fear and a sense of individual powerlessness would inhibit any individual attempt to react against what was going on. Such argument has always been hardly convincing because bad faith is an extremely individual

[48] (Waller, 2002) illustrates, with the help of several examples coming from the twentieth-century's "age of genocide", the role of dispositions, moral disengagement, self-interest, and situations in transforming ordinary people into actors of genocide.

[49] (Sartre, 1956).

and chaotic phenomenon: less-than-conscious shifts between different identities can hardly occur on a mass-scale in perfect synchronization. As bad faith is an answer to objective anxieties, it makes people's behavior erratic and often unpredictable. On the contrary, this further determines Bollas' point: there was *no* shift between the Nazi doctor, concentration-camp guard, "SS" officer and, to follow Bollas' example, the family man each of them would embody at home. Furthermore, it can be said that those double selves perfectly *coexisted* within one person.[50]

"The core element in the Fascist state of the mind (in the individual or the group) is the presence of an ideology that maintains its certainty through the operations of specific mental mechanisms aimed at eliminating all opposition. But the presence of ideology (either political, theological, or psychological) is hardly unusual; indeed it is quite ordinary" (Bollas, 1993, p. 200). I would add that this "ideology" immediately acquires the features of an *ethos*, as a cluster of beliefs and convictions that, furnishing in this special case a *total* and *rigid* explanation of human behaviors and characters, firmly and directly governs the fascist individual's actions, without hesitation, negotiations with people who do not share the Fascist state of mind, and without a serious consideration of any evidence deriving from the environment (it seems the Fascist state of mind is affected by a dominance of the so-called *confirmation bias*).[51]

5.3.1 Ideology as a Disguised Violent Ethos

Let us better focus on ideology: it is something available "out there", stored in external devices and supports (other people, books, media, etc.) of a given social collective. People readily pick up external ideological "tools" of this kind, then re-represent them internally. If we pay attention, we can indeed notice how thinking about an ideology (be it the Nazi or the Soviet communist, the catholic one or the liberal tolerant) often triggers the evocation of a complex and distinctive aesthetic and material environment. The signaling – consisting in swastikas, red stars, parades, styles of speech and even clothing fashions – does not only possess a propagandistic meaning, but it mostly enforces and empowers those who already believe in it. The historical *topos* about the difference between an army and an armed mob should be clarifying: it is the *army ideology* that allows the effective superiority of the army. It can be argued that even military training is only afforded by the presence of a distributed ideology embedded in signs, flags, words, hierarchies and, last but not least, uniforms: the very term *uniform* is significative. The uniform is the *same* for *everyone*, making every subject feel similar, both exteriorly and psychologically equal to the others.

That is to say, an ideology could be labeled as a *highly reified and sclerotized moral niche*. Morals should be present when dealing with conflicts as the presence

---

[50] It is interesting to recall how the Nazi regime's propaganda would always stress the apparent *normality* and joyfulness of the families of those very soldiers and rulers that were actuating the infamous Ultimate Solution.

[51] On confirmation bias see below section 5.4.

of a moral presupposes the presence of another conflicting one, but the Fascist state of the mind rejects this assumption and becomes the ultimate moral. Conflicts are not a matter of conflict anymore, but of *cleansing*. The adversary is not even an adversary anymore, but a mere obstacle to be annihilated and removed. This disengagement is another aspect typical of what we defined "moral bubble":[52] violence gets dissimulated because the adversaries see their active *dialectical* role refused. When you take the trash out, or when you clean the kitchen table or even when you perform the disinfestation of a building, you do not think you are performing any violence. Furthermore, you do not need to engage the opposing party as a physical or dialectical counterpart. This is sound as far as garbage, dirt or infesting vermin are concerned: we maintain that morals, the presence of an orthodoxy, can act as an immunizing factor to one's violence also when the object of violent deeds are other human beings. Hitler's "final solution" was more a necessity of hygiene rather than the elimination of an active threat. Ideologies provide individuals with a series of tools to avoid the self-emergence of violence: political, theological, biological arguments as well. If I acknowledge the dialectical role of my adversary, I would be compelled to consider myself as *his* adversary. Each of us would be just the other's *alter ego*, and the violence of the confrontation would be plain on both sides. Conversely totalitarian ideology-induced moral bubbles (by making me unaware of the intrinsic violence of my own behavior) lead me to perceive the others' deeds as extremely violent on my behalf, thus sparking a collective vicious circle.

Most of these social phenomena need a collectivization to be activated. The presence of a (large) group is what creates the difference between Adolph Hitler and Charles Manson.[53] Nazi ideals were structured so that they would be received and adopted by the largest possible number of people: constant external markers, always at hand, would make the individuals increasingly interiorize the ideology (becoming in this way less *individualized*), thus turning themselves into an externalized marker for other people.

That stated, we can further analyze this dimension that characterizes the moral bubble and consequently the Fascist state of mind. Ideologies thrive on the fact that moral bubbles are hardly an individual phenomenon but that they rest on the mutual reinforcement of moral beliefs. Uniformed individuals (the oxymoron is not meant but very significative), each in their own *individual* moral bubble, act together and combine their roughly similar moral belief in a *collective moral bubble*, which dramatically empowers its ordinary mechanisms. This collective bubble aims at systematically defusing all potential doubts, adding the action of one individual upon another in the process of self-immunization to violence, typical of the moral bubble. Within the moral bubble, the moral agent perceives his own moral principles as a given, just as much as a cognitive agent takes his beliefs as a positive, genuine

[52] Cf. chapter three, subsection 3.2.2.

[53] Charles Milles Manson is the infamous american criminal who, in 1969, commanded his adepts (known as *The Family*) to brutally slaughter the occupants of a villa in Bel Air (Los Angeles), as the climax of his bloody vengeance against a society that had made him feel rejected and unwanted. Among the victims of the carnage was Roman Polanski's wife, 8 months pregnant at the moment of her death.

truth. Ideologies project a clear coalition level, in which each *embubbled* individual assures and corroborates the beliefs of his fellows. The whole ideology-projected group becomes blind to its own violence and it is able to respond to instances of doubt with the synchronism of one organism and the power of several. The majority of the violent response is *not* a defense of the content of the questioned beliefs, but of the *tranquility* that those beliefs allow – within the moral embubblement. A violent outburst is not perceived by the agent who performs it, because it is obliterated by the unquestioned conviction of the righteousness of her own principles. That is why we are extremely aware of other agents' violence (because they clash with our own bubble) but we are virtually immunized to our own. *In extremis*, should it be impossible to suppress the corrupting belief, the solution rests in the (often physical but sometimes metaphorical) suppression of the corrupted believer. If violence perpetrated outside of the group does not succeed in its scope, it can target one of weakest members of the group itself, labeled as a deviant or a traitor.

In the case of the Fascist state of mind doubts are expelled because they immediately reverberate weakness of spirit and all weaknesses are considered as something very negative. A unidimensional mind is formed and any other external counter-views completely disregarded and eliminated. Slogans, rhetoric, violent fallacious argumentations, icons and so on substantiate (also suitably stored in external materialities) the totality of the ideology/morality at play. Bollas eloquently notes this, in a passage where semiotic and violent issues are intertwined by showing the simplification performed by the fascist implementation of the mind:

> When the mind had previously entertained in its democratic order the parts of the self and the representatives of the outside world, it was participant in a multifaceted movement of many ideas linked to the symbolic, the imaginary, and the real – Lacan's terms. Specifically, words, as signifiers, were always free in the democratic order to link to any other words, in that famous Lacanian slide of the signifiers which expressed the true freedom of the unconscious (this Other) to represent itself. But when representation freedom is foreclosed, signifiers lack this freedom, as ideology freezes up the symbolic order, words becoming signs of positions in the ideological structure. When Dukakis tried to introduce complex issues in the American presidential campaign of 1988, George Bush made the word "liberal" a sign of weakness visited upon the certain mind by doubt and complexity. To supplement his destruction of the symbolic order Bush made the American flag the sign of the difference between Dukakis and himself; sadly it signified the end of discourse and the presence of an emergent Fascist state of the mind (1993, p. 201).

5.3.2 The Fascist "Infinite Morality" as a Violent Moral Void

Of course, following Bollas and his psychoanalytic argumentations, the elimination of the "symbolic" is a kind of murder: indeed it immediately gives birth to a sense capable of murder, a sense often related to a happy feeling of an "infinite moral space [seen] as the pervert's accomplishment eliminating (at first Oedipal) opposition to desire and gaining objects without opposition" (p. 202). But the violence related to this infinite morality deriving from the killing of any opposition[54] in turn echoes a *moral void*, analogous – psychoanalytically – to the empty heart of the pervert. Quickly, the moral void created by the destruction of the opposition begins to make its presence felt:

> [...] at this point the subject must find a victim to contain that void, and now the state of mind becomes an act of violence. On the verge of its own moral vacuum, the mind splits off this dead core self and projects it into the victim henceforth identified with that moral void. To accomplish this transfer, the fascist mind transforms the human other into a disposable nonentity, a bizarre mirror transference of what has already occurred in the fascist's self experience (p. 203).

Let us consider the unconscious cognitive operations that are occurring in the process: 1) moral void is expelled by symbolically and/or actually killing the victim, 2) this destructive process is immediately denied by a "delusional narcissism" issued thanks to an "annihilation of negative hallucinations", 3) such annihilation enables an idealization of self already accomplished by the negation of any alternative self or environment (these would be just considered enviable or persecutory). We have to imagine these fascist minds as always thinking about themselves as contaminated, in need of an activity of constant purging so that a (forever) empty, aseptic, and pure self can arise (a self that does not have to have contacts with others, without past, and with a "future that is entirely of its own creation").

Bollas significantly adds that "[...] we can find this phenomenon, however, in ordinary life, whether it be spoken by those who attempt the position of pure Christianity, pure objectivity, pure science, or dare I say, pure analysis!" (pp. 204–205). The moral atmosphere generated by this phenomenon is clear and so is its immediate violent counterpart, any opposition is simply not considered, or more or less violently assaulted. The degree of the annihilation of the opposition depicts the delusional narcissistic level of the Fascist state of the mind. When the purity is seen as challenged beyond a certain threshold, a state of *perennial struggle* is implemented, like in the case of Mussolini and his followers' mentality. The words of Mussolini are very clear

> War alone keys up all human energies to their maximum tension and sets the seal of nobility on those peoples who have the courage to face it. [...] Fascism carries this anti-pacifistic attitude into the life of the individual (Mussolini, 1935).

Another sentence attributed to him is: "War is to man what maternity is to a woman. From a philosophical and doctrinal viewpoint, I do not believe in perpetual peace". A male maternity, Bollas psychoanalytically says, that reverberates death camps as deadly wombs, where adults reduced to bizarre fetuses are eventually killed!

[54] This sense of infinite morality instantly guarantees impunity, we know it is a decisive factor – together with fear – for translation to "barbarity".

Moreover, the morality of this state of continuous struggle, so to say, is no combat at all, it is not the combat of negotiations, of dialogues, it is not the continuous – for example "democratic" – engagement with adversarial opinions and with the "different" problems of diverse people: instead, these are considered on the whole a kind of illness, of cancer, to be eliminated: "The idealization of war and of the warrior is a call to a state of mind that rids itself of opposition by permanent violence" (Bollas, 1993, p. 206). A process of de-personalization accompanies the process: ideology, a leader, the state, the king are revered also to the end of murdering the self: "Thus the concentration camp, a metaphor of the psychic process of Fascism, is the place where, as the humane parts of the self are dehumanized and then exterminated, the death work is idealized in the death workers who cleanse the body politic of the undesirables" (p. 206). The leader (and the various equivalents, God, state, etc.) morally incarnates and represents the higher "cause" that the fascist mentality always uses to explain and justify what the non-fascistic minds see as mere anti-human violent behaviors; it also provides the basis for constructing that typical grandiosity "that achieves nobility" by rising above the standard human beings, rotting in the submorality of weak people. It is thanks to this grandiosity that the person that is in the Fascist state of mind thinks of himself as making happy violence, as I have indicated in the title of this section. *Happy* because it *morally* purifies everything through exciting enterprises.[55]

5.3.3 Intellectual Genocide: Killing Opponents' Cognitive Niches

In chapter two, section 2.1.4, I have said that vocal and written language is a tool "exactly like a knife". No better evidence of this role played by language is provided than by the rhetorical tools exploited by the Fascist state of mind to "kill" all opposition. I can say this is the first way of creating that more or less unconscious "sense of murdering" I have mentioned above.

Various kinds of the so-called fallacious arguments are used, also to provide those symptoms I personally consider real "alarms" through which it is possible to abduce the emergence of the Fascist state of mind: Bollas lists some of the main efficacious cases, which he eloquently considers as the tools for a possible *intellectual genocide*, which is considered as a real, new and important "category of crime". A subset

[55] Charny (2006) further illustrates the fascist mentality by analyzing the mental processes related to the complex patterns of behavior referred to as fascist and democratic. Still, fascism and democracy are not only political systems but ways of organizing the mind. They arise out of basic human needs at two very different stages of development. "the fascist paradigm", according to Charny, "is a model for attempting to solve existential anxieties that trouble all human beings" (p. 17). A democratic mind, by contrast, "recognizes that what life is all about, first of all, is the sanctity of human life. [It] takes an overriding position of caring for life and the opportunity for life as the definitive prime principle" (p. 136). Charny also suggests new principles and approaches for individual and family therapy as well, with the aim of reducing the danger of future war, genocide, and terrorism.

of these is the group of rhetorical methods which substantiate what he calls the *committive genocide*:[56]

1. distortion of the opponent's image, for example using a massive quantity of *ad hominen*;
2. decontextualization of the opposing view: this phenomenon is in turn linked to the scapegoat mechanism (which often relies on gossiping dynamics) that I have already illustrated in the previous chapters of this book. Bollas also adds that "The extreme of this act is the removal of a victim from his tribe, home (i.e. context), isolated for purposes of persecution" (Bollas, 1993, p. 208);
3. denigration, depicting the opponent's position as ridiculous;
4. caricature, which helps delineate and identify the group or the ideas that have to be considered undesirable and so "killed";
5. change of name, as an act of the elimination of proper names ("kikes" for Jews, "gooks" for Vietnamese – for two decades the current Italian prime minister has made an annoying but efficacious and aggressive use of the word "communist" to indicate simply democratic or Christian-democratic people, and in turn he has been nicknamed by his adversaries as "the dwarf" for his short size, in a bitter escalation of calling each other names). It is pretty obvious that this elimination reverberates the subsequent potential elimination of the people/ideas from the socio-political scene and in the worst case from the very "community of living people";[57]
6. categorization as aggregation, when the individual is transferred to a general category, usually with a bad connotation, in which he loses his identity and qualities: "he is a psychopath", "he is an immigrant", "he is a former alcoholic".

Cases of *omittive genocide* resort to the *absence of reference* when the life, work, or culture of an individual or group is intentionally not referred to. Of course this

[56] Salmi (2009) usefully lists the various effects of genocide under the category of alienating violence, when a person is deprived of her rights to emotional, cultural and intellectual growth, by means of racism, social ostracism and ethnocide. This has to be distinguished from repressive violence, which resorts to a mere deprivation of basic rights other than the right to survival and protection from injury, including civil, political, and social rights (Bufacchi, 2009, p. 320).

[57] Other processes of methodical "substitution" (especially active in the case of *Nationalsozialistische Deutsche Arbeitepartei*, were related to: 1) the substitution of religion by the instrumentalization of art, 2) the substitution of art by propaganda, 3) the substitution of propaganda by indoctrination, 4) the substitution of culture by monumentalism, 5) the substitution of politics by esthetics, 6) the substitution of esthetics by terror (masses – already transformed into an homogenous conglomeration where the elbow room that would enable any political or cultural relation is missing – are further weakened through the erasure of the very faces – metaphorically, aiming toward a total anonymization – and any sense of individual responsibility) (Mandoki, 1999).

is an ordinary tool that, when performed systematically and constantly against well chosen targets, can be both at the basis of the Fascist state of the mind and a simple tool in the subtle violent dynamics of everyday life, *á la* "Desperate Housewives".

To be simple and clear from the perspective of the philosophy of violence: here in this subsection I am not referring to the more or less legitimate desire of the fascist (and of all human beings) to win against others, to become richer, or to acquire power and command, but to the extremely violent and harmful tools he uses that, because of their radical character, *always* surpass in cruelty the tools used by the opponents which are less violent and so "constitutively " *weaker* if compared with those used by the fascist.

This consideration also clearly explains the rapid implementation of a "virtuous" circle of violence in the process of formation of a "collective" Fascist state of mind. All opposition, that the germinal Fascist state of the mind *violently*[58] depicts as weak and with the stigmas of doubt and complexity (like the way George Bush thought of the "liberals", as I said above), quickly *actually* becomes weak, doubtful, and vulnerable to everyone embedded in that very opposition and so perceived as weak by the collectivity dominated by the Fascist state of the mind.

In this interplay intellectual genocide may occur: usually the victims adopt few kinds of vaguely reactive behaviors 1) some people respond by not engaging in vicious gossip or trying to stop it; 2) others can simply leave the scene, no longer partaking of the group and so exiling themselves; 3) other can further express more radical views, so offering themselves up to further victimization and scapegoating; 4) some may somatize the violence received in physical or nervous disfunctions; 5) finally, others might try to form alliances with the persecutors to gain some form of protection. Bollas contends that intellectual genocide should be considered a crime against humanity and no less relevant than the sanguinary genocides. In the light of the analysis of the violent nature of language I have provided in chapter two, I cannot abstain from agreeing with him.

Finally, it is useful to remember that the persecutors who are in a Fascist state of the mind can also rapidly become object of endearment "to those who otherwise – one would have thought – would be horrified by such a behavior" (Bollas, 1993, p. 210). Usually we tend to "love" these – now cute – "monstrous monsters" (a kind of impossible object of love) because love here exonerates us from explicit and responsible opposition: for example aggression is turned into humor, or into a perverse joke, for example we pathetically say: "removed from her pulpit, she is really quite a different person, she is a very nice person". Unfortunately, in such a way the "humane authorizes the inhumane in an exchange between the Fascistic and the non-Fascistic part of the personality" (p. 221), which establishes a kind of vicious circle of justification of violence. On the contrary, it would seem more

[58] However, we have to remember that for the fascist mind this is a *moral* duty, and violence is rarely perceived thanks to the huge generated moral bubble.

non-Fascistic to try to avoid the tricks able to control our anguish and so to recuperate from the trauma caused by facing a person who is in the Fascist state of the mind, and instead frankly experience the violence received in all its force. Indeed, being a non-Fascistic listener of the words that erupt from a Fascist state of the mind *immediately* transforms us in victims, and nobody likes to clearly see herself like a victim, so she easily and immediately looks for some relief. As listeners we feel "shocked, dissociated, and deadened [. . .] we share elements in common with those who are more severely traumatized by socially operant Fascism":

> We all know how stunning it is, when discussing an issue with someone, to witness the person's vicious espousal of a doctrine that derives part of its energy from the intellectual annihilation of the other. We may be speechless. Such a rupture also occasions a sense of dissociation: we feel immediately separated out from the conversant's insanity. And following this dissociation, part of us will feel deadened by the eruption, as now it is clear to us that the other is subject to an internal Fascistic process. In a way our response is our victimage (p. 213).

It is clear: our mute reaction is already our violent victimage.

Certainly related to this fascistic aspect is the general aggressive component of the so-called "conversational dominance", which asymmetrically establishes an unequal distribution of entitlements and rights, such as the opportunity to introduce new topics, and verbally victimizes some participants.[59]

The emphasis on the role of fallacious argumentation in the formation of the Fascist state of the mind can easily explain how abusive "manipulations" in discourse interaction at the social level (through written text, speech, and visual messages) are important in totalitarian states and collectivities but also in professional settings, institutions, families, etc. They are violent tools used by dominants to establish inequalities of various types and possibly to perform intellectual genocide, when they achieve their absolute target of annihilating opponents. Van Dijk lists the major argumentative and structural tools that are involved in manipulation processes, which are almost always devoted to focusing on those cognitive and social characteristics of the recipient that make them more vulnerable and less resistant:

> (a) Incomplete or lack of relevant knowledge – so that no counter-arguments can be formulated against false, incomplete or biased assertions. (b) Fundamental norms, values and ideologies that cannot be denied or ignored. (c) Strong emotions, traumas, etc. that make people vulnerable. (d) Social positions, professions, status, etc. that induce people into tending to accept the discourses, arguments, etc. of elite persons, groups or organizations. These are typical conditions of the cognitive, emotional or social situation of the communicative event, and also part of the context models of the participants, i.e. controlling their interactions and discourses. [. . .] [Moreover, discourse

[59] (McKinlay and McVittie, 2008, chapter eight: "Dispute and aggression") also study the role of insults as precursor of physical aggression in their intertwining with arguments that can be used to establish sociability.

structures materialize suitable constraints which favor manipulations:] (a) Emphasize the position, power, authority or moral superiority of the speaker(s) or their sources – and, where relevant, the inferior position, lack of knowledge, etc. of the recipients. (b) Focus on the (new) beliefs that the manipulator wants the recipients to accept as knowledge, as well as on the arguments, proofs, etc. that make such beliefs more acceptable. (c) Discredit alternative (dissident, etc.) sources and beliefs. (d) Appeal to the relevant ideologies, attitudes and emotions of the recipients (2006, pp. 375–376).

5.3.4 Culture of Hatred and Inculcation, and the Moral "Banality of Evil"

It is clear that a component of intellectual genocide can be a culture of hatred and inculcation, that of course is not simply a characteristic of the Fascist state of mind and of the related historical counterparts, but which is also more or less violently active in many everyday ways of reasoning and feeling, for example when myths and ideologies are morally efficient at the level of the individual psyche. Following Wieviorka (2009, p. 119) we can say that Hannah Arendt's *banality of evil* refers to the fact that human beings behave in extreme violent ways without feeling or knowing what they are doing, but not necessarily being motivated by hatred or any other passion. Eichmann, she says, was "terrifyingly normal". For example, if we harm people by obeying to the order of a political authority, like in the case of the Nazis, we can do it without passion, e.g. pleasure or sadistic or cruel intents, but I believe that, even if horrible passions are present, this fact does not diminish the moral atmosphere in which the perpetrator of violence thinks her action is performed: if an agent takes pleasure in enacting something that he morally approves of, the better for him!

Servility and conformity to peers and social norms as a pervasive reason of moral transgression (or of a disengagement of morality, as I say)[60] is fully analyzed by Freiman (2010). The author also contends that servile behavior involves cognitive dissonance,[61] which can restructure or dissolve those particular desires, beliefs, and projects that fund the agent's most highly valued conceptions of herself. In this perspective it is easy to see how self-interest cannot serve as a comprehensive explanation of "immorality": humans sometimes disengage morality at the knowing expense of their personal interests, they are – often hypocritically – occasionally "too cooperative". The famous experiments devised by Milgram, which eloquently showed human inclination towards disengaging morality, are also described: "[...] few people have the resources needed to resist authority. A variety of inhibitions against disobeying authority come into play and successfully keep the person in his

[60] Cf. above section 5.1.

[61] Cf. above subsection 5.1.1.

place" (1975, p. 6). Zimbardo's Stanford Prison Experiment (Zimbardo, 2009) is also illustrated.[62]

A "normal" sadistic Nazi is not "less banal" that a non-sadistic one! Putting aside the cases of violence triggered by various kinds of madness,[63] from depressive suicidal to sadistic psychopathologic people, we can simply see that aggressiveness in both a surgeon and a Nazi is motivated by moral concerns, but we do not consider the first one violent (when both of those people consider their own aggressive behavior to be moral and the second one does not see the violence we see in his behavior! – again, "banality of evil").[64]

In a sense, the banality of evil is the supposed morality of evil, given the fact an agent acts based on a conscious self-perceived moral deliberation, possibly resorting to important and dominant ideologies and principles, like the Nazi ones. The title of this chapter is "Multiple Individual Moralities May Trigger Violence": in this perspective the life of what we label – in our philosophical framework – a violent subject, surely is "no more than a sequence of multiple, disparate, processes. Each of them is partial, and therefore ready and able to argue that it is morally not

[62] Further data on the culture of blind obedience to the orders of total institutions, which operationalize moral submission, and its impact on moral disengagement (and the consequent banalisation/moralization of violence) are provided by (Pina et al., 2010): where events in Democratic Kampuchea (i.e. Cambodia under the regime of Pol Pot) are addressed and illustrated in detail (on the related connections between demographic concerns and critical geopolitics of violence cf. (Tyner, 2009)). (Elbert et al., 2010) further stress how hunting behavior is fascinating to human beings and how it is easy to reengage it, for example in unfavorable environments (e.g. scenarios involving the presence of child soldiers), thus promoting the development of personalities which appear to be cruel "by nature": "While a breakdown of the inhibition towards intra-specific killing would endanger any animal species, controlled inhibition was enabled in humans in that higher regulatory systems, such as frontal lobe-based executive functions, to prevent the involuntary derailment of hunting behaviour. If this control – such as in child soldiers for example – is not learnt, then brutality towards humans remains fascinating and appealing. Blood must flow in order to kill. It is hence an appetitive cue as is the struggling of the victim. Hunting for men, more rarely for women, is fascinating and emotionally arousing with the parallel release of testosterone, serotonin and endorphins, which can produce feelings of euphoria and alleviate pain".

[63] There is no room for direct moral considerations in these processes of violence which just appear "for the sake of violence", because the action is out of the reach of a standard conscious deliberation.

[64] Empirical results show that, in western countries, psychopathic behavior has a fundamental impact on violence among the general population, despite its low incidence. Impulsivity is not found to be as uncontrollable as previously thought and it does not strictly depend upon an antisocial lifestyle, so that in criminal psychopaths the "instrumentality" of homicide simply prevails (Coid and Yang, 2010). On the complex relationships between psychopathy and sexual sadism cf. (Mokros et al., 2011): the authors skeptically conclude that, currently, it is unclear whether the deficiencies in emotional processing that predispose psychopaths and sexual sadists toward instrumental violence are cognitive or affective or both. According to the authors, it is still to be clarified whether these deficits represent a lack of understanding or a lack of feeling.

guilty" (Wieviorka, 2009, p. 119). It is not simply that our supposed-to-be-violent agent argues that she is morally not guilty. It is even – so to say – worse, she thinks she directly acts in a "morally" approvable way. The moral bubble that she occupies guarantees her impermeability to the possibility of facing her behavior's violent outcomes, which are morally justified so deeply that they simply cannot be "actually" considered or perceived as violent. Recently Balch and Armstrong (2009) have recovered the idea of banality of wrongdoing to explain persistent, accepted-as-normal corporate wrongdoing, particularly for high performance organizations. The establishment of a "cocoon" of a supposed to be "new morality" of the group is the tool which leads to the banality of wrongdoing. The empirical study describes five explanatory variables: the culture of competition, ends-biased leadership, missionary zeal, legitimizing myth, and the *corporate cocoon*. The conclusion is that the nature of competition generates both legitimate and illegitimate goal-seeking to adopt an iconoclastic (rule-breaking) orientation. High performance organizations are especially inclined to wrongdoing because achieving top performance commands aggressive behavior at the margins of what is normally acceptable. The way leadership reacts to competition permits various "ethical" cultures to develop: "ends-biased leadership will project strong vision, using ideology and legitimizing myth as tools to inspire and motivate. The resulting missionary zeal justifies using questionable means because of the perceived value of the end. One critical method for building strong culture is creating a sense of being separate and apart from the ordinary. This cocoon effect may create a self-referential value system that is significantly at odds with mainstream culture and in which wrongdoing is banal" (p. 292).

Other simple traits of violent states of the mind are illustrated by Wieviorka (2009, pp. 151–157): they correspond to

1. the *floating subject* (the individual experiences her negation – as a potential "desubjectivation" and loss of meaning and socialization – as something intolerable, for example because of a police "blunder", like in the case of the famous Los Angeles gang riots' reaction), a fragmentary and fleeting morality tries to reconstruct the threatened self building up – through violence – a different cultural environment;
2. the *hyper-subject* finds in ideologies, cultures, and religions plenty of "meanings", which help self-assertion and the ownership of one's destiny (terrorists, revolutionaries for example): this aims at establishing a new social and political order and justifies violence, while if subordinate and passive, this same subject can easily shift to a kind of
3. *non-subject*, who is merely obeying to authority, or to a kind of
4. *anti-subject*, who is not attached to any particular ideology or social relationship and her practice of violence is – so to say – more "animal", related to pleasure and often to a sadistic or sadomasochistic attitude, which constitutes the only way of building social relationships and of giving them a subjective moral sense. Often this "morality" is actually constructed as a kind of trained and explicit

anti-morality[65] mainly built on clusters of passions and feelings. The other's existence is typically denied as a subject;
5. the *survivor-subject* aims at self-preservation and follows the archaic-fundamental (Bergeret, 1995) violent morality deriving from an overemphasis of "the other or me?" or "survival or death", a kind of primitive narcissism. It was occurring for instance in the case of the recent Paris' *banlieues* riots, where young people's violence derived also from the failure of their adult counterparts, that proved unable to offer adequate models with which they could identify.

5.3.5 Evil and Trauma

Finally, the relationship between the disengagement of morality and the interplay between trauma/evil in the perspective of some psychoanalytic reflections must not be disregarded, inasmuch as it sheds further light on the relationship between morality and violence, also with respect to the fascist state of the mind I illustrated earlier in this book. Obviously, the psychoanalytic narratives about evil rarely clearly disclose the "moral" underside of behaviors, which on the other hand I certainly consider relevant. Psychoanalysts, used to performing therapies, abstain from moral considerations. Nevertheless their narratives, once transformed in cultural objects externalized *out there* in a non-therapeutic intellectual communication, tend to attest to a sense of fake moral neutrality concerning the agents involved in the narratives. In this way those narratives can appear in complicity with the presumption of a fake and innocent morality of any listener/reader. Briefly, the vividness of violence they often contemplate and describe is frequently relinquished, because it merely remains absorbed in the "psychic" dimension. This can reinforce the reader/listener's malignant moral bubble (cf. chapter three, subsection 3.2.2), favoring the obfuscation of the responsibility of his own destructiveness.

I said in chapter one – quoting Hannah Arendt – that human beings experience a pervasive difficulty in understanding perpetrators of evil. In spite of its clearly stated disinterest for moral judgment, psychoanalysis is very clear about this point. From a Kleinian psychoanalytic perspective the explanation of this fact is very easy and still related to what I have called, in a cognitive perspective, the "moral bubble": our unfitness to identify with a character we recognize to be evil is related to our unfitness to acknowledge our own destructiveness (Schapiro, 2002).

If, in the oedipal interpretation of *Othello*, Iago (the incarnation of Satan) (Snow, 1980) is not related to the dark impulsive *id* but rather to a punitive *super-ego*, a question arises: where is the source of this morality which compels Othello to punish Desdemona? Still in Kleinian terms, the violent punishment is related to a *catastrophic trauma*, which generates envy and aggression, as an effort to manage depressive anxiety and the related sense of loneliness caused by the trauma itself. Thus, envy and aggressions can be performed through pathological narcissistic modalities, where terror and defense work together: of course defense involves a reshaping of the good original morality to defend it (because it has been

[65] With respect to the current morality, available to the agent.

jeopardized by the trauma), through the performance of related punishments, which aim at counteracting depressive anxiety.

A similar perspective is nicely and further explored by Bollas (1995), who discusses Milton's Satan as "[...] having experienced not simply a loss of a paradisal place but a catastrophic annihilation of his position. Ruptured from the folds of nurturance, the Satanic subject bears a deep wound and good is presented now as an enviously delivered offering" (p. 184). Evil is substantiated as the nature and effect of trauma and illuminates "[...] how loss of love and catastrophic displacement can foster an envious hatred of life mutating into an identification with the anti-life" (ibid.) A new, reengaged morality works together with this identification with the anti-life, to repay the loss of the previous good morality.

Chapter seven of Bollas' book *Cracking Up: The Work of Unconscious Experience* (1995), entitled "The Structure of Evil", begins with a description of the narcissistic trauma and the related emotional abandonment/wound/deception/betrayal: it is the reorganization of trauma that substantiates the structure of evil. Genocide (which is psychoanalytically exemplified by serial killing) repeats the same structure: the serial killer distills the power of his trauma, through which – in Milton's terms – his paradise was lost.[66] A serial killer is "someone who has been allegorized: he is squeezed into an identification with one quality, evil, that obliterates other psychic qualities [...]. He identifies with the force of trauma and out of this fate develops a separate sense of the work of trauma, which, like Lucifer, he turns into his profession: squeezing others into his frame of reference" (pp. 218–219). Moreover, we humans can recognize evil within ourselves not because evil is a part of our instinctual nature but because, as Bollas argues, "we all have experienced shocking betrayals in an otherwise trustworthy parental environment" and because

> [...] we all have transformations to the allegorical plane when we identify with the force of a feeling – in the case of evil, the force of emptiness sponsored in our selves by the shock *and* its unconscious marriage with the destructive sides of our personalities. All of us have experienced this trauma, and we all know its structure. Each of us will in some respects subsequently identify with it, mesh it with mental valorizations of our own sadisms, and entertain its future in fantasy – when we are cruel to each other, or in the so-called practical joke, when we play to unfortunate (but usually not disastrous) effects on the other (p. 220).

In the most horrifying cases, the entire emotional *ego* has been lost, or "killed" (Bollas, reporting the words of a murderer, speaks of an "emotional death"), like in the case of violent serial killers, and the psychological – and at the same time moral defense (a moral defense which is often consciously perceived as right and justified by the subject himself) – is performed though violent acts. Psychoanalysis rejoins psychology, sociology and criminology, in stressing the absence of any empathy which is always seen in violent psychopaths. Repudiated love becomes

[66] Similarly, women traumatically battered by their male partners in a sadomasochistic interplay return to be victimized again and again to continually re-create and loose that comforting and caring atmosphere of their infancy which suddenly became jeopardized by a trauma, then restored, and so on. But these women are not always killed, they can "survive" this continuous remastering of an early trauma (Bollas, 1995, pp. 206–210).

hatred and resentment,[67] which triggers moral vengeance against untrustworthy people (and often also against values, objects, institutions). Here it is worth quoting Bollas again, in his contention that the serial killer "develops a separate sense of the work of trauma, which, like Lucifer, he turns into his profession: squeezing others into his frame of reference". The unwanted self, fruit of the injuring trauma, which destroyed the moral/nurturing self or the "true" self, is projected outside, to denigrate and slaughter others, provoking an equivalent failure. In fact, at first, serial killers present to their victims a kind of parental care, a charming environment of trust,[68] that – suddenly and together with a kind of catastrophic disillusion – they reverse to sanguinary violence, to reverberate their subjective drama: "The serial killer – a *killed* self – seems to go on 'living' by transforming other selves into similarly killed ones, establishing a companionship of the dead" (Bollas, 1995, p. 189). As I have already pointed out, this is often subjectively "morally" perceived as normal, just, and appropriate. The serial killers *sacrifice* the victims to their own killing trauma, and paradoxically, the emotional death due to the trauma is resurrected ("comes back to life") in a transmuted form as a new (more or less literal) necrophilic empathy with the dead body of the victim, which is often cut, raped, inspected or simply mourned. Sacrifice, as always, grants the return to a non excited state, as Girard contends.[69] Energetic moral *anticathexes*[70] are at play:

[67] This conception has been often reflected in popular culture as well, and can be thus said to be part of western cultural unconscious: consider for instance the Who's famous song *Behind Blue Eyes*, released in 1971 in the album *Who's Next*. The lyrics echo the thoughts of an emotionally dead man facing hatred and constantly dismissing his true identity and a pain for which he blames somebody else: the chorus fittingly exemplifies the idea of a will to recover from the trauma ("But my dreams / They aren't as empty / As my conscience seems to be") and yet the impossibility to fully achieve this recovery because of the compulsory need for vengeance ("I have hours, only lonely / My love is vengeance / That's never free").

[68] I agree with Bollas that it can be hypothesized that an "empty-headed other", trapped in the necessary primitive belief about the goodness of the others, so fundamental to life, that offers himself disarmed and trusting to other people, is an important part of the structure of evil. This condition is obviously *exciting* (in a kind of pathologic narcissism) to the aggressor, who erotically identifies in the other the loyal, alive child he was before the trauma, to be killed sooner or later.

[69] Cf. chapter two, subsection 2.1.5, this book.

[70] Anticathexis, according to Freud, is the energy derived from the *super-ego* to run the *ego*. The investment of energy in an object, idea, or person is known as cathexis. The function of the anticathexis is to restrict and block cathexis from the *id* for overall benefit; in such a way the *ego* can also act to block unacceptable actions from the *id*. This is known as an anticathexis and acts to block or suppress cathexes from being utilized. A typical example of anticathexis is the process of repression. Repression serves to keep undesirable actions, thoughts, or behaviors from coming into conscious awareness. However, repressing these unwanted *id* urges takes a considerable investment of energy. Because there is only so much energy available, the other processes may be shortchanged by the energy use of the anticathexes: it is in this way that, sometimes the *super-ego* disguises its reasons as anticathexes with the scope of bypassing the *ego* and presents as an *irresistible impulse* something which clearly is not.

> Some serial killers have reported the urge to kill like some horrid force that takes them over, but we may wonder if this isn't testimony to their vain effort to separate themselves from instinctual life itself, which is now mixed with its own anticathexes, forming a matrix of instinctuality and its killing, a pathologic combination of the life and death instincts. Confusing the object of desire with the source of the instinct, the killer destroys the object in order to be returned to a state of non excitation (pp. 196–197).

The psychoanalytic framework I described can be usefully applied to the case of terrorist acts, either performed internationally, internally, or by the states themselves: it seems that they "evoke anxiety and insecurity and can provoke collective narcissistic defenses. If we understand the malignity that Iago represents – the aggressive envy that attacks the good because it is good – as more responsive than instinctual, then our attention is directed to the social conditions that ignite and sustain such regressive fury" (Schapiro, 2002, p. 497). For example, state terror (and other forms of political evil, such as the hatred performed by the fascist state of the mind I have described above) undermines, with anger and rage (which relate to a sense of moral betrayal) the deep-rooted human assumptions about safety and comfort, originating in the relationship with parents, thus leading to a collapse of cooperation.

5.4 The Fascist Arguer and His Fallacious Dimensions

One of the main claims put forward by the "military intelligence" hypothesis is that arguments can be used as weapons in conflicts. More precisely, violent argumentations can certainly be part of the arsenal not only in peaceful times, but also during explicit conflicts. Indeed, words are not swords, but words can subtly support violent aggression and oppression. The connection between argumentation and violence is made explicit in the case of the Fascist state of mind. The *fascist arguer* is an example of how reasoning can serve the purpose of suppressing any opposition by resorting to some argumentative sophistication. Fascist reasoning is a set of arguments – mainly fallacious from an intellectual perspective – deliberately setting the stage for oppression and violence. As mentioned in subsection 5.3.3 above, Bollas accounted for a number of argumentative moves which fairly describe the fascist arguer: distortion, decontextualization, denigration, caricature, character assassination, change of name, and categorization as aggregation. The list he made is quite exhaustive, and it immediately puts on display the fallacious dimension of the fascist arguer, as each of the arguments listed above corresponds to a specific type of "fallacious" reasoning that we are going to deal with in detail.

Generally speaking, there are two main elements describing the kind of reasoning we called fascist. The first is a *propagandist* element and the second is a *gossiping* one. The two elements go hand in hand, indeed, as they support the same aim: destroying all opposition.

5.4.1 Negating the Truth

The first element illustrates the attitude the fascist arguer has toward truth. Indeed, an element of military intelligence is still present, even though it is more directed toward the construction of a biased truth: here we may use the term "doctrine". According to Griffiths (1990) a doctrine 1) is expressed in natural-language sentences and 2) it is a set of descriptive claims about values and things in the world. The connection with the linguistic aspects of a doctrine is relevant in order to make an interesting connection with reasoning: in the case of the fascist arguer a doctrine is reasoned out so as to oppress and suppress any information, truth or fact opposing the credibility of the fascists' claim about the world. In doing so, the fascist arguer deploys a number of arguments, which go under the broad category of a fallacious reasoning known as "the one-sidedness fallacy".

The first to highlight this kind of faulty reasoning was John Stuart Mill who wrote in his *On Liberty*:

> He who knows only his own side of the case, knows little of that. His reasons may be good, and no one may have been able to refute them. But if he is equally unable to refute the reasons on the opposite side; if he does not so much know what they are, he has no ground for preferring either opinion.

In that passage, Mill pointed out that a good argument is the one in which a person seriously considers evidence or facts that might invalidate her claim. That is, in order to strengthen her claim, she has to provide or include an effective rebuttal anticipating possible objections of the opposing part (Damer, 2000). Suber (1998) posited that the fallacious nature of this argument does not rest on the fact that the premises are false or irrelevant, but that they are incomplete, and – most of all – the incompleteness is not properly addressed, when it should and could be.

The following passage illustrates how Mill described those who commit the one-sidedness fallacy:

> [...] they have never thrown themselves into the mental position of those who think differently from them, and considered what such persons may have to say; and consequently they do not, in any proper sense of the word, know the doctrine which they themselves profess (Mill, 1985, p. 99).

The selective nature of the one-sidedness fallacy is also taken into account by the so-called *confirmation bias*. Generally speaking, the confirmation bias is the tendency exhibited by humans to overlook falsifications in favor of confirmations and, at the same time, only accept data or information supporting a certain view. So, the confirmation bias is linked to the illusion produced by the epistemic bubble, that makes us completely blind and also immune to certain facts that are inconsistent and overtly in contrast with our conclusions. The confirmation bias is usually triggered by *cognitive dissonance*[71]– the situation in which the agent upholds two cognitions that are inconsistent (Cooper, 2007). Confirmation bias is the mechanism that is

[71] Already mentioned in chapter three, subsection 3.2.1.

meant to solve an internal contradiction by suppressing one of the two conflicting cognitions, hence it also favors the process of *embubblement*.

Both the one-sidedness fallacy and confirmation bias are intrinsic elements illustrating an important aspect of the fascist arguer, and his attitude toward truth. That is, he simply negates and suppresses any fact or evidence that might contradict his convictions. The negation of any opposing fact or evidence may be of two kinds. First of all, it may resort to deception and manipulation. This is what Van Dijk (2006) called "discourse power abuse". Basically, the negation of counter-evidence is achieved by drawing attention to specific pieces of information instead of others, thus taking advantage of *information asymmetry*, which is usually due to the unequal access to knowledge.

The victims of such a manipulation lack the proper knowledge in order to debunk what they are told. The suppression of any fact questioning the accepted view is facilitated by the absence of any counterpart representing a view that would bring up the counter-evidence. It is worth noting that in this case there is no need to lie or distort the facts; those that are brought up to support a certain view are not distorted. And no lies are necessarily told. The perpetrated fraud is based upon a systematic and sophisticated *omission* that hinders understanding.[72]

It is worth noticing that in the process of hindering understanding, the fascist arguer can also make use of strategies resorting to sophisticated tools like statistics. In this case the recourse to the argumentative force of "numbers" is indeed fraudulent, as it relies on tricky manipulations. This is the case of the so-called meaningless statistics (Walton, 2008, p. 247–251). The fraudulent recourse to statistics is usually based on the use of a vague term. For instance, one may claim that "9 out of 10 patriots support the war in Iraq". The number could be true, but the word "patriot" is ill-defined or, at least, too vague to make the numerical hypothesis meaningful. The same argument may be used when statistics try to capture a phenomenon for which it is nearly impossible to have a good or reliable sample. More generally, meaningless statistics tends to produce bullshit that, however, might turn out to be a weapon, if used to discredit a certain position.

The negation of any opposing fact may be caused not only by manipulation but also by explicit suppression. Usually manipulation, and its fallacious nature, is favored by the absence of a dialectical dimension as in the case, for instance, of the politician speaking to the crowd or interviewed on TV by an indulgent journalist.

[72] As one may obviously notice, understanding can be hindered taking advantage of organized propaganda. That is, media may play a pivotal role in supporting the systematic suppression of any idea or evidence conflicting with the accepted doctrine. Manipulations and distortions perpetrated by crippled media may bias citizens' understanding and thus weaken their capabilities to make sense of what is going on and control those who hold the power. For example, ideological media may convey and contribute to spreading distorted and biased narratives about the government – especially during electoral competition. This has the main effect, for instance, of boosting a government's political agenda, covering up information that would discredit the policies implemented by the government, and preventing dissident voices from being heard. On the ideological role played by media, see for instance (Diaz, 2008).

However, the one-sidedness fallacy may also be committed in a dialectical setting in the presence of a counterpart. In this case, the evidence contrasting the accepted view is usually mentioned but just to call it into doubt and this strategy relies on the controversies it manages to generate.

In November 2008 Barack Obama was elected as president of the United States of America and Silvio Berlusconi, the Italian prime minister, – known for his blunt language and his ability in making *gaffes* – declared that he was happy about President Obama's election, as – I quote – "Mr. Obama is young, handsome and suntanned". Berlusconi was immediately accused of racism and people complained about his remark. Indeed, that of Berlusconi was a gaffe. However, he refused to apologize saying that what he said was meant as a compliment. And then he added – I quote again – "we'd all like to be tanned like Naomi Campbell and Obama". What did Berlusconi do? Of course, he could not deny that he made such a remark. He could not lie about that. He just called into doubt that telling a black person that he is suntanned is a racist remark. Even worse, he turned the thing the other way around by claiming that such a remark could actually be a compliment and black people should be proud of it.

The point here is that Berlusconi was not appointed to decide whether his remark was racist or not, if it was a compliment or just a joke. He simply refused to accept the common meaning assigned to his remark just because at that moment it contrasted with his own view.

Consider this other interesting example. In 2003 Berlusconi made quite an offensive statement about Islamic countries. I quote him:

> We must be aware of the superiority of our civilization, a system that has guaranteed well-being, respect for human rights and – in contrast with Islamic countries – respect for religious and political rights, a system that has as its value understanding of diversity and tolerance.

And then he goes on claiming:

> The West will continue to conquer peoples, even if it means a confrontation with another civilization, Islam is firmly entrenched where it was 1,400 years ago.

His statement immediately raised complaints all over the world and, of course, among the various Islamic associations based in Italy and Europe. In this case, once again, he refused to apologize claiming that his words were manipulated. This is what he said in response to his critics:

> They have tried to hang me on an isolated word, taken out of context from my whole speech.

And then he added:

> I did not say anything against the Islamic civilization. It's the work of some people in the Italian leftist press who wanted to tarnish my image and destroy my long-standing relations with Arabs and Muslims.

This case is slightly different from the previous one. He did not call into doubt that his statement was not offensive and aggressive. Actually, he could not do that.

However he did not let the incident rest, but he hit back complaining that he was misunderstood by the press on purpose. Indeed, it is possible that one's words can be taken out of context. But his case was beyond any reasonable doubt, because his statement could not be misunderstood as the reader can easily see. It is worth noting here that it is not fallacious to call something into doubt. It is fallacious to attempt at faking a controversy where there would be no grounds for it. This kind of fallacious argument can be also called "the appeal to unreasonable doubt".

We have already mentioned that skepticism is commonly considered as a rational attitude according to which a statement should be proved or tested before being accepted as true or plausible, and we have already illustrated how doubt can be a weapon that favors both the production of knowledge and ignorance. In the case of the fascist arguer doubt is instrumentally used to discard certain facts which do not corroborate the accepted view. Counter-evidence is still negated, but not because it is appropriately incorporated into the explanatory narrative of the accepted view, for instance, creating *ad hoc* explanations: it is simply rejected by claiming that it is *controversial*, therefore *false*. This form of doubting explicitly depicts the opponent's view as less credible, when, indeed, the Socratic doubt is not meant to achieve a certain conclusion, but to invite the reasoner to keep on searching for a better explanation. As brilliantly argued by Bollas, the depiction of the opponent's view as less credible is the first move towards "character assassination" (Bollas, 1993, p. 208).

The one-sidedness fallacy describes the main attitude the fascist arguer has toward truth. As argued so far, it is based on the negation and oppression of any view alternative to the accepted one, which can be fairly considered as a kind of doctrine, that is, a set of descriptive claims about the ways of the world. Before introducing the second category of fallacies describing the fascist arguer, we are going to introduce a fallacy, which is half way between the two – the so-called "straw man".

In his brief characterization of what he called "committive genocide",[73] Bollas pointed out that one of the rhetorical strategies used by the fascist mind is to decontextualize the opponent's view and to make a caricature of it. This serves the purpose, as briefly mentioned, of rendering the opponent's argument ridiculous or – at least – less credible. That, in turn, facilitates the process of character assassination. From an intellectual perspective, the argumentative move of *decontextualizing* and *caricaturing* one's view is called *straw man*.

Generally speaking, a straw man is that kind of (fallacious) reasoning based on the misrepresentation of an opponent's argument so as to be in a better position to knock it down (Walton, 2004). The misrepresentation or caricature serves the purpose of opposing a certain view without taking it seriously. Consequently, it serves the purpose of diverting the attention to another argument – a caricature or a partial interpretation – which is easier to oppose both in a rhetorical and dialectical context. It is usually considered as a fallacy of irrelevance, as it introduces information that is irrelevant for the discussion of the opponent's position.[74]

[73] See the previous subsection 5.3.3.

[74] For a more detailed treatment about relevance and straw man, see (Walton, 2004, pp. 82–84).

Talisse and Aikin (2006) introduce an interesting distinction pointing to two different forms of straw man. The first is related to the way the opponent's view is represented, whereas the second is about the selection of the reasons supporting the opponent's view. The first form captures the traditional conception of the straw man we have briefly introduced. In the case of the fascist arguer, a straw man of this kind is deployed in order to ascribe undesirable qualities to the opponent. The violent character of this form of reasoning lays on the fact that the opponent is potentially deprived of the possibility to represent herself, and her arguments, the way she really wants to. If a person cannot represent her position and have her say in a public contest or debate, then it follows that what is violated is the freedom of speech. That is, the fascist attitude emerges in the form of suppressing any representation contrasting with the fascist's one. This form of straw man may facilitate what Bollas called "change of name". Basically, the distortion of the opponent's view might lead to an act of elimination of the proper name – in our case, the original argumentative representation – which is for Bollas one step away the elimination of the person himself.

Indeed, this form of straw man is more effective insofar as it is carried out in a rhetorical context in which the actual opponent is absent. The second form of straw man is weaker than the first one, and it is more closely related to dialectical situations. In that case, the misrepresentation or caricature of the opponent's view is reached by means of refuting only the weakest reasons behind it. So, basically the debunking is made easier by avoiding to address the better arguments in opposition. This argumentative move is more defensive, but violent, as it is still promoting biased truths by distorting the original version put forward by the opponent.

Another interesting aspect worth citing about the straw man in both its forms is related to *loaded terms*. As a matter of fact, natural language is infested by ambiguities. Words are not always well-defined, and their domain of application is too broad. Words can be also interpreted in different ways or in the light of different standards. The use of a term may also be dependent on the side endorsed by the speaker. In this sense, words may serve the purpose of military intelligence: the use and interpretation of a loaded term 1) identifying membership to a group, and 2) reflecting the participation, metaphorically speaking, in a coalition on the battlefield.

One who commits a straw man takes advantage of ambiguous and vague terms to represent (and select) the opponent's argument in a way that favors his debunking and rejection or negation. Such a kind of misrepresentation may contribute to the creation of what Bollas called "categorization as aggregation". Like in the case of change of name, negation of the opposition is carried out by means of natural language. In the case of categorization as aggregation the loaded term is also used "to define the moment when the individual is transferred to a mass in which he loses his identity" (Bollas, 1993, p. 209). An example is the use of the term "anti-patriotic": anyone who dares to criticize the government is anti-patriotic. Indeed, in this case the categorization (being "anti-patriotic") is a violent attack perpetrated by a straw man for which all the due distinctions are simply wiped out at once.

5.4.2 Negating the Opponent and Moral Niche Impoverishment

As mentioned at the beginning of this section, there is a second element which illustrates the fascist character of a reasoner: the recourse to harmful and efficacious *gossiping fallacies*. What we have described so far is the attitude that the fascist arguer has towards truth. This attitude basically relies on suppressing any fact or descriptive claim opposing the accepted view. We have illustrated two types of fallacious reasoning – the one-sidedness fallacy and straw man – that are meant to constitute the core of the propagandist dimension of the fascist arguer. This subsection is devoted to describing the types of reasoning the fascist arguer deploys toward his opponents.

My main take is that the fascist arguer usually commits a number of fallacies which have been called gossiping fallacies.[75] The reference to gossip is meant to describe a common attitude among reasoners, that is the attitude to connect *ideas* emerging in a discussion with *people* who support them. In doing so, the reasoner enlarges the basis of his argumentation including information, facts, and claims that are inherently gossipy, meaning that they are about other people.

Many arguments introducing information related to other people may be easily identified as fallacious, since they contribute to the discourse with nothing but irrelevant information. I will illustrate two examples of gossiping fallacies worth citing here: the *argumentum ad hominem* (argument against the person) and the *argumentum ad verecundiam* (appeal to expert). Each of these two arguments introduces information that is not usually relevant to the matter in discussion. The *argumentum ad hominem* consists in opposing a certain idea by attacking the person or the group holding it; the *argumentum ad verecundiam* consists in supporting a certain view by appealing to somebody recognized as an expert.

The notion of gossiping fallacy clearly captures the *military* dimension of reasoning. As already pointed out in chapter one,[76] the use of language – even for argumentative purpose, as in this case – underpins an activity of coalition enforcement. That is, the reasoner tacitly establishes (or assumes) coalitions in her mind and refers to them while taking part in a discussion. Even in the most abstract cases, this military or coalitionist dimension is still present, as the reasoner always argues simultaneously both with other people and with their ideas. Indeed this is not to say that, in making up our reasoning, we as people always introduce information about those who we argue with. There are plenty of examples in which the recourse to irrelevant information or facts resulting from gossip is minimized like in the case of scientific reasoning. However, as a result of a process partly driven by evolutionary forces, human cognition is biased towards *social* problem-solving (Bardone and Magnani, 2010). That is, there is almost always "a sense of positions in the field" (Collins, 2008, p. 453) that is detected while reasoning.

The reasoner may face up to what Sperber (2010) called a *double stake*. That means that the reasoner is supposed to choose between two potentially opposing alternatives perfectly resumed by the latin motto: *amicus Platus, sed magis amica*

[75] Cf. also (Magnani, 2009; Bardone and Magnani, 2010).

[76] Cf. subsection 1.3.2.

veritas, "Plato is my friend, but truth is a better friend". I argue that this is relevant for stressing the dimension of coalition management emerging in argumentation. For instance, if a friend of mine – a learned and trusted friend – told me something I have some doubt about, I actually have a double stake: should I trust my friend? Or should I reject what he is telling me on the basis of my doubt? Indeed, from an intellectual perspective, I would agree with Aristotle saying that I should not trust what my friend told me, even though he is my friend. On the other hand, however, here a social dimension is clearly emerging that, as a member of a coalition, I should comply with – a sense of position in the field. The example may appear quite trivial. But, what if my friend and I were in discussion in a public debate and I did not want to embarrass him? Sometimes conforming to another's belief is not necessarily irrational, as it responds to the logic of coalition management. As Sperber put it:

> Moreover, participating in such a collective process involves not just an intellectual but also – and more surely – a social benefit, that of belonging, of getting recognition as a person in the know, capable of appreciating the importance of a difficult great thinker. Not participating, on the other hand, may involve the cost of being marginalised and of appearing intellectually stale and flat (2010, p. 591).

In the case of the one-sidedness fallacy, the fascist arguer attempts to destroy facts, claims or information that actually or potentially threaten the coherence and consistency of his doctrine. In the case of the gossiping fallacies, the attack perpetrated by the fascist arguer is explicit against the person, and is an example of what Bollas calls character assassination. The information spilled all over is indeed discrediting information concerning, for instance, one's private life, sexual orientation, and so forth.

Malicious gossip – furiously transmitted by the twenty-first century media – provides spectators with alleged facts about a certain person, so favoring moral conflicts and other various situations in which the disengagement of morality can be easily accomplished.[77] Gossip may contribute to rendering the quality of public discourse and of various moral cognitive niches extremely poor. Although it is not necessarily a degraded form of communication, gossiping can be recognized as one of the main factors leading to or facilitating the impoverishment of public discourse, from a moral perspective. It is worth noting here that many citizens may even learn that gossiping is the only way to be part of public discourse so that they will tend to reproduce the same impoverished patterns of reasoning (Barnett and Littlejohn, 1997).

In particular, it is also worth mentioning the recourse to a specific form of *ad hominem* called *poisoning the well* (Walton, 2006). In this case, the introduction of gossiping information – true or false, it does not matter – aims at labeling an entire category of people as not reliable and therefore to mob or destroy, metaphorically but also literally, as history tells us. The case cited by Bollas of categorization as aggression is strictly connected to this version of *ad hominem*, as the attack is generalized not only to a person as an individual, but to a person as member of a group

[77] See above section 5.1.

which is publicly mobbed and therefore victimized. Usually this attack is followed and even bolstered by a semantic shift or even change of name.

Consider the following statement made by Daniel L. Abrahamson, a member of the 9/11 Truth Movement, about Noam Chomsky:[78]

> Noam Chomsky has acted as the premier Left gatekeeper in the aftermath of the 9-11 crimes, lashing out at the 9-11 truth movement and claiming [that] any suggestions of government complicity are fabrications. The "radical" Chomsky takes a position so deeply rooted in denial that it makes the staged 9-11 whitewash commission look like a honest study.

As one can easily see, this is an example of an *ad hominem*. Abrahamson argued against what Chomsky claimed about the attack to the Twin Towers saying that the American linguist and philosopher is a Left gatekeeper, that is, a person who more or less fraudulently omits to mention certain facts just because they do not fit in with what he stands for, asserted by the mainstream media. The caricature – made by means of a personal attack – has two main components worth mentioning here. First of all, the argument aims at jeopardizing Chomsky's reputation by labeling him as Left gatekeeper. This is an example of categorization as aggression. Secondly, it is a perfect example of what we previously referred to as "a sense of positions in the field". That is, the arguer does not really pay attention to the content of his opponent's claims. Conversely, he depicts Chomsky as a member of the opposing coalition that actively takes part in the conflict boosting its own political agenda. Therefore, everything Chomsky claims is simply irrelevant, as it is interpreted in the light of military intelligence according to which what really matters is rather *why* one says something than *what* he actually says. This coalition element is captured by the well-known tagline found in the Gospel of Matthew (12:30): "He who is not with me is against me, and he who does not gather with me scatters."

More generally, the systematic recourse to forms of reasoning intimately connected with malicious gossip makes any form of acceptable debate impossible, in which the principle *audi alteram partem* is respected. Indeed, the violation of the principle may easily lead once again to character assassination, as the marginalization and annihilation of a certain voice or minority is reached by means of jeopardizing its reputation.

The *argumentum ad hominem* is one example of gossiping fallacy. The so-called *argumentum ad verecundiam* or, in plain English, appeal to authority, is another one. In the last part of this section I will introduce a special type of *ad verecundiam* that I will call *appeal to oneself.*

The *argumentum ad verecundiam* is based on the appeal to an authority acknowledged as such in order to support or boost a certain position rather than another. Consider the following example. Andrew Keen wrote a book, *The Cult of Amateur* (2007), in which he violently attacks the culture that the Internet and the Web were (and are still) nurturing. He argues that new technologies, such as blogs, social networking sites like MySpace, self-broadcasting tools like YouTube, etc., are

[78] `http://www.modernhistoryproject.org/mhp/ArticleDisplay.php?Article=NoamAsset.`

glorifying and celebrating what he calls the *cult of the amateur*: that is, through breaking up all the traditional intermediate layers between the editor and the users, these new technologies are encouraging everybody to become a source of information and entertainment about a certain issue or topic, no matter if he is an expert on the field he is writing about, or not. Conversely, he claims that a well informed public should not rely on amateurs, but talented intermediaries, like professional editors and journalists, for example. In putting forward his thesis against amateurism, he made an example, which concerns an eighty-minute movie called *Loose Change* on 9/11 conspiracy theories which sprang up on the Net some years ago:

> The "claims" made by *Loose Change* were completely discredited in the final report of the 9/11 Commission, a report that took two years to compile, cost $15 million, and was written by two governors, four congressmen, three former White House officials, and the two special counsels (Keen, 2007, p. 69).

In this case, the claim made by Keen is a fallacy based on the *appeal to experts*. It is fallacious because he does not reject the theory presented in the movie *Loose Change* by referring to evidence and incoherencies; conversely, he simply posits that the conspiracy theory presented in the movie is false, because the *9/11 Commission* reported quite the contrary. To boost this position, he simply lists some details related to the report, for example, the cost of the commission, its composition and duration. Of course, all these details are irrelevant to assess what really happened that day.

As this brief introduction has shown, an *ad verecundiam* is usually based on the recourse to a third party that is supposed and considered as an expert or an authority in a given field. The appeal to an authority is particularly effective to solve disagreement, as it may have the same function a judge has in court. Of course, the authority one appeals to might not be recognized as such by the two parties. In that case, the function of an *ad verecundiam* is more rhetorical than dialectical, meaning that it is not meant to convince one's opponent, but his audience. For instance, in a public debate concerning stem cells or abortion a catholic person may appeal to the Pope to foster his argument thereby widening disagreement with his opponent while finding favor with the audience.

There is another type of *ad verecundiam*, which is particularly interesting here: when a person appeals to himself as the ultimate authority for a certain issue. In this case, the effect is merely rhetorical and propagandist, as his opponent's position is simply rejected because it is not *his*. In this case, we consider the appeal to oneself as belonging to the toolbox of the fascist arguer. Let us now make an example.

During a recent press conference, about the upcoming Italian regional elections of March 2010 in which the PDL (right wing) candidates' list for Lazio (the region of Rome) had not been accepted due to formal irregularities, Mr. Berlusconi said:

> I want to give a documented and certified reconstruction of what happened, purified from all the untruthful and biased versions that have been provided by certain press reports, and I want to say right away that our party executives are in no way what so ever responsible for what has happened – contrary to what some people have claimed. From the reconstruction I am going to give now I'll show you that the PDL candidates have been intentionally prevented from signing up for the Lazio elections.

After describing what happened according to his reconstruction, he added:

> This report is the result of inquiries that I have personally conducted on those people who were involved in this venture.

His argument can be described as follows:

- It was me doing the inquiry.
- *Then*, the inquiry is documented and certified.

More generally:

- I say that P is true.
- I am an authority.
- *Then*, P is true.

As mentioned above, the authority one appeals to is usually a third-party one. For instance, the Pope, a Nobel laureate, a professor at Harvard University, or even a tool or a machine (i.e. Google when performing a search). By contrast, in the example of Mr. Berlusconi, he appeals to himself, not a third-party authority. Now, the question is: how could it be possible to keep the argumentative force of an *ad verecundiam* without recurring to a third-party authority?

One way is to refer to oneself in third person. For example, in the Gospel of John (9:35-37), Jesus replied as follows to a man who asked him about who the Son of God is:

> You have now seen him; in fact, he is the one speaking with you.

Another example – definitely more prosaic — is provided by the French general and statesman Charles de Gaulle who wrote this in his memoirs, telling the reader about the 1958 Algiers uprisings:

> No one really doubted that the situation could have any other conclusion than De Gaulle.

These two very simple examples are meant for pointing out how sometimes one needs to stage himself as an authority. Now, what I want to stress here is that one can do that even without explicitly resorting to the third person. In fact, what is needed is the recourse to the "virtual self", as Goffman (1961) put it. According to his definition, the idea of the virtual self describes the expectations about the character a person is supposed to have playing a given role. So, one is who he is, *and* who he is supposed to be – when playing a given role in society.

In the example of Mr. Berlusconi, he remarked that it was *him* doing the inquiry. What did he mean by that? When stressing that it was him who conducted the inquiry, he was staging his virtual self aiming to convey the impression to the audience that he was the ultimate authority over the matter in discussion. In doing so, he was involved in what Goffman called the process of "role distance" (1961). Originally,

Goffman used that expression to refer to all those cases in which a person distances himself (or who he actually is) from the role he has been assigned to in a given context. The concepts of the virtual self and that of role distance are of particular help here: I argue that in our example what Berlusconi is appealing to is – in Goffman's terminology – his virtual self or the role he plays, rather than himself. So, the process of role distance provides a valuable account about how the shift from a third-party authority to a first-person authority can actually take place. That is, my take is that it is only the process of shifting as distancing that allows the *appeal to oneself* to keep the argumentative force of the *ad verecundiam* it mimics.

It is worth pointing out that the shift is meant to create a fictional recourse to a third-party authority. At a pragmatical level, it is basically a person appealing to his *say-so* and nothing more. I posit that this is what characterizes the appeal to oneself as belonging to the fascist's toolbox, because any opposing view can be ruled out by resorting to one's say-so. The opponent is deprived of any possibility to have his say fairly discussed or analyzed simply on the basis that he is *a priori* on the wrong side. Indeed, the fascist dimension of the appeal to oneself is strengthened, or at least becomes more effective, as it reflects a power asymmetry between the arguers. In order to make this point clearer, let me go back to Goffman's idea of the virtual self. As one can easily see, in the appeal to oneself, the recourse to one's virtual self should be convincing. Otherwise, the appeal would miss the target and blow back, resulting merely pathetic. Of course, the way one's virtual self is depicted or staged is not totally at one's will. By contrast, it is a social phenomenon related to power. As far as I am concerned here a distinction should be made.

Walton (1997, p. 77-78) argued that there are two different types of authority: the *cognitive* authority and the *administrative* authority. In the first case, the recourse to authority is meant for cognitive purposes. For instance, patients trust their doctor because she is supposed to be an expert, able to read symptoms and prescribe the most effective treatment to solve the problem (i.e. treat the disease). The second type of authority – the administrative one – is usually a person that has a right to exercise command or influence. For instance, an official or a person holding a recognized position of power.

The fascist arguer usually relies on the second type of authority depicting himself not as an expert, but somebody that has a certain authority which is not derived from what he knows, but from his recognized position of power – his virtual self to use Goffman's terminology. He takes himself as an authority to exert force. So, basically, the fascist arguer is implicitly using his own power to make his statement credible so that to disagree with his view or simply to refute it would mean to question his power and more or less tacitly engage him in battle. This process is once again related to military intelligence and coalition management as I have previously contended. That is, a coalition-enforcing dimension is explicitly addressed so that agreeing or disagreeing is no longer related to the reasonableness of a certain argument or belief, but to a *sense of positions in the field*.

5.5 The Perversity of "Respecting People as Things" in Wars and the Role of Bad Faith

When exploring the problem of violent (but held as moral by their performers) *ecotage* and monkey-wrenching in chapter one of my book *Morality in a Technological World* (Magnani, 2007b) – and in chapter four of this book – I emphasized the urgency of defining some "things" as distributed "moral mediators" that can provide precious ethical information which is unattainable through existing internal mental resources and thus ascribe important new values to human beings that could not otherwise be ascribed. In the case of ecotage we face with an analogous strongly moral (and at the same time violent) process of "respecting plants as humans": indeed I observed that, in radical environmentalism, ecotage is seen as a way of defending minorities. Unprotected vegetable and animal species that are being destroyed are considered, in this construct, minorities with the same dignity as human minority groups. Moreover, if the self is ecologically extended, as many scholars contend, and it is part of a larger ecological self that comprises the whole biological community, then the defense of non human "things" is simply "self-defense". Consequently, breaking the law is justifiable for some eco-violent activists.

A similar line of thinking can help us to reframe our understanding of modern warfare and the threat it poses to nonmilitary entities. In some contemporary wars waged by Western countries, military forces have systematically violently destroyed ill-fated sets of local things, animals, and human beings, all of which have no defenses of their own and are rarely regarded by the local leaders as worthy of protection. Paradoxically, they appear on the world's moral radar screen only because of the ethical value they acquire after being destroyed. Perversely, dead bodies and bombed-out buildings have a particular kind of worth that is unavailable to living people and unscathed cities.

I contend that the traditional ethics of war has not paid sufficient conceptual and strategic attention to the problem of noncombatant immunity. This lapse has occurred, I believe, because a distorted way of "respecting people as things" has led us into assigning exaggerated ethical and pedagogical value to the violent spectacle of dead bodies and destroyed objects. Ironically, in a war setting human beings are morally "respected" as people only when they become nonliving things – that is, when they are dead – just as certain animals or sites gain value when they become extinct or are violently destroyed. This respect for violently terminated noncombatant humans echoes the paroxysm of violence in sacrificial-religious-victimary processes.[79] In a way, so-called *smart bombs* seek to minimize this bizarre "respect for people as things" (that we could label as a respect for people *when* things) by targeting only military sites and limiting civilian casualties.

[79] Again, cf. Girard's seminal work on this subject (1977), already illustrated in the previous chapters of this book.

I am not concerned here with the concept of *just war* or the question of "when to fight";[80] rather, I would like to focus on the problem of "how to fight" by revisiting the concept of "respecting people as things", I described in subsection 4.4.2 of the previous chapter. The generally accepted principle of noncombatant immunity is that civilian life and property should not be subjected to military force, but there are conflicting ideas about what counts as "civilian" and what counts as "military". I believe that we can consider unprotected things, animals, and human beings that are threatened by warfare "also" to be minorities, just as environmentalists consider endangered plants and animals to be minorities; this status would give them noncombatant immunity and accord them protection. The problem of "how to fight" would, consequently, acquire new variables and moral nuances and so help us to recalibrate the current ethics of war.

Those who refuse to acknowledge the reality of collateral damage regress into a sort of psychological refuge from the horrors of war; they find relief in such denial, but it also leads them to violently underestimate the problem of noncombatant immunity. This brings us back to the issue of bad faith: building an emotional firewall is a way for an individual to construct another self, one less sensitive to the horrors of war in this precise case. War violently kills human beings, and this fact is too horrific for some people to accept when faced with it; such people – politicians, sometimes – prefer to indulge in bad faith rather than sorting through all the complexities of war. By so doing, they maintain their ignorance about the problem of noncombatant immunity, and it is this ignorance that contributes to perpetuate human anguish and violence. In turn, more human suffering drives more people to the opiate comfort of the condition of bad faith, and the cycle continues unchecked.

Moreover, convincing oneself that collateral damage is unlikely (or, worse, acceptable) contributes to an objective ideology about war that is available "out there", stored in external devices and supports (other people, books, media, etc.) of a given social collective. People readily pick up external ideological "tools" of this kind, then re-represent them internally in order to preserve their bad faith. The condition of bad faith may be so widespread in a cultural collective that it becomes crystallized in ideological narratives shared by entire communities. Bad faith becomes, in essence, a cultural default setting. Only by continually monitoring the links between the internal and external world we can lay bare the deceptive nature of our beliefs. This process testifies to the continuous interplay between the internal and external environment, and I think it can fittingly illustrate the deceptive character of the various ideologies.

Wars compel cultures to acknowledge that they attribute greater value and respect to tanks and technological weapons (and the all-encompassing commodification of human needs) than to an intact natural community of living plants, animals, and human beings. Considering how entrenched many old ideas are, making the case

---

[80] A complete treatment of the ethics of war and peace is given in (Lackey, 1989) and, more recently, in (Coady, 2008). For an analysis of war as a complex interaction process, involving moral aspects which provide its legitimation, despite modernity's emphasis on interstate peace and cooperation, but also "immoral" aspects, when seen in the light of the dehumanization of the "other" and in the prohibition against killing, cf. (Tiryakian, 2010).

for new kinds of war or nonmilitary strategies to achieve prosperity and freedom is a considerable challenge indeed. Respecting people as things is important not only in order to ascribe greater value to others but also for preserving our own dignity and freedom. As I contend, embracing this – at first sight – un-Kantian concept requires that we develop new ways of thinking and new kinds of knowledge: if we fail to do so, one of the negative consequences is that we put ourselves at greater risk of stumbling into bad faith.

To live without important knowledge – whether we have stubbornly resisted it or we are just unaware of it – results in creating for ourselves a toxic state of ignorance that generates a vague sense of anxiety, which, I contend, is what drives us into the bad faith condition and, in our present case, neutrality with respect to violence against noncombatants. We wriggle away from unpleasantness and retreat into a kind of oblivion that just seems less harmful than confronting difficult issues. Bad faith ultimately has a deeply corrosive effect on our well-being, for being ignorant of possible choices constricts our freedom and diminishes our dignity.

But this invisible condition is so pervasive, so much a part of life in the modern world, that it is not easy to imagine how to rise above it. Bad faith may be a self-defeating coping mechanism, but it is, for many people, the best (or only) one available. My answer is to supplant it with knowledge: that which seems unbearable or impossible can often, when one has the right information, be approached differently and managed, if not easily, then at least more comfortably. Here we return to the challenge of building new forms of knowledge (and of moral knowledge) that, once developed, must be used wisely and purposefully. Acquiring a new understanding of science, technology, and ourselves is critical, but it does not complete our quest to enhance human dignity. Once we have this new awareness, we face the practical matter of putting it to use. Therefore, the most important and difficult step in the process is to identify the principles that will guide us and to develop reasoning strategies that will lead us to the best possible choices. I have never meant to be naïve and consider knowledge as the ultimate tension-solver: as a matter of fact, I have stressed throughout this book how even the most rational knowledge content can become a trigger for violent outcomes.

Chapter 6
Religion, Morality, and Violence
Faith, Violent Mediators, Overmoralization

6.1 Is Religion Violent? Faith and Its Pervertibility

The main problem I want to address here is related to a kind of prosaic paradox, which everybody knows: on one side religions are a way of explaining the genesis of violence – for example in terms of spiritual evasion, like in the case of Kierkegaard's vision – and a way of escaping it but, at the same time, insofar as religions are carriers of moral views (and take part in their "secular" construction) they constitute a great part of those collective axiologies I illustrated in chapter four,[1] which are possible triggers of punishment. Moral axiological systems are inclined to be the condition of possibility of conflicts that can lead to violent outcomes.[2] Currently we are clearly faced with the so-called "reenchantment of the world",[3] which consists in a revival of the centrality of religion in politics and media, and consequently religion appears to be at the center of a great part of people's worries and concerns worldwide. This reenchantment has also been magnified by Islamic fundamentalist terrorism. This obviously coincides with the rise of new conflicts and violence, both at the in-group (fanaticism, racism, xenophobia) and out-group levels (violent terrorist attacks and the consequent "preventive wars"). It seems that morality, more than spirituality, plays the main role in individual and group religiosity, providing stable forms of identity and axiological well-defined systems. The link between violence and the sacred has also been masterfully depicted by Girard in his well-known books:[4] Girard clearly shows that, when dealing with the sacred, we cannot distinguish between proto-religions (for example in the case of primitives or ancient people) and the great monotheistic religions. Both share similar structures of the *sacred*, even if various differences, for example in the ways of perpetration of the derived *sacrifices*, must be acknowledged.[5]

[1] Cf. section 4.5.

[2] I have already partially addressed this paradoxical situation during the analysis of "nonviolence", cf. chapter four, subsection 4.5.1.

[3] (Wieviorka, 2009, p. 167).

[4] (Girard, 1977, 1986).

[5] Cf. (1986). I have illustrated the violent/sacred mechanism of scapegoating in chapter one, page 54.

L. Magnani: Understanding Violence, SAPERE 1, pp. 235–294, 2011.
springerlink.com

This passage by (de Vries, 2002, p. 1) is extremely appropriate to introduce the problem:

> The concept of violence is both empirical and [...] transcendental or metaphysical, belonging to the realm traditionally ascribed to the a priori, to the intelligible or the noumenal (Kant would say), in short, to ideality and idealization as such. Violence, in both the widest and most elementary sense of the word, entails any cause, any justified or illegitimate force, that is exerted – physically or otherwise – by one thing (event or instance, group or person, and, perhaps, word and object) on another. Violence thus defined finds its prime model – its source, force, and counterforce – in key elements of the tradition called the religious. It can be seen as the very element of religion. No violence without (some) religion; no religion without (some) violence.

Actually, the prosaic paradox I have indicated above is one of the main theoretical questions affronted by thoughtful scholars interested in philosophy of religion and in the relationships between philosophy and religion. These philosophical approaches confirm my contention about the strict relationship between violence and morality. Any moral (religious) effort to address violence and to remain open to the "other" constitutively presupposes an exposure to censorship, punishment, and violence. For Derrida (1999) this problem coincides with the paradoxical and exoteric figure of the *á Dieu/adieu*, where turning toward God "coincides with taking leave of God, each move a disruptive doubling of the other. Any effort to address violence and remain open to the other presupposes an exposure to censorship and violence".[6] The prototype is the testimony of Abraham's sacrifice of Isaac, its religious (and "ethical") renunciation in front of God and its inherent violence.

Hence, De Vries (2002), following Derrida, links religion to violence: first of all he shows that religion contaminates philosophical and ethical thinking and for this reason it manifests itself in modern secular ideas, such as Enlightenment, democracy, cosmopolitanism and tolerance. In the non philosophical perspective of an evolutionary account of punishment and cooperation, in the light of the threat of "supernatural agents", Johnson and Bering usefully note that religious beliefs, especially because of the moralizing and sanctioning behavior they generate, may serve as a common origin of human cooperation, and so

> [...] religious traditions continue to underlie fundamental aspects of law, political discourse, appeals to public action problems, and social life, even if the modern proponents are no longer themselves believers (consider the marriage, swearing on the Bible in court, charity, many national constitutions, and calls for U.S. unity against an "evil empire" or "axis of evil" – it is not inconceivable that these norms originated and persevered because of their selective success and cognitive relevance in human history). Certainly most people today – even atheists – continue to behave in accordance with a set of values which, they appear self-evident, are directly analogous to many religious coded (and evoke the same secondary emotions of shame, empathy, guilt etc., that supervise one's own actions) (2006, p. 229).

The main author to whom De Vries refers is Jacques Derrida, together with an examination of other classical ideas by Arendt, Kant, Kierkegaard, Benjamin, Schmitt,

[6] (Reynolds, 2003, p. 480).

Habermas, and Levinas.[7] If religion is secularly intertwined with the public axiological area of human action, religion shows – exemplarily – the "pervertibility"[8] of any decision and of any responsibility, religious or not. Religion testifies something transcendental with respect to human history, which

> [...] entails an aporetic testimonial logic that makes violence practically inescapable. De Vries sets this up masterfully in a treatment of Kant's distinction between "pure" (moral, rational, true) religion and religion infected by alien, nonreligious elements, by error. Following Derrida's lead, De Vries notes how "pure" religion's sense of absolute responsibility requires the critique, censorship, indeed sacrifice, of nonreligious elements in order to mitigate heteronomy and radical evil (Reynolds, 2003, p. 479).

In this perspective "religions" appear to be indeterminate and without content, but they exist only supported by singular revealed testimonies, which "can only approximate [their] infinitely removed ideal, in comparison with which they are all not only imperfect but, strictly speaking, *false*".[9] In sum, it is in this sense that, simply, "the history of religion is one of truth and error, a nonviolence that wears the garments of violence" (p. 480). Again, a discourse about religion is at the same time a discourse about violence, as already contended by Girard. Every axiology (and so every – ineluctably secularized – religious axiology) which substantiates morality always opens up by itself to both a "potential" command to *emancipation* from violence/evil and an ineluctable *commitment* to violence/evil, for example inaugurating the space of possible threats (and so censorship of others), as I have illustrated in chapter one, in the framework of the naturalistic interpretation of morality advanced by the coalition enforcement hypothesis. Similarly:

> [...] to say "God" is to attest to an anarchic, infinitely substitutable figure ("every other is totally other") that allows no fixation or determination, bespeaking a relation without relation (as in Derrida's ideal terms: democracy, hospitality, justice, etc.) that undercuts ontotheological[10] categorical expectation and its violence. Yet such an attestation – a singular testimony – cannot escape from its inscription in public and categorical idioms of expectation that de facto betray it, that limit its purview to some situation or ideal ("this" or "that" other), inevitably excluding or censoring others. Without the self-conscious risk of this betrayal – this *á Dieu*, this violence of law, hostility, or closure – the affirmation of God, of responsibility, of hospitality, and such, turns into "an empty and even irresponsible dream"[11] (ibid.)

[7] I have discussed various aspects of some of these authors in chapter one.

[8] (Derrida, 1987). Hence, descriptive, normative, evocative, hospitable, friendly, and edifying acts are embedded in a constitutive "perverformativity", a term whose significance will be clarified subsequently. On Derrida and religion see also (Sherwood and Hart, 2005).

[9] (de Vries, 2002, pp. 88–89). It is in this sense that religions have what Durkheim (1995) called "secular utility". In current naturalistic cognitive terms, researchers speak of the "evolution of religious prosociality"; it seems confirmed, thanks to empirical data, that religious *morally concerned* thoughts reduce rates of cheating, increase altruistic behavior and trust among anonymous strangers, so favoring possible cooperation (Norenzayan and Shariff, 2008).

[10] Theology is seen as an "unreflecting faith", merely ontologically committed.

[11] (de Vries, 2002, p. 322).

Philosophically – and paradoxically – it would be only in an inconceivable "interruption for the ethical" that "responsibility, both practical and intellectual, [could] be thought and exercised at all" (de Vries, 2002, p. 10): no ethics, no violence; unfortunately, no ethics also means we just disregard or like the (natural) violence present in our world and we give up any will – supposed to be constitutively religious – to "take a stand for the best".

In this perspective, religion echoes that basic *aporia* I have many times illustrated throughout this book, when defining the relationship between morality and violence. Performing a religious act of faith – which embodies a spiritual but also "moral" openness to others – activates at the same time its immediate possible pervertibility: taking a stand for the "best" is always and constitutively opening ourselves to the "worst" or to violence. Analogously, we can say that the possibility of the worst is the condition of the best. To make an example of the same *aporia* in secular and sociological terms (and so not patently religious), we can consider how the much-celebrated modern (democratic and cosmopolitan) tolerance for "cultural difference" – a practice that surely aims to be the best – can be rapidly perverted and transformed into a dogma, which violates the "different". As a "different" I do not aim at being a different one among the many, but the specific one – positively – that counts. As different, I can feel myself violated by liberal egalitarianism or cosmopolitism (because they do not allow me to feel different) and in turn I can censor/violate their sustainers. Indeed, De Vries contends that Kant (in "Religion within the Boundaries of Mere Reason" [1793]),[12] was already perfectly aware that modernity does not "render obsolete religious categories (all figures of thought, rhetorical devices, concepts and forms of obligation, ritual practices, and so on). On the contrary, even when the 'religious' can no longer be identified as an integral and compelling system of belief or, more indirectly, as a narratively constructed way of life, it provides critical terms, argumentative resources, and a bold imagery necessary for analyzing contemporary culture successfully" (de Vries, 2002, p. 4).

At this point, one could wonder how it is possible to give a religious constructive resolution to the problem of violence, considering the strict intertwining between *taking a stand for the "best"* and *opening ourselves to the "worst"*. In the perspective of the philosophical analysis of religion I have just outlined, a positive resolution of the problem of violence is *exactly* the abstract conceptual task philosophy must *not* undertake. Any attempt to solve this problem would be to forget and betray philosophy's own contention, as just stated: the constitutive pervertibility (and its intrinsic non-philosophical character) of any religious act of faith and its related moral quality. The following is an example of this error. Criticizing De Vries' main contention about the strict relationship between religion and violence, and quoting De Vries, Jantzen says (2009, pp. 18–20):

> I do not find this definition helpful. If as he [De Vries] says, violence is involved in *every* exertion of force, even when it is justified and even when it is not physical; and if moreover this exertion is not restricted to the intentional exertion of force by persons but includes also events and even words, then nothing is left out. Everything

[12] (Kant, 1998b). See also below the following section.

> is violent. Creation is violent; so is destruction. Religion is violent; but religion is also the "counterforce" to violence.

Later on in the next page, she continues:

> By contrast, I would suggest that it is not the act of distinguishing and separating into self and others which is violent in and of itself; indeed such separation is essential if we are ever to experience the richness which respectful mutual interaction with others who are genuinely different from ourselves can bring. Violence enters, I would argue, not when difference is *defined* but when difference is perceived as *dangerous*, so that hierarchies are imposed and force is exerted to keep the hierarchies in place.

After having referred to some Biblical stories from the book of Genesis, Jantzen concludes:

> Thus, so far from this difference indicating violence, it should be understood as the very opposite. Creativity, and with it the beauty of particularity is antidote to destruction, not its enactment. Creativity invites harmony and flourishing, when the flourishing of one is interdependent with the flourishing of all.

Good! Well said, it is always heart-warming to see a positive commitment, in this case based on creativity (a theme that in the book is also related to some feminist suggestions), the problem is that this further proposal for the "best" (since it is nothing more than *yet another*) misses the philosophical point above, thus infringing and contradicting it. Following Derrida's contention, if – as a philosopher – you propose a positive universal solution to the problem of violence, it is still religion that spills over in your philosophical argumentation to suggest a good way for emancipation. But, as a philosopher, you cannot ingenuously and candidly avoid to be aware that your enthusiastic (and narcissistically "abstract") *good* will immediately presents its other perverse possible face: your "proper" philosophical space is *only* the one of the deconstruction of the ethical-religious heft. When you open *another* space for your singular act of testimony (creativity and its goodness in Jantzen's case), that marks the religious (not philosophical) capacity to risk more, you do not have to disregard, like philosophy itself also suggests, the possible hypocritical and perverted character of your "complacency of good conscience" (de Vries, 2002, p. 322).

Taking advantage of Kierkegaard's philosophy, Bellinger (2001) also repeats the spiritual gesture at the philosophical level, and stresses the role of spirituality in theological (Christian) understanding of violence. Sin is an active attempt do avoid "becoming oneself before God" and the

> [...] individual who seeks to avoid the process of spiritual maturation construes the divine command to growth as threat to the ego. The ego becomes a hardened shell, which believes that it must protect itself from the possibility of its "death", that is spiritual death, which would lead to a more mature form of selfhood. [...] Christ is at the same time the *exemplar* of the authentic selfhood and love of neighbor to which all persons are called, and the *communicator* of the call of God, which is so disturbing to the hardened ego. Thus the execution of Christ is the central revelation within human history of the nature of sin. [...] *In other words, persons become violent because they are seeking to avoid becoming more mature spiritually* (p. 7).

In the same book, chapter seven deals with Christian violence (against Jews, heretics, etc.) and contends that this violence would be in complete opposition to the central and ethical insights of the New Testament. This is of course true, because like Girard would say (in *The Scapegoat*, 1982), Jesus Christ told us that his sacrifice was meant to be, so to say, the "last one". The whole discussion in the book acknowledges the strict relationship between religion and violence and, by disrespecting its own philosophical assumption, candidly leads to "taking a stand for the best" ("persons become violent because they are seeking to avoid becoming more mature spiritually"), without remembering that that same stand, even if *said* philosophically, does not cease to be said religiously as well, and so it involves an "openness to the worst". The author thinks he is philosophizing *about* religions but he is simply still thinking *in* religious terms, he is just attesting an act of faith.

In sum, addressing the problem of a constructive resolution of the problem of violence coincides with a religious *stand for the best*, but if you candidly address this problem at the philosophical level, you just miss what philosophy clearly teaches you: that your standing for the best is in itself *religiously* opened to violence and to the worst. As a philosopher you undertake an enthusiastic religious commitment in the translucency of too much "good" conscience: given the fact that – philosophically – any religious commitment has to manifest itself empirically and historically, it is subjected to a kind of constitutive degradation, which exposes it to become chance for violence and censorship. Analogously, "pure religion" (like Kant would have said in "Religion within the Boundaries of Mere Reason"), for instance an invisible church committed to an endless emancipation from an inextirpable evil, in its secular declinations necessarily degrades into a "determinate" morality – possibly destined to generate violence, bad conflict, and censorship – that you share and practice *as a believer*. But your testimony is just "your" last mode of that "strategic negotiation that winds up naming and thus falsifying the unnameable", in that process that opens up "the possibility of ever further negotiation and substitution, even transformation, the 'becoming-true of an illusion'" (de Vries, 2002, p. 384).

Hence, religion is still fundamental in understanding the political and institutional ways adopted by Western democracies to regulate the interactions between their citizens and "others", for instance, with respect to their cultural and personal identities (legal aliens, foreigners, refugees, etc.) or in understanding the consideration to be used with fetuses, dead people, other living beings and all the artificial and technological objects (culture, tools of various kinds, and everything else that makes up the so-called cognitive niche, which very often mediates moral commitments).[13]

Hence, let us reiterate the main problem I am addressing in this section, related to the following paradox: on one side religions are always a way of explaining the genesis of violence and a way of escaping it, but, at the same time, insofar as religions are carriers of moral views (and take part in their "secular" construction) they constitute a great part of those collective axiologies which are inclined to be the condition of possibility of conflicts that can lead to violent outcomes. At the same time "religion" appears without content, but exists supported by revealed

[13] The role of external representations and resources – I call them moral mediators – in ethics is described above in section 4.4 of chapter four and in (Magnani, 2007b).

testimonies, which "can only approximate [their] infinitely removed ideal" (de Vries, 2002, pp. 88–89).

From the perspective of the individual psyche both these aspects can be explained taking advantage of the concept of the moral bubble:[14] in the first case, we solve the paradox because, thanks to the moral bubble we are in, we consciously choose a religious belief – and we perform moral actions on its behalf – because it is *inertial* and it feels correct even when we know better about its possible violent outcomes. In this sense it seems that there is a gap between the specific sensation of feeling right and being right. Paradoxically, we keep and trust a moral belief – based on an axiological conviction – even if, contradictorily, we have fresh knowledge which shows it is potentially – sometimes atrociously – dangerous. Furthermore, religions seem indeterminate and without content, before or beyond their revealed testimonies and secular expressions. Finally, religions also constitute a series of "experiences" where we go through states of knowledge that are not associated with any specific knowledge, so that the feeling in this case is a feeling of knowing, but unfortunately the knowledge is just felt – not thought – as in the case of cognitive dissonance.[15]

6.1.1 Theological Violent Argumentations

I have said above, taking advantage of some aspects of postmodern philosophy, that to say "God" is to attest an anarchic, infinitely substitutable figure ("every other is totally other") that allows no fixation or determination, that undercuts any *ontotheological* categorical expectation and its violence. I said that this attestation – a singular testimony – cannot escape from this inscription in public and categorical idioms of expectation that *de facto* betray it, that limit its purview to some situation or ideal ("this" or "that" other), inevitably excluding or censoring most of all, any others. When we turn our attention to a relatively recent book interested in comparing theology with the whole history of philosophy we can find this kind of betrayal again, which is easy to trace back in the structure itself of the argumentations performed.

John Milbank, a Christian theologian, who is among the founders of a theological current defined as "Radical Orthodoxy" (nested in the Anglo-Catholic tradition), presents an indepth knowledge about the whole "history of philosophy" in a book titled *Theology and Social Theory* (1990). But actually the title speaks of social theory, and not of philosophy. Why degrade philosophy to be just a kind of social theory? Philosophy never thought of itself as a social theory (on the whole), even less as a sociology. It is clear how, already in the title, the theologian wants to say that when he touches philosophy, philosophy instantaneously becomes something constitutively secular and *non-spiritual*. Is this not a kind of sweet violence to philosophy and its secularism? The first linguistic violence is perpetrated, Derrida would

[14] See chapter three, section 3.2.2.

[15] On the concept of cognitive dissonance see also chapter three subsection 3.2.1 and subsection 5.1.1 of chapter five. See also my analysis of (Burton, 2008) observations about beliefs in general, illustrated in chapter three, subsection 3.2.2.

say. And this outcome is immediately clear in the subtitle: "Beyond secular reason", as if sociology (read: philosophy), as a product of secular reason, is to be outclassed by something *better insofar as not secular*. Playing with the semantic ambiguity of the adjective *secular* we can say that theology is also a "secular" product of reason, it is neither something "beyond" secular reason, nor a revelation. Theology is not released directly by God but by some distinguished human beings that, to different extents, belong(ed) to particular Churches. Why should one (even implicitly) disregard the constant commitment to spirituality (and also to Christian spirituality) that is present throughout the entire history of Western philosophy? The fact that even the most materialistic doctrines take into account spirituality and religions by considering them as an opponent should clarify how, at least in the Western culture, it is nearly impossible to identify anything as purely secular, in fact the very category of "secular" intrinsically depends on what is "religious", and not vice-versa.

It does not take long before we see that the author's alternative to a world based on secular reason is the Church, which is not exactly the place where the Word of God is advocated nor where the sacraments are celebrated and the Mystery is revered, but rather a kind of "another" society.

Soon the theologian starts to deploy his gallery of intellectual subjects that must undergo a systematic work of inevitable exclusion or censorship. Let us consider some of this *theological terminator*'s victims: after the initial declaration of omnivorous cannibalism (practically all philosophical thought is totalized and aligned as a kind of target/victim to face the intellectual "slaughter" perpetrated by theology), the book indeed proceeds to eat "all" of the history of philosophy.[16] For example, Milbank draws a sharp critique of Hegel and Marx as both of them, notwithstanding their helpful analyses, still suppose a kind of "original violence" as they create their modernist myths of progress and conflict. In the case of Hegel, the violence is a "necessary" – ineluctable – moment in the blooming of the absolute Spirit toward freedom, while for Marx, linked to scientific positivism, alienation and class struggle are the "necessary" but vain tools for arriving to the socialist end. What about the so-called postmodern philosophy? Postmodernism, with its post-secular critique of modernism, tends to see the world in terms of differences. It embodies a fundamental ontological violence in its presupposition of conflicting difference since, for example for Derrida, language has no other target than itself, and nihilism is established. But the postmodern victims of theology have some merits: postmodernism was able to illustrate some aspects of the weakness of the created world and the ways in which putatively objective and rational thought can be enslaved by power.

Milbank's theological "therapy" consists in unveiling "another" ontology where the *entia* are not affected by chaos, violence, and nihilism, but rather by a "sociality of harmonious difference" in light of God's creation and the doctrine of the Trinity. Garver observes how

> [...] yet, this virtue of nonviolent Christian practice is not a practice that can be grounded in anything external to its own activity and its own narration of it. Milbank

[16] If one checks the *Index of names* at the end of the book, she will be amazed by the care that Milbank took in mentioning the majority of philosophical pantheon across the ages.

is building here upon Augustine's argument that despite the existence of difference within the created order, all creatures are related to God and so to one another and that this difference in relation is ultimately rooted within the inner life of the Trinity life together in God. In view of this kind of theological "social theory", violence (that is, any form of chaos or conflict) must always be secondary, and "peace" (that is, living in "harmony" with one another) must always be primary.[17]

In this perspective, a better title of Milbank's book would have been "Theology 'as' social theory"! Still, violence, as I have many times stressed in this book, tends to disappear as something marginal and uninteresting, which does not have to (and ought not to) be clearly depicted and known in detail. Furthermore, the supposed moral "openness to the others" one expects from theological argumentations (since they must be theoretically open to *the* Other) activates instead its immediate possible perversion and instantly evokes theoretical fundamentalism, as postmodernists are intellectually attacked because they are compromised with a "malign", which needs to be scapegoated thanks to the "unicity" of the Church-oriented vision of Christianity promoted by the author:

It will be contended that the perspective of "malign" postmodernism is the final, most perfect form of secular reason, in some ways reverting to and developing the neo-paganism of Machiavelli. For the secular postmodernists, Nietzsche became the only true master of suspicion: The present chapter, then, will show first, that a nihilistic genealogy requires an ontology of violence; second, that this ontology is only a mythology; third, that it is an entirely malign mythology (Milbank, 1990, pp. 262–263).

All the "others" lack something, and are seen in the perspective of disease, imperfection, insufficiency, also Levinas, the Jewish thinker, did not grasp the point, as eloquently shown in this passage:

This religion is not quite accurately described as "neo-paganism", because it is an embracing of those elements of sacred violence in paganism which Christianity both exposed and refused, and of which paganism, in its innocence, was only half-aware. The secular episteme is a post-Christian paganism, something in the last analysis only to be defined, negatively, as a refusal of Christianity and the invention of an "Anti-Christianity". Notice here that I am not, like Emmanuel Levinas, claiming that ontology as such is complicit with violence. If one follows this path, then one still tends to read the historical time of cultural exchange which is "Being" as inevitably violent, and to long for an impossibly pure encounter of mutually exterior subjects without mediation across a common domain, which is always doomed to infect and coerce the genuinely responding will. By contrast, I am suggesting the possibility of a *different* ontology, which denies that mediation is necessarily violent. Such an ontology alone can support an alternative, peaceable, historical practice (p. 309).

Also Plato and Girard's theories have their problems, they are found lacking and their positions need to be perfected: in the case of the first, "[...] what was lacking for it was both a notion of the interpersonal, and of the collectively corporeal as not 'debased'". In the case of the second, Milbank's argumentation goes as follows:

[17] Cf. http://www.joelgarver.com/writ/revi/milbank.htm.

> For if the victimage mechanism is culturally all-pervasive, then it could only be grasped, and exposed within culture, by one standing also outside culture. Only, therefore, by God incarnate. An abstract attachment to non-violence is therefore not enough – we need to practice this as a skill, and to learn its idiom. The idiom is built up in the Bible, and reaches its consummation in Jesus and the emergence of the Church. By drawing our attention to sacrifice, Girard helps us to articulate part of this idiom – and indeed his contribution has been one of epochal decisiveness. [...] [But] For further elaboration of the idiom we must turn back from Girard to Augustine, who by placing the Church, and not Christ alone, at the centre of his metanarrative, pays far more attention to the concrete shape of a nonantagonistic social practice [...] (pp. 376, 397, 492).

To be – I hope – moderately aggressive, I can say that, unfortunately, at least in the argumentations I am reporting, theology neither performs a "nonantagonist practice" nor tries to embrace and complement secular human ambitions. It seems that, if you just refer to Jesus in your work, and you do not mention the "Catholic" Church, like in the case of Girard, you are in the infancy of reason, an incomplete thinker. The rich arguments of the Gospels should be in a way *defused* and read in the correct way (*orthodoxy* is nothing but the right and hence righteous path) suggested by the Church, namely the ancient Fathers: "Go back to Augustine!" says the radical theologian. On my personal account, I rarely refer to the Church (and when I do, it is hardly to praise it) but luckily I can make amends for myself in a simpler way, not in a doctrinal way, please, but just as any common, lay believer: all I have to do is to take a short walk to the church of *San Pietro in Ciel d'Oro* (Saint Peter in the Sky of Gold), where the presumed relics of Saint Augustine are piously kept (*confirmed as* authentic by the visit of the Holy Father in 2007), in Pavia (Northern Italy) and where I also celebrated my Catholic marriage, hence blessed by the notorious Saint. Many intellectuals could reject the strict dualism presented by the radical theologian: as for myself, I really like Derrida's and Girard's thoughts about religion and Christianity, and this does not make me "horrified" before the Catholic Church, I appreciate many things about the Catholic Church, without such appreciation being in conflict with my appreciation of Derrida and Girard! Similarly, a (more or less) Christian thinker needs not to feel the same moralistic hatred that Augustine experienced against the Stoics, as reported by Milbank.

Indeed, the Stoics, even if positively related to Descartes' inwardness (sic!), are outdated Aristotelians, and "horrify" Augustine, who seems, at least in Milbank's account, to amazingly misunderstand the Stoic doctrine of *indifferents* as a mere nonchalant elimination of emotions. Furthermore, Stoicism was unable to conceive an unpolitical practice, that is, again, that one of the Church:

> Although the Stoic notion of the "true" ethical wisdom of the sage points in the direction of Cartesian inwardness, and an unmediated relation of the individual to total process, its conception of the everyday duty of "all" possesses elements that qualify Aristotelian inwardness in its over-emphasis on character. [...] Quite clearly, he [Augustine] rejects the Stoic ideal of *apatheia*: one cannot, he says, consider a man perfectly happy and in a state of peace if he suffers a serious physical disability. [...] Likewise, Augustine expresses horror at the Stoic ideal of unfeeling in the face of tragedies in time which are to be regarded as things "indifferent". In certain circumstances, Augustine contends, the good man should feel, horror, anger, pity and so forth;

what matters is not the having or not having of strong feelings, but whether or not they are occasioned by right desire, which alone will indicate what it is appropriate to feel. [...] Stoicism aspired towards a universal ethic, based on reason, transcending all political boundaries, and also towards a universal and ontological peace. However, it was unable to conceive of any new, non-political practice, and so the realization of peace had to remain "inward", and its political transcription could only take the form of a respect for the free space of others, and a formal acknowledgement of equality. However, virtue is now placed in a new and positive relation to difference, and, like stoicism, Christianity starts to validate liberty and equality (pp. 419–422).

Finally, what Milbank calls the self-torturing circle of secular reason can be interrupted so that humanity can finally go *beyond* it, as suggested by the title of the book: "Both nihilism and Christianity decode the inconsistencies of this position. And the Catholic vision [I say, only the Catholic one?] of ontological peace now provides the only alternative to a nihilistic outlook. Even today, in the midst of the self-torturing circle of secular reason, there can open to view again a series with which it is in no ontological continuity: the emanation of harmonious difference, the exodus of new generations, the diagonal of ascent, the path of peaceful flight" (pp. 419–422).

To summarize, Milbank's entire faith seems to be "I believe in the Church", a slightly disturbing actualization of *extra Ecclesiam nulla salus*. God the Father, the Son and (especially) the Holy Ghost play all very peripheral roles in his theology, everything else in Western culture has to be scapegoated, victimized, or, in some cases, therapized and "completed" (read defused and turned into something else and especially something harmless). Is this an example of the non-antagonistic practice proposed by Augustine? Finally, the theologian's most radical spirit emerges and generously, but also a bit violently, concedes a conclusive suspicion, concerning the fact that the Church did not do much – if anything – to realize its scope, given the fact we *actually* face with dangerous post-modernism, violence, and secularity; maybe, then, the (historical) Church is not much better than secular institutions and culture:

In the midst of history, the judgement of God has already happened. And either the Church enacts the vision of paradisal community which this judgement opens out, or else it promotes a hellish society beyond any terrors known to antiquity: *corruptio optimi pessima*. For the Christian interruption of history "decoded" antique virtue, yet thereby helped to unleash first liberalism, then positivism and dialectics and finally nihilism. Insofar as the Church has failed, and has even become a hellish anti-Church, it has confined Christianity, like everything else, within the cycle of the ceaseless exhaustion and return of violence (pp. 419–422).

It seems that Milbank, with these last words, paradoxically comes back to Derrida and his contention that performing a religious act of faith – which embodies a moral openness to the others – activates its immediate possible pervertibility, taking a stand for the "best" is always and constitutively opening ourselves to the "worst", to violence. Now, at the end of the book, the Church is clearly attacked.

With this conclusion, the author also seems to come back to Kant, who says something similar – in a gentler way, in "The End of All Things" [1794], as he envisages the possible perverted end of all things in a moral respect:

> If Christianity should ever come to the point where it ceased to be worthy of love (which could very well transpire if instead of its gentle spirit it were armed with commanding authority), then, because there is no neutrality in moral things (still less a coalition between opposed principles), a disinclination and resistance to it would become the ruling mode of thought among people; and the *Antichrist*, who is taken to be the forerunner of the last day, would begin his – albeit short – regime (presumably based on fear and self-interest); but then, because Christianity, though supposedly *destined* to be the world of religion, would not be *favored* by fate to become it, *the* (perverted) *end of all things*, in a moral respect, would arrive (1998a, p. 205).

6.2 Religion as a Natural Phenomenon

In the previous section, devoted to the relationship between religious faith and its pervertibility, we have seen that it is philosophically consistent to consider religion a basic mediator of morality, even if religion is much more than just morality, of course.[18] Scholars from anthropology, psychology, evolutionary biology, neuroscience and philosophy are developing an extremely rich and new cognitively comprehensive study of religion. In this perspective religious beliefs and behaviors are seen as a "natural outgrowth of basic cognitive functioning" (Teehan, 2009, p. 233).[19] Indeed religion beliefs and behaviors can be seen in the light of mental processing (Boyer, 2001), and/or as the fruit of evolutionary processes (Atran, 2002; Broom, 2009; Tremlin, 2006) that have shaped our emotional and cognitive predispositions: Darwinian evolution favors the human brain's inclination toward religion.

Hence, the consideration of religion as a natural phenomenon is made possible by cognitive science and bio-neurological research intertwined with some empirical data, more or less justified and correctly interpreted. Sometimes the results are infested by poor speculations about the overall significance of religion, an attitude sadly typical of scientists who want to philosophize. I think that considering religion as a natural phenomenon interestingly leads to the following main consequences:

1. if religion is seen as a natural phenomenon, it is not a *special* kind of human experience: its moral function is central because morality favors the – supposed to be good – bio-cultural evolutionary adaptivity of cooperation;
2. religion can be scientifically seen a) as a natural phenomenon in a Darwinian or b) in a non (strictly) Darwinian evolutionary perspective: in the first case

[18] On the strict relationship between morality and historical religions such as Confucianism, Hinduism, Buddhism, and Christianity, cf. (Peters, 2003). The author further emphasizes that what is good or bad depends in part on the customary morality that develops in order to manage inner-group conflicts while at the same time allowing the group in question to be competitive in conflicts with other groups.

[19] Cf. also the recent (Teehan, 2010).

a. some researchers (for example (Wilson, 2002a)) highlight its direct or indirect Darwinian adaptive function – which contemplates both moral and violent aspects,[20] others address religion's merely negative overall outcome as evolutionarily maladaptive (Dennett, 2006);
b. in the second case many authors (for example (Atran, 2002)), even if engaged in studying religion taking advantage of the cognitive, semiotic, anthropological, and historical methods deriving from the "arsenal" of human sciences, are more skeptical about the possibility of a strict evolutionary encapsulation of both religion and of the whole development of human culture. Some authors think that religion is an "incidental by-product" of biological evolution (and does not play a direct role in strict Darwinian evolution), an hodgepodge of genetically based mental mechanisms designated by natural selection for mere mundane purpose (Wright, 2009), such as for example confirmation bias (believing what your peers believe), explaining events in terms of personal agency, and interest and inclination in remote control (a bias toward beliefs that promise influence over predators, disease, and bad weather);

3. religion is easy, because it is grounded – like magic and myth – in very rudimentary and simple human cognitive capacities (that is, when compared to higher "rational" ways of thinking). As a matter of fact, religion provides some cognitive tools – rooted in basic human cognitive capacities – for hypothesizing supernatural agents and building moral confabulations [21] which provide acceptable stories about people and life, typical of religious believers;[22]
4. the violent aspects of religions are usually either

   a. disregarded, favoring the secular prosocial function; or
   b. overemphasized, to underline that – being a natural phenomenon – religion is nothing special, to eventually support atheistic outcomes.

A possible implicit consequence of the above cognitive research, for example in media reports, basically consists in two opposing and very simple *epistemological moralities*, which unfortunately favor more or less violent conflicts between believers and non-believers: 1) religion and Gods are necessary and ineluctable, given the fact they also are, so to say, "wired" in human biology and evolution, as *science* teaches; 2) religion is simply something natural, like many other human activities, endowed with positive (for example cooperative altruism or psychological relief) and negative (violence, moral conflicts, waste of time and high costs, magic mentality, etc.) aspects. The rough ideological results would contemplate (and still

[20] So to say, the violent fruits of morality and thus of religions are bad, but nevertheless adaptive.

[21] In the previous chapter I used the expression "moral confabulation" to refer to a pathological dimension concerning psychopaths. Here I use the same expression to indicate narratives whose plausibility should be questioned from an intellectual perspective.

[22] For further reading see (Bertolotti and Magnani, 2010).

do) two more or less violent – and pathetic – conflicts in Christian countries: 1) Darwinian evolution – and in general, science – absorbs religion and, by explaining it, annihilates its specificity; 2) Biblical narratives absorb scientific results and try to delegitimize them. I cannot abstain from adding a further comment, which derives from a personal epistemological-philosophical attitude, that I am perfectly aware is in turn endowed with a potential mildly violent outcome: it was and it is atrocious for a philosopher to see credited intellectuals, politicians, prelates, journalists, and also intellectually decent everyday people, endorsing in such a dogmatic way one of those two conflicting – each of them supposed to be absolute – "truths", candidly engaging in a barren controversy which could hardly find a proper home even in the most commercial and vulgar pop philosophy.[23]

Let us illustrate some aspects of the recent intellectual studies on religious tradition I have just introduced.

6.2.1 Religious Moral Carriers, Religious Violent Carriers

The cognitive analysis of religion provides a wealth of results. (Søresen, 2007) gives an interesting synthetic illustration of the cognitive and semiotic structure of religions and of their rituals in the light of individuals' basic cognitive capacity to understand actions (and intentions, beliefs, desires) performed by people in their own group, and to ascribe agency to natural and physical events. In this perspective religious beliefs – such as seeing natural objects as instances of an intelligent design pointing to a responsible hidden non-observable agent – can be easily seen as the natural outgrowth of basic cognitive functioning.

Religious (but also magical) rituals are seen as usually monopolized and performed by people ordinarily endowed with special qualities, who act on behalf of those hidden supernatural agents. Thanks to rituals, these people are able to violate domain-specific causal expectations, for example attributing special symbolic virtues to some substances: the act of touching a child's head with water is exactly a performative act of baptizing by means of the symbolic attribution which relates to the actions performed in mythical narratives of Christ being baptized, and to ideas of essential change with sanctified water. Rituals facilitate the ascription of superhuman agency responsible for representations of both the ritual efficacy and the relationship – inside the ritual – between the proximate and the ultimate intentions.

[23] A rigorous analysis of the controversies concerning the possible adaptive Darwinian, non Darwinian, and loosely Darwinian role of religion is furnished in (Proctor, 2005), (Richerson and Newson, 2009), and in other contributions contained in the interesting (Schloss and Murray, 2009), for example (Murray and Goldberg, 2009). Religion as an "evolutionary accident", as a by-product of certain highly structured systems that have evolved for understanding the social world, is illustrated in (Bloom, 2009). A collection of studies which aim at correcting some of the misconceptions which give rise to and help perpetuate what are here labeled "myths" about science and religion is the recent (Numbers, 2009).

The stability of ritual actions is guaranteed by the construction of doctrines that fix the general meanings which can be performed in the specific instance.[24]

Neurological correlates of religious and mystical experiences have been studied by using technology like PET and SPECT scans, and fMRI: temporal lobes, parietal lobe, and the thalamus areas resulted extremely important in the case of deep religious and mystical experiences (Newberg et al., 2001). This kind of research is typically interpreted by the media (but also by sophisticated intellectuals, both atheist and religious) in terms of a (very rudimentary, philosophically speaking) reductionistic perspective. Indeed, if the religious experiences are nothing more than by-products of neuro-chemical processes and so established by Darwinian evolution, the specificity of religion, so to say, disappears. It is obvious that religious believers can feel themselves violated by such approaches and can possibly in turn retaliate with more or less violent suspicion and hostility toward science. Other more peaceful agreements – I have to note, still philosophically rudimentary – between science and spirituality can be built for example arguing that God designed the brain so that it is capable of experiencing him (ibid.)

On the other hand, as I have anticipated, many evolutionary processes that have shaped our emotional and cognitive predispositions which favor human inclination toward religion have been studied. In this last case, moral functions are the fundamental issue, seen as evolutionary strategies that of course are involved in religious practices of various kinds. Morality carried by primitive or ancient religions – or proto-religions, if we aim at distinguishing them from modern monotheistic ones – is at the center of interest insofar as it promotes cooperation/altruism in the terms I have already depicted in chapter one, section 1.3, when I described the coalition enforcement hypothesis. Various theories of cooperation have been proposed so far, it can be useful to provide an extremely basic account of the different approaches, which all present consequences for the analysis of religion-based morality. According to the *kin selection* hypothesis (Hamilton, 1964), it is important if we morally immolate ourselves for the benefit of our children because this way we are protecting our genetic investments (inclusive fitness). Cooperation based on *reciprocal altruism*, instead, can be both direct (in which altruistic acts are returned sooner or later (Trivers, 1971)) and indirect (in which one's reputation for cooperation is rewarded indirectly through the favor of third-party observers (Alexander, 1987)), and it also explains why we morally sacrifice ourselves for someone genetically unrelated to us as well. Another theory explains cooperation as a form of *costly signaling*, in which generosity serves as an advertisement of high fitness to would-be mates or allies (Gintis et al., 2001). Finally, *group selection* theory sees the whole groups, as constituting a further level of selection, engage each other in a struggle for survival where morality and punishment (and so religion) play a crucial role (Wilson, 2002b).[25]

[24] In chapter four, subsection 4.8.3, I have illustrated an example of religious rituals in the framework of Jungian psychoanalysis, showing its related psychic, cognitive, moral, and practical role.

[25] On cooperation in the perspective of economics, anthropology, evolutionary and human biology, social psychology, and sociology cf. also the collection of studies contained in (Gintis et al., 2005).

Indeed, an articulated defense of the consistency of the evolutionary interpretation of many (proto-)religions is given by Wilson (2002a), which especially addresses the relevance of the controversial *group selection* theory, that I mentioned many times in the first chapter of this book. This theory proposes the reformulation of a strictly Darwinist view of the development of human culture. Religions can be seen as approximate mechanisms that motivate adaptive behaviors at the group level.[26]

A skeptical hypothesis about the adaptive role, advanced by group selection theory, of human (costly) engagement in cooperation is proposed by Burnham and Johnson (2005), as they contend that the costs of such cooperation cannot be recovered through kin-selection, reciprocal altruism, indirect reciprocity, or costly signaling. They contrast the main tenets of the evolutionary account, which states that cooperation is (a) a newly documented aspect of human nature, (b) adaptive, and (c) evolved by group selection: instead, they argue that the phenomenon is: (a) not new, (b) maladaptive, and (c) evolved by individual selection. They conclude "[...] because humans cannot be relied upon to work for the good of the group, we must craft social, economic, environmental and political interactions to ensure cooperation against selfish temptation. If the human propensity to cooperate were shaped by group selection, why is punishment so essential to promote sacrifice for the group?" (p. 131).

Alexander (1987) stresses the role of altruism as a way of gaining reputation for being a cooperator (and to attract other cooperators) so that altruistic individuals can gain reward by their social group. These theories see cheaters who do not cooperate as targets of a more or less violent punishment, which is part and parcel of the supposed moral behavior of the cooperators. It is obvious to note that religions are immediately involved, insofar as they are fundamental moral carriers. To distinguish between potential cheaters and potential cooperators we need a shared (religious) morality, which would be able to demarcate the "others". Therefore, (religious) morality is related to the possibility for a community to prosper and flourish.

The following passage by Teehan (2009, p. 239) provides a clear example of how religion can provide a solid basis for establishing the building blocks of a morality:

> Pascal Boyer points out that we conceive of other humans as "limited access strategic agents" (Boyer, 2001, p. 144). That is, we assume that others do not have access to complete or perfect information relevant to social interactions. Our information is limited, our ability to discern another's intention faulty, and this limitation is mutual. [...] People all over the world, however, represent gods as "full access strategic agents" (Boyer, 2001, p. 155). That is, they view their gods or ancestors not necessarily as omniscient but as having access to all the information relevant to particular social interactions. The gods have access to all that is needed for making a sound judgment in any particular situation. They know my child is healthy and at school and that actually I intend to spend the day watching a baseball game. Now, not all gods may be represented as possessing this quality, but the ones that do have special significance. And beings that possess such a trait are in a particularly privileged position to

[26] See also chapter one, section 1.3. On the evolutionary approaches to the analysis of the development of human cultures cf. (Norenzayan, 2006).

assume a moral role. Gods, as full access strategic agents, occupy a unique role that allows them to detect and punish cheaters, and to reward cooperators. In moral religions such gods are conceived of as "interested parties in moral choices" (ibid.) They are concerned with social interactions and fully cognizant of the behavior and motives of those involved. [...] Communal belief in such beings, therefore, lowers the risk of cooperating and raises the cost of cheating by making detection more probable, and punishment more certain. Religion then becomes the vehicle for the moral code of a society required for that society to continue to function as a coherent unit as it grows in size and complexity. Furthermore, religion not only supports evolved moral mechanisms by providing supernatural oversight, it also powerfully functions as a signal of willingness to cooperate. As noted it is imperative to be able to discriminate between potential cooperators and cheaters.

Here some examples of the moral functions of religion: to justify norms humans always took advantage of the threat of supernatural punishment (Johnson and Kruger, 2004; Johnson and Bering, 2006). Religious rituals provide "signs of commitment" in the members of the community, to favor the in-group separation of cheaters from cooperators and the discrimination of out-group individuals who do not share the same rituals. The history of the Patriarchs in Judaism approximates the logic of kin selection. The history of Christianity presents a more complex situation because of a more fluid and dynamic community with respect to Judaism – even if Christianity, from a moral point of view, seems filled with teachings that encourage kin selection: applying to it the idea of tribal morality no longer appears appropriate (Teehan, 2009, p. 243). Following Teehan, even Jesus' message to "love your enemies" (Matt, 5.44) would not refute the evolutionary account of religion which incorporates the "moral" commitment to the group which the Christian religion considers central. Teeham (2009, p. 244) is inclined

[...] to point out that as moral exhortation Jesus' teaching may move beyond evolutionary logic, but as a guide to behavior it is evolutionary logic that often holds sway. While this is an admittedly contentious point, I would claim that the history of Christianity is filled with examples (e.g. the crusades, the inquisition, the persecution of heretics and Jews, etc.) that speak to the power of the underlying evolutionary logic to overwhelm attempts to develop moral attitudes contrary to it (e.g. "turn the other cheek"). The response of Christians in history to enemies and to attacks has often been much more in line with the psychology of evolutionary morality than with these particular teachings of Jesus. This is not so much a condemnation of Christianity as it is a lesson on the difficulty of moving beyond these evolutionarily ingrained moral predispositions.

It seems to me that this passage can be interpreted in the light of the strictly philosophical considerations I made in the previous section: performing a religious act of faith – which embodies a moral openness to others – activates at the same time its immediate possible pervertibility; taking a stand for the "best" is always and constitutively opening ourselves to the "worst", to violence. Violent outcomes of religions should not astound the cognitive scientist (and common people), because when morality, exclusion, prohibitions, and punishment (physical or – less sanguinary – "spiritual") are central, possible threats and related violence are probable:

> Morality develops as a tool to promote within-group cohesiveness and solidarity (Alexander, 1987; Wilson, 2002a; Johnson and Bering, 2006). This cohesiveness functions as an adaptive advantage in competition with other groups [...]. Morality is a code of how to treat those in my group; it is not extended, at least not in the same way, to those outside the group. Since these others are not bound by the same code they must be treated as potential cheaters. Those outside the group are in fact a potential threat to my group survival (Teehan, 2009, p. 246).

In this perspective, for example *killing* an out-group is not generally considered *murder* in the sense of contravening the moral norms of the group/society.

6.2.2 Rendering Human Behavior Predictable through Religion and Morality

As I have illustrated, religion as a moral carrier provides knowledge that directly shapes human behavior, thus making it predictable and favoring cooperation inside groups. In my previous book on morality and technology[27] I contended that cognitively tracking the external natural and artificial world provides the "elbow room" necessary to build a free deliberative agent, and that, unfortunately, one of the main obstacles to free choice (and thus to making free will effective) is not only the lack of suitable predictive knowledge about the processes of the natural and artificial world, but also about the behavior of other human beings. From this perspective, other people are "natural things" whose behavior is a priori difficult to predict: how can we track human intentions? Consequently, human behavior poses a very different sort of challenge, with respect to the constitutive incompleteness or lack of trustworthiness of available knowledge about natural and artificial phenomena. Indeed, when we "morally" seek ownership of our own destinies – in our case through religion – we expect to be able to reach objectives through consciousness, free will and intentionality, thus undertaking a stand for the "best". We can only obtain the desired results if we can count on some consistency and predictability in the behavior of other human agents. If, in an attempt to "author" my destiny, I consider *merit* as a way to achieve a desired position, I must be able to assume that other human beings of my collectivity value it similarly. I contend that many objectified entities like common and religious morality, moral philosophy, human and social sciences, and of course ethical knowledge, are clearly connected to our existing need to operate at our highest level of conscious activity, as is the case when we seek to exercise free will and to claim ownership of our destinies: religion is central in providing ethical knowledge appropriate to achieve this aim.

Amazed, the cognitive psychologist contends that "we find it enormously seductive to think of ourselves as having minds, and so we are drawn into an intuitive appreciation of our own conscious will" (Wegner, 2002, p. 26): I do not find it "seductive" simply to think that we have minds with conscious wills, as we are clearly seeing. How can I fruitfully employ my brain's free will mechanisms if I cannot trust other human beings? How can I work on a personal project or participate in

[27] (Magnani, 2007b).

a social project if not by relying on the commitment of other human agents? How may I "author" my life and reach my goals if I am unsure which actions to choose because I cannot be assured that others share my values and support my intentions? Religion, morality, moral knowledge and teaching enhance and permit free will because they impose order on the randomness of human behaviors, giving people a better chance of owning their destinies. In the perspective of the relationship between morality and violence it has to be said that recent research[28] has provided evidence that "lack of control" increases illusory pattern perception (for example images in noise, illusory correlations in stock market information, perception of conspiracies, and development of superstitions), thus increasing chances of moral conflict and possible subsequent violence.

There are many human actions that affect others' free will and ownership of destiny; among them, as recently observed by some scholars, is the practice of gossiping.[29] These authors affirm that the practice is not just an exchange of information about absent people, which can of course be a form of indirect aggression: it is also, they contend, a form of sociable interaction[30] processing others as "moral characters" (Yerkovich, 1977). The narratives created by gossiping become a possible source of shared knowledge about evaluative categories concerning (moral) ways of acting and interacting. Gossip need not be evaluative, but it is "moral" insofar as it describes behaviors and presents them as interesting and salient and, consequently, as potentially or *de facto* sanctionable. Gossiping could play an important role in morally – more or less violently – "policing free riders" (Dunbar, 2004), that is, those who enjoy the benefits of sociality but refuse to pay their share of its costs. Commenting on the behavior of such people, or casting aspersions on their character, helps us to control their potentially destructive effect on societies.

I have contended above that moral practices protect the ownership of our destinies because ethics renders human behavior more predictable, and when we can count on shared values in dealing with other "moral" human beings, we can better project our future. Consequently gossip helps safeguard the ownership of our destinies as it constantly shapes our narrative constructions of morality: empirical data have shown, for example, that gossip works as a form of low-cost (moral) social cognition that conveys valuable information about culture and society. The act of gossiping can allow us to recognize that others are at risk of exploitation by moral free riders even though we ourselves are not (Dunbar, 2004, pp. 106–109). Kant said that the "Kingdom of Ends" – that is, the moral world – "is a practical Idea used to bring into existence what does not exist but can be made actual by our conduct – and indeed to bring it into existence in conformity with this Idea" (1964a, p. 104).

[28] (Whitson and Galinsky, 2008).

[29] I have already mentioned it in chapter three, subsection 3.2.2.

[30] In the previous chapters of this book I often emphasized the role of gossip. Dunbar (2004) originally gives scientific cognitive dignity to gossip explaining it in the framework of the so-called "social brain hypothesis". Posited in the late 1980s, this hypothesis contends that the relatively large brains of human beings and other primates reflect the computational demands of complex social systems and not just the need to process information of ecological relevance. See also the footnote at page 87.

Hence, the kingdom of ends is a kingdom of possible free choices created by (and contingent upon) human beings, for it is only their reliability that make free will, and thus responsibility and freedom, possible. Dennett, when discussing the status of "self-made selves", makes the following comment: "Kant's famous claim in *Foundations of the Metaphysics of Morals* that the law we give ourselves does not bind us suggests that the selves we become in this process are not constrained by the law we promulgate because these selves are (partly) constituted by those very laws, partly created by a fiat that renders more articulate and definite something hitherto underdone or unformed" (1984, p. 90). Moreover, human aspects that are the underpinnings of the kingdom of ends – such as religion, in our present case – must be successfully and appropriately activated; their being in good working order is a basic condition for exercising morality and allowing free will to become "good" will. As I already illustrated in the previous chapters, I am a sustainer of the importance of knowledge in constructing a new ethical commitment that embraces the idea of "respecting people as things", and I always stressed the fact that particular kinds of technologies can threaten the growth – and even the existence – of freedom, responsibility, and the ownership of our destinies.

How could we use our free will without the constraints of objectified morality, religions, laws, and institutions that impose regularity and predictability on human behavior, and that, in turn, bolster people's trustworthiness? In this sense, we are responsible for our own free will (and, therefore, our freedom) because its existence and its perpetuity seem not to be an inalienable given, but they depend on our intellectual and practical choices about knowledge, religions, scientific and moral institutions and the related techniques and technologies, and on their use in everyday settings, work environments, education, communication, and economic life. For example, environmental imperatives are matters of principle that cannot be economically bargained away because they represent a kind of paradox of liberalism. Indeed, in matters of conservation, one could maintain that neutrality is necessary to preserve the rights of the individuals involved, but this notion is obviously outweighed by the fact that the freedom to destroy natural goods and things today will, paradoxically, inhibit freedom in the future, when people will have as a result fewer options when choosing among competing ideas of the good life.[31]

In the following subsection I will illustrate the moral/social functions of religions and their effect in maintaining various levels of cultural stability in a human collective.

[31] A naturalistic and interesting approach, which is incompatible with the view (I have described in this section) that morality basically derives from religious teaching, is contended by Thagard (2010), who focuses on the importance of scientific knowledge. The whole treatment is in tune with my emphasis on "knowledge as duty" in morality. He proposes a theory of ethical intuition derived by a neural account of emotional consciousness: scientific evidence informed by philosophy provides a better source of knowledge than religious faith or philosophical "pure reason" to favor happiness, morality, and living meaningful lives.

6.2.3 Religion Is Easy: Religion Is Violent

Some cognitive scientists, for example Norenzayan (2006), hypothesize that cultural narratives such as myths and folktales are more likely to achieve cultural stability (that is they are culturally "selected") if they are characterized by a *minimal* counter-intuitive cognitive structure which includes clear intuitive concepts combined with a minority of counterintuitive ones. They are easy to be recalled, favor the greatest number of inferences with the least cognitive effort and sometimes they afford psychological relief, as in the case of the narratives informed by supernatural beliefs: in sum they are attention arresting and cognitively rich.

Cognitive research basically stresses that religion is "easy". I have already illustrated at the beginning of this section 6.2 that religions can be scientifically seen as an adaptive natural phenomenon in a Darwinian or non Darwinian evolutionary perspective: in the first case some researchers (for example (Wilson, 2002a)) address their direct or indirect Darwinian adaptive function, others their merely negative overall maladaptive outcome (Dennett, 2006); in the second case many authors (for example (Atran, 2002), prefer to study religion taking advantage of the cognitive, semiotic, anthropological and historical "arsenal" of human sciences, and are more skeptical about the possibility of speaking of a Darwinian evolutionary explanation.

The seminal analysis given by Boyer (2001) looks for an explanation in cognitive terms of some basic accounts of religion such as the explanatory, comfort enhancing, moral, as well as the illusive cognitive characters. He provides an analysis of counter-intuitive religious concepts, which are seen as parasitic upon intuitive ontology present in normal human brain, so that they are easy to acquire and communicate. In this innovative book religious practices are framed in terms of 90's cognitive neuroscience research in the "modularity of mind". For instance, Boyer contends that widespread beliefs in "supernatural agents" (e.g., gods, ancestors, spirits, and witches) result from the same operations involved in agency detection, evolutionarily rooted both in many animals and humans: in humans these beliefs lead to the intuitive modular process of assuming intervention by conscious agents, regardless of whether they are actually present.[32] Atran (2002) further deepens the above perspective on religion by analyzing basic human folkmechanics (tracking objects and movements), folkbiology (categorizing and reasoning about living things), and folkpsychology (attributing to others intentions, beliefs, desires, and "minds"). He also highlights and investigates their role in explaining and depicting the mental origin of bodiless supernatural agents, their stability across cultures, the birth of beliefs and their transmission and optimality for memory, the role of emotions (especially anxieties), metarepresentations, and deceptive capacities in the rise of "moral" religious commitments. It is important to note that the author is convinced that religions

[32] Barrett (2009) calls "intuitive inference systems" those elementary mental tools – such as naïve physics, naïve biology, agency detection device, theory of mind – which are able to generate basic human beliefs, and which can in turn provide the condition of possibility of those *minimally counter-intuitive* religious beliefs, so fruitful in their moral and social problem solving capacities and so easily communicable.

are "not" adaptations in any Darwinian sense, even if they conform, so to say, to an evolutionary landscape.

In a further analysis of the basic cognitive insights about religion I have just illustrated, Bulbulia (2009) notes that it is the link between religious beliefs in norm-supporting gods and "emotions" that provides a good advantage to individuals facing cooperation dilemmas. He also tries to explain how religious people are able to experience superhuman agents and to use them as justification of morality while remaining otherwise functionally engaged with distal realities – ecological and social. His conclusion is that religion is a kind of "mental time-travel", one that relies on specially contrived fictions and encapsulated self-deceptions over their reality:

> The most important and obvious difference is that religious conviction is not self-consciously represented as fiction. It is self consciously represented as true. [...] So religious commitments must be surrounded by cognitive firewalls similar to those that prevent our counterfactual and fictional musings from collapsing into a split-minded failure to distinguish actual from non-actual representations. [...] We must consciously believe but we must also unconsciously not believe that religious representations are true. Needless to say, this division of a religious mind imposes strict demands on the metarepresentational system that controls and isolates religious fictions (pp. 62–65).[33]

Furthermore, it is through fictional religious counterfactual representations of sacred agents that human beings are able to manipulate specific representations for special consideration off-line, for example to represent non-actual but possible scenarios. Reality is conceived as god-infested through a conviction that resorts to a kind of confabulation (Hirstein, 2005), which in turn permits to overcome the causes of their judgments, emotions, and behaviors. Through motivation and moralistic confabulation humans easily fulfill their urgent and continuous needs, thus favoring that ownership of their destinies I have illustrated in the previous section: "In expressing our personal pasts we invent reasons and fill missing gaps, supplying a coherence and control to our autobiographical selves we do not actually possess. [...] The experience of some religious agents seems to be confabulatory in the present tense" (p. 68).

McKim (2008, p. 389) contends that religions exhibit what he calls an "extremely rich ambiguity", given the fact there is an abundant evidence in history and in human experience that is relevant to them. Ambiguity does not jeopardize the prosocial function of religions:

> People may also feel a sense of obligation to be faithful to their religious community, or to their tradition, or to the ways of their ancestors, and so forth, and it may seem to them that they have every reason to take these feelings seriously. They may have a strong sense that doing so is extremely important. They may also be aware of themselves as members of a particular historical community of like-minded individuals, whose way of life appears to be valuable and worthy of preservation. This may in fact be

[33] Of course the "division" of the religious mind is not psychopathologic, for example schizophrenic, because the religious experience is *encapsulated* and so the division does not dangerously affect other mental functions.

the community in which they have acquired their evaluative criteria and their outlook on life, so that they cherish it on that account. Membership in it may be partially constitutive of who they are. And all of this can be so even if there is rich, or extremely rich, ambiguity, and even when this is acknowledged to be so.

Other researchers, who still approach religion as a natural phenomenon, particularly stress its prosociality function. For example, Norenzayan and Shariff (2008, p. 62), studying the prosocial effect of religions, also acknowledge that

> [...] although religions continue to be powerful facilitators of prosociality in large groups, they are not the only ones. The cultural spread of reliable secular institutions, such as courts, policing authorities, and effective contract-enforcing mechanisms, although historically recent, has changed the course of human prosociality. Consequently, active members of modern secular organizations are at least as likely to report donating to charity as active members of religious ones. Supporting this conclusion, experimentally induced reminders of secular moral authority had as much effect on generous behavior in an economic game as reminders of God, and there are many examples of modern, large, cooperative, and not very religious societies (such as those in Western and Northern Europe), that, nonetheless, retain a great degree of intragroup trust and cooperation.[34]

Finally an evolutionary account on beliefs in supernatural agency is provided by the so called Error Management Theory (EMT). EMT is a theory – deriving from signaling theory – according to which certain errors (including, for instance, biases, misbeliefs etc.) have been selected during evolution insofar as they allowed humans to avoid *worse* kinds of errors. Usually, such errors appear whenever there is an asymmetry between the costs related to false positive and false negative. To make a very simple example, it is better to mistake a stick for a snake (false positive) than the other way around, a snake for a stick (false negative). Interestingly, Johnson (2009) applied EMT to religion and the idea of supernatural agents, claiming that God might be considered as an error that, however, would allow humans to avoid worse errors related to punishment and the evolution of cooperation. More precisely, he claimed that belief in a supernatural agent would make humans feel constantly observed, so that they would consequently tend to desist from any behavior that could be sanctioned and punished by the deity (but actually by the group). Such a belief, on the one hand, would constrain one's freedom; on the other hand, it

[34] Also the beneficial effect of meditation on mental health has been studied taking advantage of psychological and neurological experiments: it seems that meditation affects neuroplasticity and thicker brain regions associated with attention, *interoception* (i.e. the sense of the physiological condition of the body) and sensory processing, like the prefrontal cortex and the right anterior *insula*. Cf. for example (Mohandas, 2008). Recent, more skeptical results are illustrated in a research by (Harris et al., 2009), that takes advantage of better nonreligious control conditions with respect to the previous ones adopted in other experimental studies on this topic. They refer that even if both religious and nonreligious beliefs "differently" engage broad regions of the frontal, parietal and medial temporal lobes, the differences between belief and disbelief appear to be independent from the processing of the lobe.

would permit him to avoid infringing social norms, especially when the probability of being caught and punished by the group is high.[35]

Given the subject of this book it is also important to mention other research, directly addressing the violent fruit of religious commitment, especially providing sociological and historical data. Scott-Appleby (2000) richly describes various historical and cultural conditions during wars in which religious actors choose the path of violence (providing religious "arguments" or "justification", for example linked to ethnicity and nationalism), while others seek justice through non violent means and works of reconciliation and forgiveness, thanks to educators, advocates, intercessors, reconcilers, scholars, NGOs, media, and inter-religious and ecumenical organizations. Stressing the ambivalence of the sacred, which parallels, from the socio-historical point of view, the philosophical problem of religion and its pervertibility (that I have illustrated earlier in philosophical terms), Scott Appleby for example recognizes that suicide bombers are no less "religious" than the persons who renounce violence and give themselves entirely to compassion and healing. Thus, if a suicide bomber is motivated by this power of the sacred – in his interpretation and understanding – and considers that act to be an expression of devotion to the sacred, then it is a religious act.

Juergensmeyer (2000) in turn explores the use of violence by marginal groups within five major religious traditions: Christianity (reconstruction theology and the Christian Identity movement, abortion clinic attacks, the Oklahoma City bombing, and Northern Ireland); Judaism (Baruch Goldstein, the assassination of Rabin, and Kahane); Islam (the World Trade Center bombing and Hamas suicide missions); Sikhism (the assassinations of Indira Gandhi and Beant Singh); and Buddhism (Aum Shinrikyo and the Tokyo subway gas attack). The theological justifications for violence and the bases for its authorization are illustrated, together with the analogies between the strategies of the performed violence, the religious rituals, and the problem of cosmic war, seen as a transcendent and absolute conflict.[36]

---

[35] EMT is also relevant in discussing the rationale of the so-called fallacies I described in chapter three. For instance, from an intellectual point of view "hasty generalizations" – just to mention one of the most well-known examples – are considered fallacies or errors, as they lack specific formal features belonging to more sophisticated forms of reasoning; for example, they are not truth-preserving and they usually rely on samples that are statistically meaningless. However, hasty generalizations are errors in reasoning that might turn out to be successful for the person who actually employs them. For example, a toddler who got his finger burnt on the stove might infer that, from that moment on, it is better to keep his hands off the stove. In this sense, hasty generalizations can be considered – from an eco-cognitive perspective – a resource that enters in a sort of human survival kit (Magnani, 2009, chapter seven).

[36] Juergensmeyer concludes *Terror in the Mind of God* with some suggestions for the future of religious violence that are certainly controversial, arguing that "the cure for religious violence may ultimately lie in a renewed appreciation for religion itself" and in a new acknowledgement of religion in public life. Hall (2009). proposes an exploratory typology to sociologically characterize the range of "cultural logics" that support the possibility of religious violence. He discusses the importance of such factors as nationalism, colonialism, the presence of religious regimes, interreligious competition, and establishment repression of countercultural religious movements.

Selengut (2008) as well systematically documents how religion should occupy a central role in understanding many forms of violent acts and convincingly argues that its importance should not be routinely dismissed or ignored. He outlines many convincing arguments and examples that further jeopardize the image of religion as a simple pacifist institution; Selengut argues, giving plenty of examples, that religious beliefs and institutions have a darker side which triggers violence – or at least works as a dominant precursor allowing violence to intensify or burst more fully. Finally, in *The Tenacity of Unreasonable Beliefs* Schimmel (2008) analyzes fundamentalist versions of Judaism, Christianity and Islam, illustrating their respective claims about the direct divine revelation and origins of their sacred scriptures. Adopting a rational perspective, the author shows how claims that some religious people make for some of their beliefs – and the theological concepts they endorse and invoke – do not meet the criteria of coherence and consistency.[37]

It is not my intention to undervalue all the precious and sophisticated results of the cognitive research I have just illustrated. My main point here is to stress that when cognitive oriented philosophers and scientists interestingly stress the moral and violent nature of religion, basically collocating it in a historical-sociological or naturalistic evolutionary framework, they unhappily result trivial from a philosophical point of view, because – as I already suggested – the specificity of religion simply disappears. Philosophically, religions can result violent just because they simply "originally" and "fundamentally" address the problem of overcoming violence/evil, as Derrida's interplay *á Dieu/adieu* clearly illustrates.

These researchers mainly stress the moral and/or violent aspects of religion, which certainly are astonishing – so to say – from an empirical point of view, but they also miss the philosophical point: the core aspect of all religions is that the acts of faith contain in themselves – that is in their empirical and historical actualization – their own pervertibility:

> The idolatry of the ontotheological is both indispensable and perverted, a closure disrupted from within, demanding its own correction in a rhythm of alternation between reason and history, the general and the singular, that keeps each perpetually off balance, thereby empowering critical leverage and resistance as an absolute task. This gives deconstruction an ethical-religious heft, generating space for the singular act of testimony, that marks the capacity to risk more than "the indifference of indecision or, worse still, the complacency of good conscience" (de Vries, 2002, p. 322).

Thus, to summarize, it is not that "neither philosophy nor theology has been able to adequately reconcile these two aspects of religion [morality and violence]" (Teehan, 2009, p. 250), rather, the paradox is at the roots of religion, and does not undermine its value and specificity. Either dismissing one side of the paradox 1) religion is just peaceful, people who make violence on behalf of religion are psychopaths, etc., or the other 2) religion is just violence, it is a theoretically wasteful and philosophically empty task. Furthermore, these dismissals are not devoid of

---

[37] Goldstein (2006) provides a rich collection of Marxist-oriented studies in sociology of religion, also addressing the analysis of the so-called "redemptive violence".

aggressiveness, in the first case "against violence", which is disguised in its "intellectual" dignity, reasons and roots, and in the second case "against religion", which is reduced to something merely violent, or maladaptive,[38] or uneconomic, thus forgetting its founding "commitment to the best".

The philosophical framework I have illustrated above in this section also sheds light on the following complaint: the idea that religion has a dangerous tendency to promote violence (the so-called "myth of religious violence") is part of a conventional wisdom of Western societies, and it also seems to underlie many of our institutions and policies, from limitations to the public role of religion to the anti-religious efforts to promote liberal democracy in the Middle East. For example Cavanaugh (2009) contends that religions can become victims of that myth, because what counts as religious or secular in any given context is a function of political configurations of power. Even if there is much truth in this observation, the author, unfortunately and candidly thinking that there is a transhistorical and transcultural concept of religion which depicts religion as non-rational and prone to violence, seems to contend that religion could instead be something uncontaminated by violence. It is just the bad "social context" which orients religion to violence. I think this is a wrong conclusion, which disregards the philosophical efforts and results I am illustrating in this chapter. Absolving religion from its relationship with violence (and so with morality) simply sterilizes its main concern.

I do however agree with the author and his opinion that various institutions and policies of the liberal and tolerant Western societies "violently" aliment and exploit the "myth of religious violence" for example to legitimate excessive neocolonial violence against non-Western *others*, particularly as far as the Muslim world is concerned. If the "myth of religious violence" is the assumption there is a notable difference between what is classified as religion – Judaism, Christianity, Islam and so on – and what is classified as a secular system – marxism, capitalism, humanism and so on – and the former is thus inherently and *a priori* considered more violent than the latter, of course the author is right. Similarly, it is certain that very hypocritical liberal States demand their citizens to *sacrifice*, even unto their death, reminding them that violence in the name of the nation is necessary and honorable and worthy of medals and grandiose funerals (*Dulce et decorum est pro patria mori* still echoes today), while those same liberal States target violence in the name of religion as something to be feared and labeled as "fanaticism". Unfortunately, this acceptable consideration also tends to undervalue the point I have illustrated earlier at the beginning of section 6.1: the constitutive link between religion and violence shows that to religiously – morally – "take a stand for the best" contaminates philosophical and ethical thinking and for this reason it manifests itself in modern secular ideas, such as Enlightenment, democracy, cosmopolitanism, and tolerance. However, I say,

[38] Cf. the critique of religion as a "cognitive niche" (cf. also this book chapter four, section 4.1.1), supported by the discussion addressed by Bulbulia (2008) to the Dennett's work (2006). Dennett argues that religious ideas endure because they are adapted for their own survival, not ours: they are just good replicators. Sharing his evolutionary picture of religion, Bulbulia tries to contrast what he calls Dennett's cultural maladaptivism.

why the violence of secular institutions should have to attenuate or to absolve the violence of religions, and vice-versa?[39]

It is from the philosophical perspective I have endorsed in the previous sections that we can analogously criticize research which addresses the problem of the "biological" root of violence and morality, such as in the case of the recent (Smith, 2007), concerned with the origins of war. Smith contends that both our warlike nature and our aversion to war are part of human nature. To support this hypothesis, Smith illustrates the behaviors of various animals, from ants to wolves to primates, with special attention to chimpanzees, our closest living (evolutionary) relatives. Other covered topics range from human beings' socio-cultural evolution to bacterial studies, as the author also addresses the development of body-and-mind systems (evolutionary biology and psychology), to the aim of showing how we have always been conditioned by biological factors, for war, and against wars. It is important to stress that the emphasis on the biological roots of violence/war does not have to involve an endorsement of their ineluctability, as the author himself notes:

> Lesser beings blindly, rigidly, and mechanically act out their biological programming, whereas we humans *choose* how to live. This supposedly makes us the authors of our destinies, and this opens up a chasm between the human world and the realm of nature. Free will makes us morally responsible for our actions, but explaining human behavior biologically robs us of our responsibility and reduces us to the stature of mere animals. [...] Human beings wage war because it is in our nature to do so, and saying that war is just a matter of choice without taking into account how our choices grow in the rich soil of human nature is a recipe for confusion (p. 26).

A similar consideration has to be applied to the philosophical awareness of the paradox about faith and its pervertibility. That awareness does not exclude hope and the openness to the best, but rather offers a rich knowledge on how hope itself is very very often easily, candidly, and unawarely betrayed and perverted.

6.3 From Magical Thinking to Religion: Bubbles and Cognitive Firewalls

Further knowledge on religion as a natural phenomenon can be gained taking advantage of two constitutive architectures of human cognition, the epistemic bubble and its moral counterpart.[40] Many human phenomena can indeed be framed within those two concepts: let us consider here the case of religion.

Religion is a complex object of investigation because it consists of two separate parts: as a matter of fact, any religion contemplates a core of ontological beliefs that is coupled to a set of social and moral implications. It is very important that this be

[39] The complicated problem of religious pluralism in democratic theory and practice is analyzed in the recent (Mookherjee, 2011): various approaches about a political theory of modus vivendi to religious conflict are proposed and analyzed.

[40] Cf. this book, section 3.2.2 of chapter three (Woods, 2005).

always kept in mind as those two faces can be analyzed separately, but always considering religion as a whole. Sociobiologists, group-selection scientists, and evolutionary psychologists, for instance, provide us with brilliant insights about the social implications of religion,[41] describing it as a series of fortunate adaptations, but they fail to stress the ontological relevance of religious beliefs. A religion that does not offer positive ontological statements about reality and who/what populates it is not a religion, but more of a skeptical philosophy, or a mere atheistic morality. Of course, religions provide us with a set of instructions regulating our behavior and informing our moral judgements, but this is derived from the commitment to other ontological beliefs.[42] Wilson's churches-as-organisms hypothesis is meaningful, but many groups can be told to behave as super-organisms: modern armies, for instance. Still, there is some constitutive difference between the group of believers in Jesus Christ, Dead and Risen, and those who endorse Communism, Liberalism, or the fans of any rockstar or book writer.

Furthermore, the label "religion" indicates a phenomenon not only complex in its structure, but also in its diachronic development. They all go under the name of religions, but a great shift occurs between the beliefs of our ancestors in the spirits of Nature, the faith in the Egyptian gods, between the religion of ancient Israel and the cults to the Greek pantheon. Not to mention the difference with contemporary post-secularized Christianity. Each particular religion contemplates different social and ontological commitments (from which different moralities descend), and one cannot necessarily individuate a general trend as far as those commitments are concerned. We will try, though, to separate the phenomenon into two major areas: properly called religion and its ancestor, magical thinking.

We can shed further light on this subject by taking into analysis the formation and the crystallization of beliefs concerning what is supernatural. To make a case, North and South American native populations would experience a cognitive irritation concerning thunderstorms: this is not to say that American Indians were meteorologists, as it is not to say that we are detectives just because, if we hear something falling on the ground, we start wondering who dropped that item. Our basic cognitive unreflective beliefs about the world (*folkphysics*) tell us that an effect must always have a cause. They were shaped by common cognitive past experience with natural objects and everyday events. To our mind-brain system, cause means an agent doing something. If we experience an effect in our ecology, then an agent must have caused it. The bigger the effect, the more powerful the agent *must* be. It is important to stress that these universalizing inferences from the cause-effect situation do have an abductive nature (that is they are referred to our capacity to guess hypotheses): if any cause has a discrete effect, then anything perceived as a discrete effect hints at a cause.

---

[41] See for instance (Wilson, 2002a).

[42] John Locke's famous claim about atheists, in his *Letter Concerning Toleration* (1689), can be read in this perspective: "[...] the bonds of human society [such as oaths, promises and commitments], can have no hold upon an atheist" insofar as he has no superior being, i.e. God, to guarantee for his behavior (2003, p. 246).

Thus, agent-detecting abductive processes would pick up some of the signs originated by the storms as if originated by an agent, and the resulting *super*agent would be elaborated as a Thunderbird, whose enormous wings stirred the wind and whose powerful cry was thunder itself, as the South American tradition held:

> According to the Ashluslay Indians of the Paraguayan Chaco, thunder and lightning are produced by birds who have long, sharp beaks and who carry fire under their wings. The thunder is their cry and lightning the fire which they drop over the earth. They were also the owners of fire and their enmity against mankind began after they had been deprived of that element (Métraux, 1944, p. 132).

Another Indian myth, North American this time, provides a very similar account of the Thunderbird. The cultural relationship between the two supernatural agents is more of an anthropologist's matter, and hence we will not deal with it; conversely, it is important to notice how in both cultures the same kind of signs – related to the weather and the phenomena of sky and air – are processed *as if* symptomatic of an avian super-agent.

> When it is stormy weather the Thunderbird flies through the skies. He is of monstrous size. When he opens and shuts his eyes, he makes the lightning. The flapping of his wings makes the thunder and the great winds. Thunderbird keeps his meat in a dark hole under the glacier at the foot of the Olympic glacial field. That is his home. When he moves about in there, he makes the noise of thunder there under the ice (Reagan and Walters, 1933, p. 320).

Both these accounts can be taken as fair examples of epistemic bubbles (Woods, 2005): they concern ontological expectations, placating a condition of doubt about the world's reality. We might also say that those particular individuals' commitment to their beliefs was complete. As already noted, to them the Thunderbird was as real as any other ordinary agent in their ecology.

Processes such as those described by anthropological and cognitive studies,[43] explain why certain supernatural inventions, especially those including minimally counterintuitive elements, were favored by cultural selection and managed to become extremely popular and resilient, thus acquiring a shared social dimension.[44] The bubble originated with the first unconsciously creative explanation seems prone to transcend the limits of individual cognition, and become transmittable to a whole community.

It could be hard to understand how such elements could be grounds for agreement and not primarily for argument. Curiously enough, the caricatured answer to this question is uttered by a fictional character in the popular animated series *The Simpsons*, as Chief Wiggum mocks the young, knowledgeable, scientific and therefore irritating Lisa by observing: "[...] Yes, everybody's heard of angels, but who's heard of a 'Neanderthal'!" (from the episode *Lisa the Skeptic*). The idea behind this sentence is that concepts representing supernatural agents are inferentially rich and simple, as opposed to the precision of scientific ones. For instance,

---

[43] Cf. above, section 6.2.

[44] Cf. (Norenzayan, 2006).

> A tree that hears the thoughts of human beings [is] an inferentially rich concept; upon learning of such a tree, one wonders what the tree knows about one's own thoughts and what the consequences of such knowledge might be. Such a concept invites *individual and collective* elaboration on the possible consequences of the existence of such a tree (Barrett and Lanman, 2008, p. 115, italics not in the original text).

This leads to their characteristic of displaying a reduced number of fixed features shared by believers, and a comparatively high number of features that may vary according to cultures, groups and single individuals; on the contrary, the meaning and denotation of scientific concepts is not arbitrary, and when it occasionally is, a speaker must be careful to specify which one she is referring to. Let us consider the very concept of "angel": the core of the belief represents a winged human being, whose role involves filling the gap between God and his Creation. This core is immediately activated in anyone who knows the meaning of angel: all the subsequent traits, such as their sex, corporality, the number of wings, the color of their skin and their garments, their size, and so on, may vary but this does not compromise the mutual understanding of the word and the concept "angel". This allows supernatural concepts to play the role of *cognitive wild cards*: to avoid futile proliferations and facilitate transmission based upon mutual recognition, different prerogatives can be attributed to the same agents, making them available for further explanations[45] at different and well-separated levels.

As a further example, once the aforementioned Thunderbird-concept is ready, it can be inserted in another explanation, which will make sense to those who are already comfortable with the original thunderbird concept. In the following case, the Thunderbird is used to make sense of the avalanche as well:

> Some men were hunting on Hoh mountains. They found a hole in the side of the mountain. They said, "This is Thunderbird's home. This is a supernatural place". Whenever they walked close to the hole they were very afraid. Thunderbird smelled the hunters whenever they approached his place. He did not want any person to come near his house. He caused ice to come out of the door of his house. Whenever people came near there, he rolled ice down the mountain side while he made the thunder noise. The ice would roll until it came to the level place where the rocks are. There it broke into a million pieces, and rattled as it rolled farther down the valley. Everyone was afraid of Thunderbird and of the thunder noise. No one would sleep near that place over night (Reagan and Walters, 1933, p. 320).

The problem with this argument is that it seems to fit for our ancestors, for the very *inventors* of the supernatural: their commitment to the supernatural was indeed an epistemic bubble, sustained by the "feeling of knowing" (Burton, 2008) that such beliefs allowed. Their embubblement displayed no sign of self-deception or confabulation, they believed intrinsically in those objects, not only in a contemplative dimension, but also in a practical one. They behaved *as if* such entities existed because, for them, they really existed.

---

[45] Angels and devils, for instance, may appear in cosmological narratives but also in considerations concerning the behavior of single human beings, to the point of becoming moral mediators for the development of children's moral character, if used in appropriate narratives.

What about the kind of religious belief we are accustomed to? For most of contemporary Western human beings, positive assumptions about reality proposed by religions are hardly as true as anything else. How could we indulge in sin (and use violence against our brothers) if we sincerely believe in an all-knowing and almighty being always watching over us? How come that one, though firmly convinced of the existence of Heaven and of her own righteousness, still prefers life over death? How come that religion, and its proper holy domain, is most often constrained to particular and well-determined dimensions? Bulbulia's studies (2009) shed a very interesting light on this matter, stressing how a great part of religious convictions can be described as a kind of split-belief which, we contend, might help characterizing religion as yet another kind of bubble.

As said, an analysis of religion cannot be complete if the ontological or the social aspect is excessively neglected. In a recent article[46] I attempted to maintain that the "invention" of a supernatural being showed how the inferential abductive basis for the generation of such concepts is constitutive of humans' brain-mind systems, and can hence provide good arguments for the universality (and *easiness*, as I just showed in the previous sections) of religion. Anyone can easily understand religious or magical discourses because everyone is prone to make such inferences, thanks to the way our brain seems to be wired as a result of the co-evolutionary pressures due to the interplay between ecological and cognitive niches.

Such universality proved to be a powerful element to favor the enhancement of social dynamics among hominids, which at least partially explains the persistence of religion over the centuries. This is not to say that religion *per se* is an adaptation, especially in the strict Darwinian meaning: given its complex nature and the impossibility to operate a reduction to only the social or ontological commitments, it is hard to label it as an adaptation. Furthermore, it is hard to consider a Darwinian adaptation a cognitive misjudgment that causes our brain to interpret ordinary natural signs as symptomatic of the presence of a super-agent.

If religious beliefs were really put this way, even sociobiologists would have it hard to present religion as a loosely adaptive trait: paraphrasing Scott Atran, we might say that it is highly unlikely for any population to augment the fitness or welfare of its individuals in an intergroup struggle by taking to be true what is materially false and to be false what is materially true. Joseph Bulbulia displays similar concerns:

> For naturalists, religious beliefs are at least partially unfounded. Agents attribute beliefs to gods; we have assumed the gods play no causal role in activating these beliefs. How could religious persons go so badly wrong? More puzzling: how can they go so badly wrong without injuring themselves, in the way schizophrenics do? For there is no evidence that adult religious agents are cognitively impaired, globally irrational, or immature. Unlike schizophrenics, religious agents flourish. Whatever one thinks about the sanity of religious belief, religious persons are sane (2009, p. 45).

No matter how benefiting to a group religion can be, the cost of a sincere commitment to religion's ontological prescriptions could be more than what one is willing

[46] (Bertolotti and Magnani, 2010).

to sustain. That is to say, if we consider religion as a mere cognitive bubble providing us with easy satisfactory explanations, the whole enterprise would look like a mass-suicide. Our mental flaw, from which the invention of the supernatural originated, would seem merely a cumbersome impairment. Nevertheless, if we take a look at reality, we can witness how religion is still persisting millennia after its origins, and religious human beings, though decreasing in number because of ever-growing secularization, seem rather healthy in body, mind and social life.

Bulbulia tries to provide us with the tools to solve this delicate issue, by pointing out how "[...] when religious conviction seeps into practical domains, an otherwise adaptive trait becomes damaging. Religious agents face an encapsulation constraint" (2009, p. 57): that is to say, the core of a religion is a self-limiting matter. Furthermore, the ability to fit supernatural concepts into narratives is a top achievement of our linguistic abilities: not only language affords decoupling and reference to things and agents that are not present at the moment, but it also allows different levels of fictional narratives.

Fictional narratives play a crucial role in human being's social interactions, because not only do they allow the sharing of moral templates[47] by means of parables (such as the one's in the Gospels, or the Gospels themselves), short stories, fables, etc.; they also provide a counterfactual (yet not necessarily counterintuitive) environment in which one can make predictions and experience emotions according to the same inferential patterns as in reality. As an example, let us think of when we read a compelling novel or watch a thrilling movie: we *know* it is fictional, still if the character is badly wounded we *predict* he might die and *grieve* as if it was happening to one of our beloveds.

Considering religion, we can witness how magical-supernatural concepts, once they become part of the cultural patrimony of a group, constitute its religion by crystallizing into a set of more or less fixed narratives. These narratives play a role in the society as far as they support norms and commitments – so that they are also related to conflicts, punishment, and violence – which partially explains the persistent success of religion.[48]

For religion to be effective, though, believers are expected to sincerely believe and commit themselves to the positive moral and ontological core of their *credo*. The problem is that, as mentioned several times, a positive commitment to ontological oddities such as supernatural beings, in spite of the social advantages, can be rather dangerous: as for this matter, Bulbulia claims that religious beliefs are characterized by a mental ®IMAGINE marker, thus illustrated:

®IMAGINE [Zugroo is Lord Creator]
practical inference: NOT TRUE [Zugroo is Lord Creator]
workspace inference: CERTAINLY TRUE [Zugroo is Lord Creator] (2009, p. 63).

[47] I have described the concept of "moral template" of behavior in (Magnani, 2007b).

[48] On narratives and religion cf. the innovative contributions contained in (Boyer, 2001; Wilson, 2002a; Atran, 2002; Bulbulia, 2009).

The ®IMAGINE marker allows us to introduce a further kind of cognitive embubblement, that is the "religious" bubble. If the simple cognitive bubble had an illusional dimension, the religious bubble has a necessary self-deceptive dimension, which acts to reinforce the ontological commitment onto the belief itself, but simultaneously it *defuses the practical inferences* about our ecology, *though enhancing the moral ones*. To clarify this with an example, let us consider Europe in the middle ages: religious belief in demons, succubi and similar creatures was very widespread: but if a party had to set out for a journey at night, they would not (only) protect themselves against demons but rather against bandits and highwaymen. The latter could nevertheless be considered and feared *as if* they were evil spirits and daemons (many wrongdoers acquired in fact a supernatural *aura* in old legends!) Nevertheless, their belief in demons was utterly sincere, as much as their fear, and – among many other entities – demons would *warrant* for their moral behavior and that of their fellows.

As we said, objects of belief in the supernatural display an inferential richness, so they can be easily shared within a community. Though, the community cannot be a warrant for the religious bubble as much as it is for the moral bubble: moral bubbles rest entirely on a coalition dimension, while the *cognitive firewalls* preventing religious beliefs from being dangerous seem to arise from an individual base.[49] Such ideas suggest that if we want to benefit from religion's powerful social possibilities, we have to cope with the trade-off between managing a functional society and the costs of dealing with a distorted vision of reality that informs our expectations about the world.

In the next section I will contend how an active solution to this conflict resulted in the potential *confinement* of the supernatural to the dimension of the *holy*. Thus, religion maximized the benefits (for the group) of (individual) commitment to supernatural agents reducing the ecological risks caused by an adoption of magical thinking as a "strong" cognitive bubble. The confinement of what is *holy* to a well-defined dimension of social life, a delineation that is conveyed through education as well, is reflected in the rise of specific cognitive firewalls that prevent magical-supernatural concepts and inferences to flood the areas of our mind-brain system dedicated to ecological survival.

As a matter of fact, different people display different strengths of cognitive firewalls, and even modern religions may comprehend believers that still belong to a magic dimension – that is, who are virtually "devoid" of any cognitive firewall. Let us consider suicide bombers engaged in a *Jihad*, or holy war. It must be pointed out how, even if such individuals remain in an ancestral magic and violent dimension as regards themselves (i.e. they really believe that by choosing martyrdom they will be granted an afterlife of perennial bliss, surrounded by virgins eager to satisfy

[49] The notion of "cognitive firewall" belongs to a computational representation of the mind: Cosmides and Tooby define them as "computational methods for managing the threat posed by false, unreliable, obsolete, out-of-context, deceptive, or scope-violating representations. Cognitive firewalls – systems of representational quarantine and error correction – have evolved for this purpose. They are, no doubt, far from perfect. But without them, our form of mentality would not be possible" (2000, p. 105).

their needs and so on), for the whole religious-group they act as a powerful costly-commitment, further enforcing the beliefs of the rest of the groups, but without affecting their individual way of coping with ecological material reality.

6.4 Sacred and Sacrifice: The Violent Boundaries of Magic

As I contended in the previous section, religion could be seen as a cultural barrier developed over the ages as a protection against the overflowing of magic in every domain of natural and psychic life. That statement, though, begs the *why* question. The solution I provided before is not exhaustive. We shall hereby follow our intuitions about the necessity to fence the magic in a religious framework by further developing them with respect to the intrinsic bounds between religion, magic and violence.

In a recent article,[50] in line with a major trend in evolutionary psychology and paleoanthropology, it is contended that the invention of supernatural beliefs is connatural with human beings, yet the supernatural immediately transcends the act of its production, by becoming something *other*. This region can be defined as supernatural, magic or divine and what is common to all these labels is the intrinsic *otherness* compared to the natural domains. Can the violent dimension of religion originate from this fundamental ontological rupture?

Religion is commonly perceived as having something to do with violence, namely with sacrifice. Religious violence scares us more than many other kind of violence. No matter how trivial this may seem, religion is considered violent because it has often made people *hurt* other people, physically or psychologically. The very word "religion" makes us think about altars, prayers, priests and temples, but also about sacrifices, crusades, inquisition, *jihad* and *mujaheddins*, penance, sects and other elements all conveying a strong connotation of violence. Here we are not committing a hasty and insulting generalization: it might be claimed, on the contrary, that many forms of religion and spirituality, for instance Buddhism,[51] do not cause suffering *per se*. As I will try to show in the following section, the peaceful nature of oriental religions might be a wishful myth, as opposed to the violence experienced by Westerners regarding their own religious history, or ultimately a dangerous misunderstanding generated by the juxtaposition of the oriental framework with the Western-Christian one: as observed by Žižek, (2009) commenting on Buddhism, a blindly benevolent and peaceful attitude towards the world could be phenomenologically impossible to distinguish from universal indifference.

I hereby suggest that violence and hurting others should be considered in the wider sense of *limitation* of potentialities, of freedom to act differently with no foreseeable harmful natural consequence. From this perspective, self imposed

[50] (Bertolotti and Magnani, 2010).

[51] We do not feel it proper to adopt the canonical objection of pop-theology on the matter, which typically consists in denying that Buddhism and other Asian religions are actually religions, but rather philosophies: Buddha, *karma* etc. are clearly supernatural concepts, which allow us to label as religions those doctrines that deal with them.

vegetarianism is a mitigated form of violence, for instance, as well as forbidding oneself from killing a cow when it would provide a satisfying and protein-rich meal: vows and the celebration of chastity are embraced by the same category. The Jewish observance of the *Sabbath* can be considered as the "negative sacrifice" par excellence. To further stress my point, I are not referring to the moral and possibly violent consequences following from the intentional or unintentional infringement of such religious rules; we are referring to the very acts of abstention from things that would naturally be in our power to do.

Religious violence is overwhelming because, being an expression of the magic, it occupies a zone beyond the reach of our possibility of understanding. To quote De Vries' (2002, p. 288) rich imagery, the otherness of magic entails a necessarily violent dimension insofar as the ontology of the natural world becomes a "hauntology"[52] where, in the tradition of Derrida's writings, the object of belief is so intrinsically *other* that it becomes the Other of the Other, lest it loses the Alterity that defines it.

Magic is potentially highly violent because of this shift which allows the degeneration of wishful thinking: *ex magico, sequitur quodlibet*. Magical correlations, by definition, do not need to comply with any procedure of empirical causation. Magic, and consequently religion, allows and compels us to a "leap of faith, [...] and in the act of taking a stand for the best we open ourselves to violence, to the worst".[53] Divinity, magic, absolute alterity etc. embody a dual significance of violence, and it is both a cognitive violence, as far as it halts and makes powerless our usual comprehension heuristics, and a physical, pragmatic violence because of the consequences it triggers in our behavior with respect to other beings in our ontology.

The whole idea underlying my analysis is that religion, as a complex cultural phenomenon, developed mainly as a cognitive fence around *magic*, to prevent it from overflowing into other domains than the peculiar domain of magic itself. Religious doctrines, rituals, symbols, artifacts play the role of setting a boundary before what is *Other*, *divine*, *magic*, and *what is not*. More precisely, the element allowing this juxtaposition and contraposition between something understandable – such as culture – is the *sacred*, the holy.

The holy can be said to partake of both the cultural, intelligible dimension and the magical domain. Holiness may concern a wide range of objects, and can be more or less reified: a physical item can be sacred, but also a place, a person, an image, a word, a phrase, a ritual. Holiness, in fact, seems to embody not so much an ontological hiatus but rather a cognitive, gnoseological one: in our natural ecologies, sacredness is a self-transcending property as far as intellection is concerned.

Holiness almost always involves a subsequent *feeling of privations*: what has been instituted as holy usually loses its natural role, or keeps it but in a metaphorical sense. This is clear, for instance, as far as animals in religious settings are concerned.

---

[52] Cf. also the first section of this chapter. The word *hauntology* is a blend of the verb *to haunt* and the suffix *-ology*, originally coined by Derrida (1994): it aims at labeling the region of what *neither is nor is not*, the typical condition associated with the specter. It constitutes a kind of suspended, fluctuating and unintelligible region that *haunts* our usual ontologies.

[53] (Reynolds, 2003, p. 480).

> The animal nature of these creatures is striking and significant. At one and the same time, it points away from itself and is mingled with humanity and even with divinity. In fact, the animals in the procession either do not behave like animals or are not real animals at all. This characteristic animality suspended and reinterpreted is typical rather than peculiar when animals or images of animals appear in religious settings. Thus the religious significance of these animals does not lie primarily in their inherent animal nature but in that to which it gives added meaning. There is a synergetic effect between the actual animal and the being with which the animal is combined or connected be it a human or a god (Gilhus, 2006, p. 94).

That is because the institution of the sacred as *the* meaningful hiatus requires an act that replicates in a mimetic fashion a distinctive characteristic of the level it connects with, namely the absolute Other: this effective act is sacrifice. If divinity is the absolute other, which cannot be understood, sacrifice must be an act that transcends its own understanding, replicating the cognitive and physical violence entailed by the *gnoseo-onto-hauntological* (to use De Vries' terminology) region of magic.

René Girard illuminates this matter by reflecting upon the very etymology of the word *sacrifice*: it originates from the latin word *sacrificium*, which is in turn composed by the union of the verb *facere* and the attribute *sacer, sacrum*, that is, *to make something/someone holy*. But once again, the very word *sacer* has a violent connotation insofar as it recalls the act of ritual slaughtering (1977, section X, for instance). To sum this up, a sacrifice is the violent act by which something is made holy, that is partaking of yet another original violence. Sacrifice is oriented towards magic, but every time one directs himself towards magic, towards divinity, he also directs himself towards the violence intrinsically embedded within it.

I could condense this conception of sacrifice, and consequently of religion, borrowing from De Vries (2002, p. 287) his interesting neologism "perverformativity", a fusion of the words performance and perversion. The "pervertibility" of any sacrificial religious performance rests in its original constitution: turning to divinity is conceived as *good* in the performer's mind, yet this turn can happen only via a turn to violence. Again, we could be challenged on this assumption by proof of non violent or bloodless sacrifices but violence does not need to be addressed towards the victim of the sacrifice. Furthermore, it seems short-sighted to see only one victim in the sacrifice, that is, the creature being slaughtered. Let us consider a sacrifice consisting in the offering of the harvest's firstlings to the deity: do the firstlings suffer? Of course not, no great violence is perpetrated against fruit and vegetables, still a residual violence is committed by the very offerers against themselves, as they would have benefitted from eating those products. Several parts of the Torah prescribe strict rules about the consumption of sacrificed meat: even if a part of it would be actually consumed by the offering party, they would have gained greater benefit from it consuming it all by themselves. In the natural world, what would happen to an observant Jew should he eat pork meat? Would he get poisoned or struck by lightning at once? Surely not. What if he took care of his garden during Shabbat: would anything happen? Not in our natural, physical ontology.

If religions are about commitment to something supernatural, this commitment must have a cost because otherwise it would be virtually irrelevant, and there would be no commitment at all: there would be no difference between uttering "I believe that Jesus is the Christ" and "I think Casablanca is the best movie ever". That kind of commitment, which generates from different kinds of sacrifice and substantiates the border region of the sacred, should originally have been a *disruption* of cognitive habits determining our expectations about the world, but it might have become invisible because of its sedimentation in new habits.[54] We claim that sacrifices can be harmless because we think, for instance, about the Sunday mass. The violent and restrictive dimension of the Commandment that reads "Remember the Sabbath and keep it holy", is perceived not as much by the adults (to whom Sunday mass is a habit) or the elders (who may even consider the Mass as an escape from boredom) but rather by children: they understand very well that those fifty-something minutes could be used in another and probably more pleasant way, i. e. watching Sunday morning cartoons, playing with friends and siblings, etc.

Still, it is important to bear in mind that there is yet another kind of violence – more subtile and even easier to obliterate – present in every sacrificial scenario: that is, every sacrifice is a *sacrificium intellectus* insofar as it is a violence towards comprehension, in a mimetic fashion that reproduces the violent cognitive puzzlement induced by magic *per se*. Again, Derrida's (1999) figure of *à Dieu/adieu* offers a powerful insight as far as this aspect of sacrifice is concerned.

Any act of Faith, any kind of sacrifice, be it violent, bloodless or merely intellectual, implies saying *adieu* – farewell – to the very object of the act, because it is being dedicated, consecrated in this turn towards God, in the moment of the *à Dieu* – to, towards God. The violence of total privation (i. e. death or irreversible departure) of the object can be mitigated by a partial privation of some characteristics of the object, which becomes something else from what is was before. Yet, on a similar ground, the *adieu* moment concerns the executor as well, as long as she is saying farewell to the object and to her possibility of understanding: not only her cognitive habits are disrupted by the object becoming at once the same and something else, but also the very end of the process transcends the possibility of her intellection.

Sacrifice, as we contended before, originates in our natural ecology but tends toward the Other. This cognitive *adieu* is absolute, total: "[...] to say adieu, if only for an instant, to the ethical order of universal laws of human rights by responding to a singular responsibility toward an ab-solute other – for example, the other par excellence, God – implies sacrificing the virtual totality of all innumerable others" (de Vries, 2002, p. 159). The *à Dieu/adieu* figure then is not a bi-polar one: furthermore, not only the farewell can be interpreted in different, yet coherent, ways. The second pole is polysemic as well, since the French preposition "à" can be a locative and dative proposition, but can denote possession as well. French maintained the *dativus possesivus* from latin: *c'est à moi* means "it belongs to me", hence *à Dieu* signals what belongs to God. Sacrifice makes an object to partake of *what belongs to God*. But if it is up to God, it is *not* up to me anymore. The object does not belong to me, to my ontology, to the gnoseological regions I am empowered to navigate.

[54] About this cf. the collectivization of the moral bubble and self immunization to one's morality in section 5.3.1 of chapter five.

To recollect our thoughts, a sacrifice could ultimately be understood as the famous cuts in canvas performed by the Italian artist Lucio Fontana: it is a violent cut into our natural, intellectual and cognitive habits directed towards the magic and at the same time trying to confine the Otherness to its proper domain, and achieves this result by a mimetic reproduction of violence intrinsically contained in that Otherness, by bringing out and reproducing the hiatus between the ontology we are accustomed to and the *hauntology*, the sphere of the magic which is produced by our minds but violently transcends them at the very moment of its generation. Sacrifice and hence the Sacred, the Holy, play a liminal role by setting a cognitive and cultural framework around the region of magic: this frame behaves as a selective cognitive portal in both ways, as far as it allows human beings to direct their gestures towards magic and divinity, but at the same time prevents that very Other from flooding natural domains.

6.5 Overmorality and Wisdom: Morality or Indifference?

Before we get into the matter, I feel I should inform the reader that the argumentative discourse of this section is slightly different from many other parts of this book, insofar as the initial part reflects some extremely personal considerations about me as a human being and what I think to be my own *philosophy of life*: the thoughts I will expose are absolutely crucial and acted as the more or less explicit core that sparked the *need* to commit myself and try to write a philosophically comprehensive book about violence.

Justice (legal and/or moral) is historically and geographically variable, as everyone knows. I can say that, of course, I have my personal morality, which aims at minimizing both moral commitments and violent outcomes, but it is just an "aristocratic" (in the sense of intellectually aristocratic) sophisticated concern, which, even if shared with others, is far from the reach of uncultivated human beings. Moreover, I can just plausibly hypothesize that if I try to teach a kind of Stoic morality to those "other" non cultivated people they will perceive my teaching as something violent and incomprehensible (of course, this is the violent effect of any intellectual aristocracy). There are *preferable* moralities (for example for me and for other intellectual people). I prefer life imprisonment to the death penalty, because I see a weaker violence in it: yet, it is clear that the success of such a preference depends on the struggle in the objective life of groups and their cultures, and on the fact that the option I endorse is the established one in my legal framework (sadly, I often notice that the established habit is the most violent one). I am almost sure that many people here in Northern Italy, who actively militate in the "Lega Nord" (*Northern League*), basically embedded in a xenophobic state of the mind,[55] consider the death penalty to be less violent than life imprisonment (because – I guess – from their own inner "moral" perspective they think that killing a killer diminishes the global violence of a society and that it is the "just" punishment).

[55] Cf. above, section 5.3. of the previous chapter.

Is perceiving injustice "natural"? Obviously I do not think so. Perceiving injustice is very variable: do not you see how different the feeling of injustice is in different people? I like capitalistic social-democracies and I am very surprised to see how many people do not perceive at all the global injustice of this horrible (at least for me) neo-con/neo-liberal capitalism. So, perceiving injustices consists in a practice heavily affected by cultural and axiological factors: here in Italy there are people who lost their jobs or their private enterprise/business – and became poor (or poorer) – because of the neoliberal policies, but they still morally think neoliberalism is the best and do not perceive its atrocious outcomes like I do. I think they are not coherent and ignorant but they think I am a stupid dreamer and/or a horrible and verbally violent member of a presumptuous intellectual social-democratic élite! I like equality, but I do not have any reason to think that equality is something special. I am aware that, if we want to impose and establish equality, we have to perform what Walter Benjamin calls a law-making violent act (consider for example the French revolution): indeed many people and groups do not like equality, and do not think equality is a positive aspect of a strong morality, even now in the twenty first Century: they consequently feel constantly violated by the modern civil idea of "égalité". So, why think that equality is void of any relationship with violence? The egalitarian groups usually hate non-egalitarians and vice-versa, as I think they are both convinced they are dealing with a "pure/good morality" (which justifies any related violence, thanks to the moral bubble they are in).

As a philosopher – adopting a naturalistic perspective – what I want to avoid is to establish a final and stable truth about violence, that is a dogmatic and "locked" moral philosophical perspective about "what is violence?" Of course I also want to avoid answering questions like "how can we get rid of violence". Answering these questions *inside philosophy* seems to me the perpetration of a high degree of intellectual violence, disrespecting the *banality* and – so to say – *moral dignity* of human violence. However, the reader must not misunderstand me! When I say I want to provide a "moral dignity" to violence I am referring to the fact that we have to respect it, as a human behavior that cannot be neutralized with an abstract, narcissistic, emancipating, conceptual philosophical theory (too "low cost", from both the intellectual and emotional point of view). This would be a kind of violence, a merely abstract terminator machine, "written in more or less complicated books", which just fakes a perverse atmosphere of an almost empty moral "militancy". I just wanted to increase philosophical and cognitive knowledge on violence's multiple aspects, to show how violence is *de facto* intertwined with morality, and how much violence is hidden, and invisibly or unintentionally performed. In a few words, I think it is mandatory, in our times, to stress the other face of (presupposed) good things and beliefs. Gogol, for example, was perfectly aware of the fact that knowledge, inclination, and sensitivity to good is always inextricably bound to knowledge, inclination, and sensitivity to evil.

As an individual I have of course my own (evolving) morality, but – as I already said – I keep it as something aristocratic that I do not intend to "teach" to anyone. Here are some elements of what I consider to be my morality: 1) I do not like the *overmorality* (cf. below in this same subsection) that I witness everywhere; 2) in my behavior I always try to "lower emotions" , to avoid *a priori* conflicting situations where morally-dependent conflicts (and conflicts of other cultural perspectives) can arise; 3) I try to treat people according to their "nature", like Zeno of Citium says: "The goal of life is living in agreement with nature"; 4) I do not like revenge but I try to transform it, when it is possible of course – better to avoid revenge if it involves too violent an outcome – into a moderate, non retaliatory, didactic reaction (if this is not possible or feasible I give up and give in to revenge). As you can easily see this is not a morality in the common sense of the term (like my inner Catholic morality, which I learnt when I was a child, and anyway I still love and try to follow); it is something more personal and also characterized by meta-moral aspects (individual, customized to me and through my history), related to a possible good construction of myself.

As I hope I have made clear in this book so far, I am convinced that there is a spontaneous generation of violence through morality. As a further example, imagine that you are a good Stoic (as I would like to be), and you are *morally* intelligent because you prefer to prevent violence. Other people that entertain relationships with you – and that instead prefer and like the behavior you perceive as "violent" – can perceive "you" as violent! Moreover, we must not forget that people often "like" violence (emotionally, or, more rationally, because they are morally convinced of its "moral" function: punishment, purification, revenge, edification, etc.), and it is difficult to dissuade them. When we approach violent people to persuade them about the badness of their behavior, we are in turn very often considered as non-violently violent![56]

Let me address the problem of what I can call a "moral epistemology" (which comprehends the intrinsic "morality of sound reasoning" and is concerned with a somehow moral "commitment to the truth"), supposed to be clever in a pure way and able to foster good moral outcomes for everyone. The following is an example of how this morality can be a severe conflict-maker: you like logic and trust logical reasoning, and the role of empirical evidence, so you explain to a person what *modus ponens* is and its wonderful capacity to preserve truth, and therefore you aim at transferring what you consider to be pure logical information – that you candidly think devoid of any harmful potential. Your interlocutor can nevertheless feel violated maybe because she prefer magic and does not care about correct reasoning, which she instead regards as dangerous because too rational and "frigid"! The dramatically ridiculous point is that you were considering yourself so intellectually pure and non-violent, being so candidly embedded in your own moral bubble!

I have said above that the dimension of wisdom (Stoics, etc.) is a good morality but it is just aristocratic. An aspect of this preference of mine can be further explained by illustrating my ideas about what I call *overmorality*. I maintain that overmorality (that is the presence of too many moral values attributed to too many human features, things, event, and entities), is dangerous, because it furnishes too many opportunities to trigger more violence by promoting plenty of unresolvable conflicts. I recently realized that overmoralization is analogous of the problem of

[56] On this interplay violent/non-violent cf. also the subsection 4.5.1 "Nonviolent Moral Axiologies, Pacifism, and Violence", chapter four.

overcriminalization, when I found the book by Husak (2008): overcriminalization presents similar discontent with respect to overmoralization. For example, Husak contends that the state lacks a good reason to punish drug users, and that, thanks to overcriminalization, injustice (consequently) increases and it is pervasive throughout the criminal domain: the results of criminal justice in presence of overcriminalization are often perverse and "unjust", with the consequence of an exceptional and expensive quantity of people in prison. Provocatively, Husak contends that a *right not to be punished* should be implemented, given the fact that, like Jeremy Bentham already contended, any punishment *is* a violence. Why would this right deserve less protection than free speech, freedom of association, or liberty of conscience? Too many people are more or less violently legally punished because of the infringement of mere *mala prohibita*, and not because they also did *mala in se*. Thus, often those punishments *are not deserved*, not even as mere didactic examples to be presented to other humans. A description of the main assumptions indicated by Husak is simply given by Donoso (2010):

> Husak exposes with clarity how a system characterised by overcriminalisation puts at stake basic principles of the rule of law by making people unaware of what types of conduct are criminally proscribed; precluding them from having adequate notice of some of their legal obligations; and, ultimately, by undermining one of the main goals of the system of law, namely, to guide people's behaviour (p. 11). Overcriminalisation also breaks principles of legality by making the criminal law outsource from non-criminal branches of the law (p. 13), which runs the serious risk of making the criminal law even less intelligible for the layperson and making its limits dependent on the limits of spheres of law that are beyond its proper domain. Despite these and other reasons to be worried about overcriminalisation, Husak emphasises – guided by "the peculiar American penal context" (p. 14) – that the principal reason to be troubled by having too much criminal law is that it produces too much punishment. That is the most urgent source of injustice on which the book focuses.

Analogously, I think that an excess of morality coincides with an excess of punishment and conflict, and this explains my sympathy for Husak's ideas. Moreover, in the present book I have often stressed the fact that our era is characterized by a huge quantity of fragmentary, often contradictory, moral values and allegiances, that affect human behavior in confusing and conflicting ways. This complexity often makes people simply ignorant of *basic* moral rules which would be instead useful for their practical life in a community, to avoid potential violent conflicts. Indeed it can be contended that fragmented pieces of morality can corrupt and transform more basic and fruitful tenets. The Stoics always emphasized the need to limit the over-expansion of morality. They contended that humans could recognize that many of the values they attribute to events, behaviors, artifacts, and so on should be considered *indifferent*. On closer analysis, many things are indifferent and to take excessive moral care of them is wrong and pernicious:

> Since such things as health, wealth or reputation could not affect virtue, it made no difference to the wise man whether he had them or not, and he could not consider them good or evil. Zeno termed them all "indifferents", but he called such things as health and wealth "according to nature" and the opposite of these "contrary to nature".

> [...] But even if the virtue of the wise man was not affected by the loss of property, it was necessary for him to earn his own living and to support his family. Zeno called actions of this kind "duties", "acts of which a reasonable account can be given". A man could be virtuous in sickness or poverty but from the practical point of view he had to pay enough attention to health to be a good soldier in the defence of his country and enough attention to money to earn his own living (Reesor, 1951, 152–153).

Duties of this kind are of low or no moral value, seen in proportion to how they aid the natural instinct for self-preservation. Hence, from the point of view of this book it seems clear that

1. if we deprive "things" of their excessive moral value and reduce them to "preferable" or "not preferable" targets, which are related to merely "practical" duties, it is less likely they can trigger deep passion and unmanageable intra-personal and interpersonal conflicts, and we will certainly be less inclined to use them as a way of punishing ourselves and other human beings for not respecting our too many unquestionable moral commitments; less wrongdoing would help us, like Coeckelbergh (2010, p. 243) says, "[...] to set up institutions that prevent 'interminable generations of prisoners' and pay more attention to those who do good in spite of, and in response to, the tragic character of human action and human life";
2. unfortunately, to espouse the doctrine of the indifferents is still a strong "moral option", which can generate conflict with people who think those things you do not consider to be valuable, are instead worthy of being endowed with some positive moral values, and certainly not to be neglected from this point of view.

We have said that the disciples of the doctrine of the indifferents are less inclined to punish themselves and/or others because they do not endorse an overabundance of moral commitments. Analogously, it is less likely for them to experience resentment and envy. Furthermore, some basic discontent is present in the doctrine of the indifferents: Žižek (2009, p. 46) usefully warns us about the wisdom and innocence of attitude where the increasing of indifferents is at the same time a possible carrier of disengagement of empathy, taking advantage of a critique of the Buddhist ethics of solidarity with every living being, also involved in learning to be indifferent with respect to many thus "unjustified" moral commitments: "After all, what Buddhism offers as a solution is an universalized indifference – a learning of how to withdraw from too much empathy. This is why Buddhism can so easily turn into the very opposite of universal compassion: the advocacy of a ruthless military attitude, which is what the fate of Zen Buddhism amply demonstrates" (2009, p. 54).[57]

6.5.1 Escaping Overmorality?

As I have noted above in section 6.1, philosophically – and paradoxically – it would be only in the "interruption for the ethical" that "responsibility, both practical and intellectual, [could] be though and exercised at all" (de Vries, 2002, p. 10): no ethics,

[57] On this aporetic question concerning universalized indifference cf. also (Žižek, 2003).

no violence; but, unfortunately, no ethics means we just disregard (or like) the violence there is in our world and we give up on taking any "stand for the best". It would seem good to lessen our commitment to morality, still saving its effect of cooperation and of promotion of the ownership of our destiny, to the end of diminishing conflicts and so chances for violence.

I have already asserted – going beyond my previous emphasis on the positive value of the ownership of our destinies – that, once a human collective is engaged in an *ethos* (which preserves both that individual ownership and coalition enforcement) unfortunately also all the puzzling consequences of moral commitments derive from that ethos, such as conflicts between views, infringements and violations, which are seen as such just because basically they jeopardize the ownership of the destinies we care about, and the related cooperation too. In turn, these processes are accompanied by more or less obvious violent outcomes, with the possible complication of an increase of the ineluctable states of resentment and envy.

A further problem arises, related to the importance, for human psychological health, of fate. Paradoxically, humans beings care about cooperation (i.e. coalition enforcement), morality, and thus the ownership or their destinies, but they also take advantage of the idea of fate. Here an eloquent example. Quoting Friedrick Hayek, Žižek contends that is easier to accept inequalities "[...] if one can claim that they result from an impersonal blind force: the good thing about 'irrationality' of the market and success and failure in capitalism is that it allows me precisely to perceive my failure or success as 'undeserved', contingent" (2009, p. 76). From this perspective the supposed – illusory – "a-morality" (and so the unjustness) of capitalism appears to be the feature that renders it acceptable to the majority. After all, the scarce power that capitalism reserves to individuals as regards the control of their destiny is a guarantee of exculpation when they experience a failure (and also envy and resentment are weakened), and at the same time it is a guarantee of hypocritically self-attributing the merit in case of success. On the contrary if I have failed to possess my destiny because of others' (and my) violations of a moral code, which lead to a lack of cooperation and to a disaster in the ownership of my destiny, I certainly cannot call it misfortune, and the space is opened up for punishment, violence, envy, and resentment against myself or other human beings. I failed to preserve the ownership of my destiny and so to gain a desired object: this triggers the libidinal change which generates the shift of the libidinal investment from the object to the obstacle itself, like Žižek usefully observes.

Let us examine the problem of silence, which can be of help in increasing knowledge about the issue. Silence can be seen as a way to morally and stoically disregard a situation – basically we do not express our moral opinion or judgment – in order to avoid the possible rise of conflicts. Unfortunately, silence can also favor violence: Zerubavel (2006) observes that conspiracies of silence pose serious problems, and we therefore also need to examine their negative effects on social life (p. 16). For example taboos, euphemisms and tact are certainly related to the act of weakening our – so to speak – excessive "expressions of morality" (a kind of overmorality) but:

> Aside from the pressure to see and hear no evil, there is also a strong social pressure not to acknowledge the fact that we sometimes do indeed see or hear it. Not only are we expected to refrain from asking potentially embarrassing questions, we are also expected to pretend not to have heard potentially embarrassing "answers" even when we actually have. By not acknowledging what we have in fact seen or heard, we can "tactfully" pretend not to have noticed it (p. 30).

From this point of view breaking the silence is a way of controlling violence and diminishing it. Unfortunately, often silence triggers violence. Politics of denial and sex repression are clear examples: repression is often operated as an injunction to silence, an admission that there was (and is) nothing to say about such things, nothing to see, and nothing to know, as it has already been clearly analyzed by Foucault (1979; 2006). And we know perfectly well that – unfortunately – *silence breakers* are often ridiculed, vilified, and even mobbed/ostracized, thus becoming victims of other kinds of violence. Furthermore, various groups consider silence breakers as threats to their existence and safety. We also face with a paradoxical effect when, in an effort to preserve group cohesion, conspiracies of silence can undermine that cohesion by jeopardizing the development of trusting relations presupposed by overt communication. Silence is also embedded in conflicting moral situations. With the "moral" scope of "protecting" her family, a woman who suspects that her husband is molesting their daughter may thus pretend not to notice it. In such a way she perpetuates the inflicted violence, but by breaking the silence she would attack the integrity of the family as a whole, which is always seen as a value. Conspiracies of silence may also trigger feelings of loneliness.

On the other side, the moral non violent function of silence has to be emphasized: "[...] denial also helps protect others besides oneself. Being unaware that the person with whom I am talking is constantly yawning may indeed be self-protective, yet pretending not to notice it so as not to embarrass him is clearly motivated by altruistic concerns" (Zerubavel, 2006, p. 75). Being tactfully inattentive and discreet helps save the public image of our neighbors and avoids conflicting with them and hurting their feelings. Of course, the conspiracy of silence does not protect only individuals but also the group collective image and safety.

Silence is also used to help people in crossing the line into committing violence: they are kept in ignorance of what they are doing for as long as possible. Keeping people in the dark is often performed by manipulating ambiguity and misunderstanding in order to promote wrongdoing (Baumeister, 1997). Other related ways of encouraging violent behavior – which in multiple ways comply with lack of information and/or silence – are related to:

1. diffusion of responsibility (the larger the group the less responsible and informed any individual member feels),
2. division of labor (that helps conceal responsibility and even the possibility of tracking and individuating the single responsibility as regards harm perpetrated),
3. exploitation of trust and loyalty (when for example people at the bottom trust the decision-makers at the top to do the right thing, and unaware, become evildoers),

4. suppression of private doubts and inhibitions thanks to the immersion in groups (whatever their peaceful private feelings, people tend to express only the "correct" view imposed by a group they belong to, for example of strong hatred against an enemy).[58]

6.5.2 *The Ostrich Effect: The Limits of Docility*

The relationship between silence and violence has to be further deepened. The moral problem of silence can also be linked to the concept of docility. In the previous chapters we have been dealing with docility as something virtuous and undocility as something vicious (cf. chapter one, section 1.3). Indeed, we have implicitly assumed a normative stance on docility, meaning that docile behaviors would be somehow "moral", and undocile ones immoral. Indeed, we do not want to deny that this assumption has been made. However, in depicting who the undocile person is, it is worth considering the limitations docility has. This is to say that, even if docile behaviors are *somehow* good, there are some occasions in which they are not desirable, worth seeking, or beneficial.

Docility is that attitude underlying a number of activities concerning the sharing of information and cognitive resources. As just mentioned, it assumes a kind of ethical attitude that involves, for instance, avoiding information asymmetry, granting the access to cognitive resources to as many people as possible, being committed to the general principle of clarity, and so on. All these prescriptions, however, do not take into account the simple fact that sometimes it is better to leave some things *unsaid*. Sometimes it is better to be undocile: that is, it is better not to disclose and then share information. As we are going to discuss in the following, the attitude of keeping silent – even for good reason – is indeed "undocile".

We refer here to the so-called "ostrich effect".[59] Basically, the ostrich effect is the tendency to ignore unpleasant information by means of avoidance and/or denial. The result is to keep silent on certain matters, thus blocking the possibility that a certain piece of information is made publicly available or acknowledged. This is also described – still in metaphorical terms – by the English expression "the elephant in the room", to refer to an object that everybody is indeed aware of yet no one wants to publicly acknowledge (Zerubavel, 2006).

There is a number of reasons why people prefer not to publicly acknowledge that they have certain problems. Consider, for instance, the case of some holocaust survivors who refused to recount their violent experiences passing on the same attitude to their children and grandchildren. What they experienced in the concentration camps was so violent and horrific that they decided not to disclose it even to their closest relatives. And even when asked to explicitly mention some experience from that time, they still resisted by resorting to the use of euphemisms, for example, in

[58] These cases are typical of what is called *structural* violence, which I have described in chapter one, section 1.1, and reverberate some aspects of the kinds of disengagement of morality described in the previous chapter, section 5.1.

[59] Cf. (Karlsson et al., 2009; Brown and Kagel, 2009).

the case of holocaust, "unmentionable years" or "the war". Paradoxically, the recourse to euphemisms (that basically change the names of certain events) allows victims to refer to brutal experiences without actually mentioning them. Holocaust survivors are "silent witnesses", who prefer not to share their experiences because it would be extremely painful and traumatic to disclose them.[60]

Conversely, one may opt not to talk about certain events not because of a trauma, but because it might involve fear or lack of confidence. An example similar to the one related to domestic sexual abuse illustrated in the previous section is the following: in a family, members may decide to keep silent about a member's drinking problem just because they are afraid of the consequences and conflicts or because they are afraid not to be able to address the problem or tackle it. Active avoidance is meant as a surviving strategy, when, for instance, people lack the resources to cope with a problem.

On some other occasions, denial is just a matter of *tact*. We purposefully avoid noticing a certain detail in our friend just because we are supposed to ignore it. This may regard more or less trivial things like, for instance, bad breath, weight gain, or hair loss. Disclosing and communicating certain information may irritate people or make feel them embarrassed. We would rather prefer not to make people *lose face*. This might also be related to privacy and its ethical underpinnings. For instance, we may omit to say that our colleague's husband cheated on her, thereby causing depression, when the boss is going to assign a new and important task, because that could influence his decision. Here what is implicit is that the communication of certain information is not ethically neutral, as it can promote malicious gossiping or, even worse, discrimination or mobbing. More generally, even in the absence of a legal duty to privacy, silence can be a protective measure related to the respect of people's ability to develop and realize their goals.[61]

Denial may also acquire a social dimension, especially when it regards something that might turn out to be threatening for a group. For instance, some documents are *classified* by a government because they contain information pointing to some vulnerabilities that might be harmfully exploited, for example, in a terrorist attack against inert citizens, or information having strategic value, for instance, in the stock market. In this case, the interests protected are the ones of the investors.

Silence or active avoidance may also regard the communication within a particular institution like the army. Recently, US congresswoman Loretta Sanchez has called attention to the stories of women who have been victim of sexual assault while serving in the US army. Thanks to her initiative, a Sexual Assault Database has been set up and developed in order to encourage women to break the silence. Interestingly, a new "restrictive reporting option" has been introduced so to allow women who have been victims of sexual assault to get help (i.e., counseling and other treatment) without their command having to be notified.[62] In this case silence

[60] Aspects of violence pertaining to the Holocaust in the perspective of Hannah Arendt and feminism are dealt with in (Bar On, 2002).

[61] Cf. (Magnani, 2009, chapters three and four).

[62] The whole story is reported here:
`http://news.bbc.co.uk/2/hi/americas/8511010.stm`.

and active avoidance is explicitly supported by the law to protect the offended. This is an interesting example, since it shows how – paradoxically – silence can be usefully deployed for breaking the silence.

The ostrich effect can be interpreted as a particular case of self-deception. More precisely, it stresses the relationship between language and self-deception. Indeed, silence is not an inattentive attitude, according to which we simply do not pay attention to something. Instead, it is a sort of *negative selection*, meaning that we simply prevent something from being transmitted to the public arena. As already argued, this is basically the undocile character of silence and denial. There is something more to say. In active avoidance, we simply refuse to extend our cognition and to consider some potential chances that are *out there* – distributed in the various external artifacts available in the cognitive niche. In the case of silence, we simply refuse to use language as *the ultimate artifact* to enhance understanding or to share potentially relevant information. As a matter of fact, language reconfigures a variety of cognitive tasks as it is a medium for "non-domain specific thinking". It also stabilizes certain experiences enabling important meta-cognitive abilities like having *thought of thought*. Conversely, avoiding the use of language enables what we call *obliteration*. Let us make an example in order to clarify this point.[63]

As argued by Hirstein (2005), self-deception always involves tension or nagging doubts. Self-deception is the *result* of a process in which two inconsistent cognitions appear in one's mind. For instance, consider a man whose wife cheated on him. In this case, the self-deceptive process is activated or triggered because he has some evidence making him think that his wife has an affair. He certainly experiences a sort of tension between the unhappy proof and his belief that his wife *is* "faithful". The *initial awareness* of such a conflict is fundamental for the process of self-deception to be instantiated. It is only by means of manipulation that he is able to resolve the tension he is experiencing. Such a manipulation is not performed "internally", but it occurs in a "distributed" framework and in a hybrid way: that is, the initial awareness of a tension between the two conflicting cognitions is appropriately manipulated so as to favor obliteration. In the case of silence, it is the active avoidance of using language that helps to resolve the internal conflict. The initial awareness is literally switched off like a lamp promoting the process of embubblement. This process is clearly undocile, because it is based on covering up certain information that would have been potentially public.

This undocile behavior more or less explicitly acknowledges the violent dimension of language. Silence and denial can be explicit responses to this. Language – which is indeed the product of massive cognitive delegations made possible by docility – turns out to be like a weapon able to harm people. As I contended in the first two chapters of this book, language is rooted in a kind of *military intelligence* as a *morality carrier* (and therefore potentially a *violence carrier*). Not only language, but also silence can be a "tool exactly like a knife". Basically, language – even in its own peculiar structure and syntax – transmits vital pieces of information that, once externalized and made available *out there*, acquire a kind of *semiotic agency* affording certain

[63] On language as the "ultimate artifact" cf. (Clark, 2006) and (Magnani, 2009, chapter three).

moral behaviors variously related to *expanding* and, at the same time, *constraining* our action possibilities. Due to this creative role, language (but also silence) may promote or even create new in-group and inter-group conflicts and violence.

To sum up, undocility also underlies those situations in which a person simply refrains from using an external object as a product of cognitive delegation basically for protection. As already mentioned, this is a form of acknowledgment that cognitive delegations do come at a cost. As we have just pointed out, the unskilled or irreflexive use of external artifacts like language, for instance, can cause negative or unhappy outcomes that could have been prevented or avoided. In this sense, undocility may be an option to re-appropriate and re-gain control over the cognitive meanings that people have lavished on external things and objects.

6.6 How Can Forgiveness Be Violent? The Ideal of Mercy and Its Discontents

The Golden Rule is an ethical code that states one has a right to just treatment, and a responsibility to ensure justice for others. It is also called the ethics of reciprocity. The Golden Rule has its roots in many world religions and cultures, and is of course a tool which different cultures use to resolve conflicts. The concept of the Golden Rule originates most famously in a Torah verse: "You shall not take vengeance or bear a grudge against your kinsfolk. Love your neighbor as yourself: I am the Lord".

It is important to keep in mind how such a disposition of the mind is not a natural or instinctive one. The Judeo-Christian culture did not portray the original state of bliss in the Garden of Eden as that of the *good savage*: as a matter of fact, history starts off from that situation with a violent act of insubordination, immediately followed by a quarrel in which each party tries to put the blame on the other (Adam blamed his wife, Eve blamed the Snake), and life outside Eden is immediately defined by Cain murdering his brother Abel.

Consequently, the Commandment to love our similar as another human being enters our cultures as an injunction, a law, or "the" Law itself, because it is extremely complicated to deal with the Other human being without letting oneself fall within the temptation of annihilating the other. As Žižek points out, elaborating Levinas' thought,

> [...] when the Old Testament enjoins you to love and respect your neighbour, this does not refer to your imaginary semblable/double, but to the neighbour qua traumatic Thing. In contrast to the New Age attitude which ultimately reduces my Other/Neighbour to my mirror-image or to the means in the path of my self-realisation (like the Jungian psychology in which other persons around me are ultimately reduced to the externalisations/projections of the different disavowed aspects of my personality), Judaism opens up a tradition in which an alien traumatic kernel forever persists in my Neighbour – the Neighbour remains an inert, impenetrable, enigmatic presence that hystericises me. The core of this presence, of course, is the Other's desire, an enigma not only for us, but also for the Other itself. For this reason, the Lacanian "Che vuoi?" is not simply an inquiry into "What do you want?", but more an inquiry into "What 's bugging you?" (2004)

The Jewish culture, summarized in the multiple facets of the Talmud, seems well aware of how troublesome and therefore fundamental this injunction is, to the point of considering it as the pivotal point of all God's revelations, as shown by this story presented in *Shabbos 31*:

> Upon one occasion an unbeliever approached Shamai and mockingly requested the Rabbi to teach to him the tenets and principles of Judaism in the space of time he could stand on one foot. Shamai, in great wrath, bade him begone, and the man then applied to Hillel, who said: "Do not unto others what you would not have others do to you. This is the whole law; the rest, merely commentaries upon it."

Rabbi Hillel's utterance is of course nothing but yet another formulation of the Golden Rule, "Love your neighbor as yourself".

Furthermore, personally I cannot give meaning to the concept of forgiveness if not in connection with love, mercy, and – especially – compassion: in *Leviticus* 19:18[38], the "Great Commandment" in the Jewish mystical tradition, is particularly clear about this. One rabbi has put it this way: "Kindness gives to another. Compassion knows no other". It seems that only thanks to compassion can you overcome the ultimate cause of violence in the narcissistic *fear of the Neighbor* and his consequent potentially *inhuman* dimension: simply, thanks to compassion, the other is not *other* anymore, because "compassion knows no other". Hence, compassion and love are seen as the main means for avoiding resentments, as clearly re-stated by Jesus Christ: Mark 11:25 (NIV) "But I tell you who hear me: Love your enemies, do good to those who hate you, bless those who curse you, pray for those who mistreat you. If someone strikes you on one cheek, turn to him the other also".[64]

Forgiveness is naturally specified as the process of concluding resentment, indignation, or anger as a result of a "perceived" offense, difference or mistake, and/or ceasing to demand punishment or restitution. From the point of view of the analysis of violence I have given in this book forgiveness seems the best way to avoid moral conflict and so to interrupt the chain. How can you take vengeance if you feel compassion? From the psychological view point forgiveness would have to correspond to a redirection of negative motivations, which would aim at being accompanied by more conciliatory motivations toward the transgressor. A restructuring of emotions is expected thanks to a replacement of negative, unforgiving ones with positive, other-oriented feelings. The positive emotions would have to neutralize some of the negative ones, resulting in a decrease in the overall experienced negativity. Once the negative emotions are substantially eliminated, positive ones can be built (Worthington Jr., 2005a, p. 4).

Forgiveness may be conceded without any expectation of restorative justice, and without any response on the part of the offender (for example, one may forgive a person who is unknown or dead). However, forgiveness can be highly constrained: it may be necessary for the offender to offer some form of acknowledgment, apology, and/or restitution, or even just ask for forgiveness, in order for the harmed person to believe himself in any condition to forgive.

[64] On forgiveness and Christian ethics see (Bash, 2007).

I hope this book has clearly presented that people have a propensity to offend and be offended. Violence has to be primarily emphasized as a philosophical problem: other traditions, for example religions, tend to avoid the direct use of the word "violence", preferring words like "suffering", "evil", "testimony", etc. which inevitably tend to obliterate the fact that what basically matters – when for example speaking of suffering – is just violence, and violence perpetrated directly by human beings, much more than violence derived by natural events or God. In the following subsection I will illustrate some aspects of forgiveness that are very important to further comprehend the morality/religion/violence interplay.

In the following sections I will mainly stress the actual multidimensional material complexities of cognitive dynamics underlying forgiveness, thus paying minor attention to the important philosophical considerations by Arendt, Jankélévitch, Derrida, and Améry, that I will try to acknowledge whenever possible.[65]

6.6.1 Forgiveness Is Multidimensional

The practice of forgiveness is much more complex than expected. If we try to increase knowledge about forgiveness, by showing its multidimensional aspects, soon we curiously see that forgiveness can also be a trigger of violence. The recent book edited by Worthington (2005b) provides a lot of scientific, philosophical and theological knowledge about forgiveness, also showing a wide range of related problems and discontents.

Despite the particular inclination of human beings everywhere to adopt the ancestral "morality" of retaliation and revenge, there is a widespread and common conviction that the potential benefits of forgiveness are localized in four areas: that is, those of physical, mental, relational, and spiritual health. And of course the unwillingness to forgive is stressful and makes people feel hostile toward transgressors. Is this true?

6.6.1.1 Positive Aspects

First of all, among the positive aspects of forgiveness, even research on non-human primates has produced evidence for so-called reconciliation and consolation, that is, post-conflict contacts that serve to repair social relationships and comfort the distressed, a kind of animal variant of forgiveness. Furthermore, forgiveness and lack of forgiveness might affect mental health and well-being. I have said that people have a propensity to offend and be offended. Unforgiving behaviour naturally

[65] On these provocative and complex philosophical speculations about the "paradox of forgiveness" and in particular about the "unforgivable", I direct the reader to (Arendt, 1998), (Jankélévitch, 2005), (Derrida, 2001a,b,c), and (Améry, 2009). Cf. also the multifaceted contributions contained in the collection (Caputo et al., 2001). Some very detailed and specific literature is available, which is related to the role of forgiveness in the political sphere, as a form of politics of apology – also in absence of a religious view and with respect to the aim of reconciliation – cf. for example (Levy and Sznaider, 2006) and (Moolakkattu, 2011).

spreads out from hurtful offenses. Research has shown that both forgiveness and justice may be beneficial in reducing victims' unwillingness to forgive, "which is associated with prolonged physiological activation, which in turn is theorized to have more cardiovascular health implications than short-term stress reactivity" (Brosschot and Thayer, 2003). In sum, a vast amount of empirical research provides support of the notion that forgiveness may have a salutary effect on mental health.

For many years, psychologists have studied the impact of life traumas on the psychological, social, and physical functioning of people and have soon acknowledged that life's traumatic events also affect people spiritually. Offenses and distressing life events of "desecration" can take on even greater power when interpreted within a spiritual context. More than a stressful life event, desecration can become a violation of the "deepest, most precious aspects of a person's life" (Mahoney et al., 2005, p. 69), as violations of one's soul. Here we face with a basic challenge to forgiveness because the process of forgiveness shifts from dyadic (self-perpetrator) to triadic (self-perpetrator-sacred) in nature. That is, analyzing the relationship between the victim and the perpetrator is not sufficient in the case of desecration, but also the relationship with the sacred is at play. It is for this reason that researchers should also "extend their studies of trauma and forgiveness to examine the spiritual character of life's most painful events and most powerful resources for resolution" (2005, p. 69). Anger toward God is another amazing case of aggressive behavior that is overcome through forgiveness with real difficulty.

6.6.1.2 Problems

The multidimensionality of forgiveness is even clearer if we acknowledge it has to be extendedly and complicatedly

> [...] mapped out in an incomplete three (offer, feel, seek) by three (self, others, God) table that yields seven distinct dimensions of forgiveness that should be investigated. They are: (a) forgiveness of oneself, (b) forgiveness of others, (c) forgiveness of God, (d) feeling others' forgiveness, (e) feeling God's forgiveness, (f) seeking others' forgiveness, and (g) seeking God's forgiveness. We leave feeling and seeking forgiveness from oneself undefined at this point (Toussaint and Webb, 2005, p. 358).

These seven dimensions of forgiveness may relate differentially to mental health. Research has demonstrated that forgiveness of the self and others were associated with less distress and greater well-being, but it seems that feeling forgiven by God was not associated with these outcomes, and seeking forgiveness from others was indeed associated but in the opposite direction. Hence, it seems that forgiveness does not necessarily provide benefits to the forgiver or can even harm him, in the case of being forced to forgive.

Furthermore, a more complicated issue in forgiveness is the extent to which retributive motives in the victim reflect cold emotions of revenge or retaliation driven by self-interest, or reflect a genuine societal concern for the welfare of others and/or commitment to an ethical standard of fairness: "To be sure, it is far too simplistic and inaccurate to equate the desire for retribution with unforgiveness, because a retribution motive can not only potentially lower the injustice gap and therefore reduce negative emotions [...] but also reflect concerns beyond the self" (Hill et al., 2005, p. 483).

Finally, forgiveness is often dependent on the offender's attitudes. Once the victim has forgiven, the transgressor has to consider a response both intra-psychically and interpersonally. A problem arises which can have a feedback on forgiveness. Will forgiveness be accepted? The transgressor's experiences, thus, are complicated and become intertwined with the experiences and responses of the victim. The value of forgiveness for the victim himself becomes strictly related to the victim's further reaction. Here it is clear that usually forgiveness does not only depend on the victim's psychological attitudes and moral convictions, it also presents an interpersonal aspect which is occurring in the context of ongoing relationships, which point to the critical roles played by both victim and perpetrator in promoting (vs. impeding) forgiveness and relational repair. For example, research into causal connections between forgiveness and marital outcomes has provided evidence that they do not depend on single source reports, but overlap with global marital satisfaction (Fincham et al., 2005).

6.6.1.3 Negative Aspects

The practice of forgiveness can have violent effects, insofar as forgiving paradoxically produces further violence to the victim. The following is a clear and eloquent example. There is evidence that reconciliation (one important aspect of forgiving) is potentially harmful for victims of sexual abuse. Hence, reconciliation should be encouraged only with the care:

> Those engaging in forgiveness efforts with sexual abuse victims should consider alternatives to reconciliation that do not necessarily require the victim to repair the relationship with the offender but that simply culminate in some level of empathy or acceptance of the offender's flaws and failures. The degree to which one can forgive depends on the degree of damage to basic life assumptions. The violation of basic trust and protection of incestual abuse may be particularly difficult to forgive. Further, incest victims may need to engage in aspects of intrapersonal forgiving of the self [...] if, as is often the case with incest, there is any guilt or self blame associated with being abused. Making sense of a trauma and finding benefit in the experience are among the most effective ways to cope with victimization. Learning to forgive or acquiring a forgiving disposition may be perceived as potential benefits of having gone through a life-changing traumatic experience (Noll, 2005, p. 371).

Furthermore, it has to be remembered that some level of anger begins to be maladaptive only beyond a certain threshold, which depends on many factors, individual and context-related. Not all anger is counter-productive. Not everyone is ready to forgive and not necessarily, lacking forgiveness, is he inclined to perform retaliation. This leads to a skeptical conclusion: forgiveness, "only" when properly understood and practiced (Freedman et al., 2005), can be good at least for the victim (if not – much better case – for the victim "and" also for an improvement of the tranquility of his human environment and for a real reconciliatory outcome).

Often philosophers have prosaically but knowingly noted that vindictive passions have some positive value, especially if not related to cruelty or malice. Going beyond vindictive passions – and not merely controlling them and making them appropriate – can foster, both at the individual and collective level, the danger

> [. . .] that something of value is being improperly sacrificed. This sacrifice may be of proper self-respect, sacrifice of respect for the moral order, or – if hasty forgiveness leads to hasty reconciliation-sacrifice of reasonable self-protection. Even as we rightly preach the virtues of forgiveness, we should recognize that victims deserve to have their vindictive passions to some degree validated. Even if these passions should generally not be the last word, they have a legitimate claim to be the first word. Even when they should not control, they should be listened to with respect instead of met with pious sermons and sentimental, dismissive clichés. In short: even if one subscribes to a brand of Christianity that must ultimately reject the legitimacy of Elie Wiesel's (1995) prayer at Auschwitz or is committed to forgiveness for purely psychological reasons, one must surely still have some sympathy with that prayer and not see it merely as a symptom of irrational illness or evil: God of forgiveness, do not forgive those who created this place. God of mercy, have no mercy on those who killed here Jewish children (Murphy, 2005).

Alas, everyone knows that reconciliation after mass violence and genocide is a difficult task. Hannah Arendt (1998) shares with Jankélévitch the belief that forgiveness is an act of love that permits the creation of a new future relationship – according to her, forgiveness is always a promise and plays the political role of reestablishing a sense of community – thus avoiding cycles of revenge. Forgiving enables us to come to terms with the past and liberates us to some extent from the burden of the received harm and its irreversibility. However she also contends that forgiveness can only be understood within the realm of human affairs. Forgiveness can be conceded only to violators that we can comprehend, of course forgiveness for the unforgivable is impossible and incoherent. For Arendt radical evil – for example in the case of the ultimate solution – is by definition beyond the realm of forgiveness or punishment: we cannot forgive what we cannot punish. The cases of Jean Améry (2009) and Esther Mujawayo – who claim the unforgivable does exist – deviate from the common imaginary of the agents of *un*forgiveness. Neither of them is consumed by hatred or out for sanguinary revenge: they are suffering and a moderate amount of rage suffuses the writings of Améry in particular, but their aims and arguments are not deluded or morally humiliating. Améry thinks that victims of torture, like the survivors of Auschwitz, have lost their ability to trust in human beings forever - such a subject is a "dead man on leave". He acknowledges, in a way, the irreducibility and irreversibility of unforgivable violence which cannot be taught otherwise: the problem is that the Other can be absolute, and can exercise this absolute power by inflicting suffering.

At this point I have to stress that some suspect about the excessive ideality of forgiveness is justified. Forgiveness is the ideal of the spectacular morality of "Love your enemies, do good to those who hate you, bless those who curse you, pray for those who mistreat you. If someone strikes you on one cheek, turn to him the other also". Unfortunately forgiveness very easily *de facto* gives rise to moral and intellectual confusion and so becomes a very difficult concept to be grasped and very difficult to be properly performed by people: it, *paradoxically*, offers a perverted underside.

Jankélévitch also (2005) clearly sees how many actions, that may appear to be forgiveness, in the end amount to nothing more than a kind of pseudo-forgiveness, full of instrumental and utilitarian aspects (such as reconciliation, overcoming the anger, "moving on", etc.). However, he gives philosophical room to a noble, ideal, positive idea of forgiveness: forgiveness as a pure moral choice, an act of gift giving or grace that reverberates true freedom, beyond reason – there are no reasons to forgive – and a way of recreating a new future, reshaping our relationship with the past and the violator. It is moral and yet beyond the ethical imperatives, because one is not obliged to forgive. Furthermore, still according to Derrida forgiveness is not a system of exchange (Derrida, 2001a,b,c): he put forgiveness aside as something utterly separate from politics and articulates it in the ultimate aporetic formula "forgiveness forgives only the unforgivable". For Derrida, forgiveness is unconditional and we must forgive the guilty as guilty without a reference to a request for forgiveness, and without transforming the guilty into the innocent.

What is important to note from our perspective is that forgiveness delineates a kind of threat to "every" morality, as I will further explain in the following subsection. How can this be? Does forgiveness not contain – as a spectacular *moral* ideal – a germinal challenge to the appropriateness not only of revenge but also of any other moral *punishment*? In this book I have illustrated that (more or less violent) punishment is the very constitutive aspect of every moral system. In a sense, we can say that morality of forgiveness – at least in its ideal and abstract version – "obscenely" aims to go beyond, to skip punishment, eventually leaving it to the legal system or directly to God, but fundamentally proscribing it. Consequently, how can we preserve moral orders without punishment? I have often said that morality generates punishment, conflict, and so violence: morality is strictly intertwined with violence, so forgiveness certainly aims to overcome violence, but, at the same time, presents the confused face of a difficult and ambiguous human practice, which seems to jeopardize every morality.

Some thinkers literally contend that emphasis on forgiveness is often used by leaders to perpetrate cycles of ongoing abuse. Kramer and Diana (1993, p. 295) clearly state that "[...] to forgive without requiring the other to change is not only self-destructive, but ensures a dysfunctional relationship will remain so by continually rewarding mistreatment". They also observe that faith-based ideals of forgiveness, while appearing selfless, contain implicit selfish aspects: "when forgiving contains a moral component, there is moral superiority in the act itself that can allow one to feel virtuous" (p. 292). And an even more perverse outcome is that it is often unclear who benefits more from forgiveness, the one doing the forgiving or the one being forgiven.

The Stoics were aware of the paradox of forgiveness. Some basic tenets of Stoic doctrines are usually misunderstood: for the Stoics humans are perfectly allowed to feel emotions (for example, rage) of any kind, and *do not* have to eliminate them.

They have only to control them so that no *moral* outrage is felt. Nevertheless humans cannot give up revenge and punishment. How? We cannot forgive if we forget,

> [...] for forgiving requires remembering the moral injuries we suffered, and making a conscious decision to forswear for the future (or at least the immediate future) our resentment or rage. To forgive is to move beyond or overcome punishing anger for *moral* reasons – because the wrongdoer has repented, or because we see other redeeming features in the person untarnished by the wrongdoing, or because we are ready to readmit the shunned wrongdoer back into the moral community. In forgiving, we neither renounce nor repress the protest that our anger registers; instead we decide that it is time to move beyond it on moral grounds. Forgetting, in contrast, is something that happens to us over time, and as therapeutic as it may be, it is not something we *do* for moral reasons (Sherman, 2005, p. 85).

Forgiving is not condoning, like Kant "Doctrine of Virtue" says: "Men have a duty to cultivate a *conciliatory spirit* (*placabilitas*). But this must not be confused with placid toleration of injuries (*mitis iniuriarum patientia*), renunciation of the rigorous means (*rigorosa*) for preventing the recurrence of injuries by other men; for in the latter case a man would be throwing away his right and letting others trample on it, and so would violate his duty to himself" (1964b, pp. 459–460).

Following the Stoic wisdom, instead than forgiving (or punishing in a retributive way), we need to laboriously adopt the "kindly gaze of a doctor viewing the sick". Seneca says "So the wise man will be calm and fair to the errors which confront him; a reformer of wrongdoers, not their enemy, he will start each day with the thought 'Many will meet me who are given to drunkenness, lust, ingratitude, avarice, many who are disturbed by raving ambition'. All this he will view with the kindly gaze of a doctor viewing the sick" (1995, *On Anger*, 2.10.7). Retaliation, vengeance, or retributive punishment are a business of cure rather than a *moral* (and thus violent) protest. The reaction that focuses on the moral mistake of the wrongdoer – and so that defends the value that makes the action wrong – is wrong and misguided. In sum, we face with a difficult and general *didactic* role of the actions that have to help us to go beyond the social, physical and psychological damages that were caused to us by other people's violence, through educating the wrongdoer and his emulators, a task that is always related to teaching and making use of knowledge and to the formation of a resilient reason-based responsibility: it would be the only non retaliating way the victim of violence possesses to manage his resentment and indignation.[66]

6.6.2 Moral Viscosity and the Ultimate Paradox of Forgiveness

In 1887 Friedrick Nietzsche composed and published one of his most polemical yet insightful books, *On the Genealogy of Morality*. The second essay, entitled "'Guilt', 'bad conscience' and related matters" provides an interesting discussion about the forgiveness (moral and judicial) of wrongdoers, which is seen as something "beyond the law". Nietzsche says:

[66] Further philosophical aspects of "genuine" forgiveness are studied in (Griswold, 2007).

> As the power and self-confidence of a community grows, its penal law becomes more lenient; if the former is weakened or endangered, harsher forms of the latter will re-emerge. The "creditor" always becomes more humane as his wealth increases; finally, the amount of his wealth determines how much injury he can sustain without suffering from it. It is not impossible to imagine society so conscious of its power that it could allow itself the noblest luxury available to it, – that of letting its malefactors go unpunished. "What do I care about my parasites", it could say, "let them live and flourish: I am strong enough for all that!" [...] Justice, which began by saying "Everything can be paid off, everything must be paid off", ends by turning a blind eye and letting off those unable to pay, – it ends, like every good thing on earth, by sublimating itself. The self-sublimation of justice: we know what a nice name it gives itself – mercy; it remains, of course, the prerogative of the most powerful man, better still, his way of being beyond the law (2007, p. 47–48).

Nietzsche correctly acknowledges how forgiveness is essentially a context-dependent matter: it rests on a number of factors but the *conditio sine qua non* is that the forgiver must be able to sustain, in terms of psychological, social and physical resources, the act of forgiving. As for this, there is not much difference between forgiveness and tolerance (which conveys more of a passive idea): both forgiveness and tolerance have a cost. The feeling of tolerance and forgiveness presupposes an emotional cost because the very ideas of tolerance and forgiveness entail the fact that one is not *automatically* supposed to tolerate and forgive a certain thing or event (Smith, 1997).

Nietzsche, though, contends that once the members of a community can afford it, they will necessarily and uncritically perform all of the forgiveness they can: this is, needless to say, the exact nemesis of *morality* and justice. Not only, through forgiveness morality becomes its own self victimization – *self-sublimation* in Nietzschean words: a morality that forgives *automatically* is a morality that engages in a sophisticated oeuvre of self-destruction.

Let us consider this violent yet interesting claim: it is sensible because an act of forgiving, not as mere tolerance but as an active practice (something like embracing and lifting someone back up to her previous status within the moral group), always involves – from a theoretical point of view – a second infringement of the norm inasmuch as the forgiver *rehearses* in her mind the action of the wrongdoer.

The forgiver must consider, as if they were both before her, the (infringed) norm and the wrongdoer. She can decide to support the reasons of the latter (that are after all the reasons for forgiveness) against the norm: this is obtained by the *rehearsal* and hence by the second violation of the norm. By forgiving, thus, the forgiver as well places himself *outside* of morality to join the wrongdoer. This physical exit of both actors from the original state is well exemplified by the parable of the Prodigal Son in the *New Testament* (Luke 15:11-32): the parable is strictly about forgiveness and it is important to notice how, at the end of the episode, when the prodigal son comes back to his original group, the forgiver/father does not wait for him but runs *out* to meet him. Thus he makes explicit the act of placing himself out of the original morality as well! This clearly shows through the words of the "good" son, who perceives his father's forgiveness as a violence against him, the one who had always behaved according to the morality of the family. As Tara Smith puts it, "[...] forgiveness is the judgment that a person's immoral action should not be treated as proof of a grave moral defect or an irredeemably bad character. The person offering forgiveness in effect says: 'I will not write you off on the basis of this incident'"(Smith, 1997, p. 37). In other words, the forgiver's action is indeed a second attack on the moral norm: the forgiver says, "I care about the norm but I care more about you, about our relationship, your relationship with the group, etc."

This is precisely what we referred to in chapter five[67] as a double stake. The person who is receiving apologies from another person has to choose between two opposing alternatives: on the one hand, forgiving immediately affects one's sense of *position in the field*, which means that forgiving immediately projects an imaginary coalition conflicting with the one the forgiver belongs to. This implies – as I have just stressed – a sort of temporal disengagement from his morality. On the other hand, failing to offer forgiveness literally means cutting off a social bond established with the person to be forgiven. As we will see in the following, *moral viscosity* is a major factor in defusing the various double stakes one might face.

In this perspective, Nietzsche's account seems exact: *systematical* forgiveness is indeed the ultimate self-mortification of morality. One must take a step backwards, though, and understand if such a systematic nature of forgiveness is really the case: in the philosopher's *Genealogy*, morals originate from a *rebellion* of the slaves to their ancient, mighty masters.

This conception of the origins of the moral is often advocated by denigrators of morality as a mere instrument of thwarting and repression, yet, as Žižek points out,

> The ultimate irony is that this "critique of ethical violence" is sometimes even linked to the Nietzschean motif of moral norms as imposed by the weak on the strong, thwarting their life assertiveness: "moral sensitivity', bad conscience, guilt-feeling, as internalized resistance to the heroic assertion of Life. For Nietzsche, such "moral sensitivity" culminates in the contemporary Last Man who fears the excessive intensity of life as something that may disturb his search for "happiness" without stress, and who, for this very reason, rejects "cruel" imposed moral norms as a threat to his fragile balance (2004, p. 2).[68]

As we read in the *Genealogy*, morality is the rule of the weak, who – out of resentment – label as *evil* the prerogatives of their former master. Weaklings are not *able* to contrast those who do them wrong, so their resentful withstanding of any abuse is "given good names such as 'patience', also known as the virtue; not-being-able-to-take-revenge is called not-wanting-to-take-revenge, it might even be forgiveness" (Nietzsche, 2007, p. 28).

The systematical character of morality seems to be overstressed in the Nietzschean account. If it were, forgiveness as a moral norm would indeed have

---

[67] Cf. subsection 5.4.2.

[68] On another account, Nietzschean reconstruction is strikingly similar to the result of recent findings in cognitive paleoanthropology: according to such research, the rise of egalitarian moralities was permitted by the development of remote killing artifacts and techniques, that would make it harder for alpha males (better hunters) to bully feebler individuals and administer the game to their own liking (Bingham, 2000; Boehm, 2002). Cf. also chapter one, section 1.3.2, this book.

annihilated morality itself long since: yet the moral groups embodying Jewish and especially Christian thoughts on the goodness of forgiveness have been thriving for more than three thousand years, now. This is the *paradox* of forgiveness: on one hand forgiveness reiterates the attack upon the moral norm and sets the forgiver *out* of the group as the wrongdoer, on the other hand both the forgiver and the wrongdoer are found back in the group, and morality survives in spite of the double attack and ultimately forgiveness receives moral praise.

It is important to take into the account how forgiveness is hardly ever a moral injunction like the other norms. If a set of moral norms were accompanied, *ceteris paribus*, by a final norm imposing forgiveness of any wrong deed, then this would really mean a short-circuit of the whole moral institution! One must not confound *forgiveness* with its, alas, widespread degeneration supported by *do-goodism*: an indiscriminate exercise and recommendation of general tolerance that is as self-pleasing as it is self-deceiving. This is often the accusation directed at Christian/Catholic eagerness to forgive, and probably the target of Nietzsche's polemic, but it is a degenerated interpretation of the recommendation to forgive.

Forgiveness is a *plus* rule that situates itself on another level with respect to the rest of moral norms, if only for the fact that it *may* refer to any other moral norm. It is wrong to steal and yet a thief can be forgiven, it is wrong to lie and yet a liar can be forgiven, it is wrong to commit adultery and yet an adulterer can be forgiven, and so on. Forgiveness is a *possibility*, however encouraged it must remain in the realm of what *can* be: forgiving requires a complex emotional progression that could never assume the characteristic of compulsion. Besides, our cultural backgrounds are full of characters that are deemed morally righteous even if they opposed severity and blind justice to a will to forgive. Not to forgive, as a matter of fact, is usually morally deprecated only as far as it betrays incoherence: it is *wrong* for me not to forgive one who has indulged in the same wrong deed that I indulged in, especially if I have been forgiven for that wrong myself. Apart from in this case, one can ironically assess that whereas violence – that is morality through its related punishments and generated conflicts – is always a duty (and its sublimation may lead to further violence and perversion), forgiveness ... is no such duty! Forgiveness is *not* a duty and it is not usually perceived as such.

The paradox of morality, worth further investigation, consists in the fact that in all of the cases we just mentioned, in spite of the double attack on the rule (from the wrongdoer and the forgiver), the rule survives and is still *morally* believed by the forgiver (and probably by the wrongdoer): stealing, lying, and being unfaithful is *still* wrong. A situation of *do goodist* indiscriminate tolerance and it is important to state this once again, cannot even be analyzed under this schema because the wrongdoer knows from the beginning that her deeds will not be recriminated, let alone punished. Forgiveness, on the other hand, implies that the forgiver will achieve a deeper level of understanding, with respect to the moral infringement that has been perpetrated: it should have generated resentment and a will to perform a just punishment of the violator but, just because of forgiveness, the forgiver will not recur to punishment and will attempt to placate resentment.

The solution to this paradox is foreshadowed by a term introduced by Lahiti and Weinstein (2005), *moral viscosity*.[69] Connected to the concept of *moral character*, that is the sum of (usually positive) moral traits one person builds as a result of her ethical education, moral viscosity represents the tendency of moral beliefs to *stick* to their holder in spite of the changing contexts. Morality is not a solipsistic matter but always rests on a shared and distributed dimension. The issues we have just debated regarding the violence of forgiveness (forgiver and wrongdoer both exiting the community) have an extreme relevance for the whole moral group, where every individual is intertwined with the others in a holistic fashion. From this point of view, the offense to the norm perpetrated by the forgiver is felt by all other individuals who partake of that moral. Yet, the *viscosity* of morality prevents the whole group-morality from being excessively jeopardized by any attack on the shared norms.

It is tempting to draw a sharp juxtaposition between moral viscosity and forgiveness: if we start from the individual moral agent, we can argue that viscosity is essentially about self-forgiveness. We occasionally behave in a way that does not comply with our own moral requirements, yet we manage to get over it. Even if we consciously break a norm or *disengage* it,[70] that viscous norm remains bound to our moral conscience and we can maintain our commitment to our group's morality thus preserving its effect in coalition enforcement. Clearly, this is not the case of a schizophrenic or split personality, neither we are referring to a permanent or reiterated state like *bad faith*,[71] nor with plain hypocrisy. Viscosity is what allows us to deal with and overcome single moral inconsistencies: "I cheated on my partner once but I still think that cheating is wrong", "My coworker dropped his ten-dollar bill and I took it but I still think that stealing is wrong". It is important to bear in mind how this is an extremely cultural issue and not a global psychological one. In far-eastern culture, for instance, practices such as *seppuku* (the Japanese ritual form of suicide) inform us of a different sensibility as regards personal failures and self-forgiveness.[72] Other cultures would relate to a less viscous nature of morality with a symbolic elimination of the wrongdoer: this is the case of the *scapegoat* ritual, as described in the *Bible* (Leviticus 16), in which the guilt of wrongdoers would be *passed on* to a third creature, namely a goat, that was then to be expelled from the group and chased into the wilderness.[73]

If we get back to our analysis of the interpersonal dimension of forgiveness we see how the dynamics of viscosity can help us understand and solve the initial paradox. If viscosity represents single beliefs sticking to an agent, in a distributed framework of moral beliefs viscosity is what keeps single agents stuck to morality, and hence to the groups projected and supported by the moral in question. Of course the degree to which viscosity can cover for an abrupt modification in the observance

---

[69] I have already introduced this term in chapter three, subsection 3.2.2, this book.

[70] Disengagement/reengagement of morality has been described in the previous chapter, section 5.1.

[71] On bad faith see also chapter five, section 5.3.

[72] Further details on this topic are illustrated in (Fuse, 1980).

[73] On scapegoating and its effect in the violence that characterizes religious rituals cf. (Girard, 1977, 1986). Cf. also subsection 2.1.5, chapter two of this book.

of a norm depends on the "power and self-confidence of a community" (Nietzsche would agree). If the offense is too dire (there are plenty of unforgiven wrongs) or the number of perpetrators is too elevated there can be only two solutions: either the wrongdoers are definitively expelled from the group or the whole moral institute crumbles with the dispersion of the same group. It is important to stress how the viscosity of morality is not the theoretical counterpart to simple hypocrisy, it does not allow morality to withstand an unlimited number of violations without displaying actual impoverishment: as a matter of fact, the level of viscosity is an extremely contextual factor so that an offense for which viscosity can act as guarantee on one occasion, can prove to be fatal (for the wrongdoer) in another situation. Furthermore, there are also cases in which the violation is so significant that forgiveness cannot be granted without the forgiver permanently joining the wrongdoer(s) *outside* of the original group.

When the *viscosity*[74] of morality allows and supports forgiveness, the whole process can be thus summarized: 1) as the wrongdoer infringes the moral rule, she immediately places herself outside of the moral group; 2) by the theoretical dynamics of forgiveness which I exemplified before, the forgiver reenacts the violation of the norm and for an instant (though this cannot be transposed into actual, physical time) takes a stand with the wrongdoer against the norm, exiting the moral coalition as well; 3) morality, due to its characteristic of *viscosity*, acts as a blubbery substance and reconstitutes its structure before the rest of the community feels the *wounds* as *their own*; 4) viscous moral remnants upon the forgiven wrongdoer merge with those of the forgiver so that they can both be re-absorbed into the main viscous unity, that is the original moral group.

[74] This rough parameter, which is more qualitative than quantitative, can be traced back to different influences. The religious one is fundamental: it is easy to argue that Western society's eagerness to forgive is heavily dependent on the importance of the theme of forgiveness in early Christian writings (McCullough and Worthington Jr., 1999). The philosophical contribution, later absorbed into an *idéologie spontanée*, plays a crucial role as well.

Conclusion

In place of a formal conclusion, I offer here a sort of final comment about what I see as the most important elements of *Understanding Violence*. I hope that I was able, through my arguments, to encourage a serious commitment to increase our knowledge about violence, both as a cognitive and philosophical topic. The philosophical breakthrough that I tried to operate, focused on the analysis of violence *as it is*, as I abstained from explaining it *away* by any reduction, can improve our chances to maintain the theoretical commitment towards the study of human "violent" attitudes, thus keeping research lively and fruitful.

At the same time, given the importance of violence in actual human behavior, studies on this topic help to maintain an intellectual focus on moral commitments, hopefully, also beyond the strictly intellectual community. The knowledge about violence I passionately endorsed in this book aims also at supplying researchers with the rational poise required to handle controversial issues regarding religion, morality, science, and language, now and in the future.

Violence, harm, and aggressions are so pervasive, so much a part of human (and animal) life, that it is not easy to imagine how to live without stumble upon them. The violent attacks we bear on other human beings, animals, and nature depend on complicated biological and cognitive mechanisms, fundamentally intertwined with the role of moralities. People can make use of different moralities as weapons of violence, meaning to harm others – both for defensive and aggressive scopes – as I described along the various chapters of this book. Stressing this issue has been a further extension of the analysis of morality, transcending some precious and sophisticated – but candid, in the perspective of violence – results of moral philosophy. Acknowledging our "condition" of "violent beings", and increasing knowledge about our stupefying capacity to harm, is a way of accepting our responsibility: it could hopefully be of help in individuating cognitive firewalls that prevent violence from always escalating, to further weaken errors and damages and to improve our freedom and a safer ownership of our destinies.

Nikolai Vasilievich Gogol was perfectly aware that our knowledge, our inclination and our sensitivity to good are always inextricably bound to knowledge, inclination, and sensitivity to evil. First of all, to understand violence does not mean to forgive or justify it, but – possibly – it can be of help for example to avoid falling in a situation when you must forgive or you need to be forgiven. Seeing evil from the point of view of the gray morality of its perpetrators – where violence appears, clearly, so value-driven – can help to grasp violence both as a common and almost ineluctable event, which cannot be disregarded. To understand violence I had to suppress my own personal moral judgments, but, for example, I have gained a new metamoral awareness about (at least my) human condition. The spectacle of human violence taught me that I have to live with it and I have to clearly monitor it (instead of silencing it, either in my own conscience or in my intellectual activity). I also learnt that trying to avoid it both as a passive and active agent requires something much more complicated that the mere candid acquisition, adoption, refinement, and practice of a traditional morality, whatever it is.

Let us begin.

References

Abbott, M.: The creature before the law: notes on Walter Benjamin's "Critique of Violence". COLLOQUY text theory critique 16, 81–96 (2008), www.colloquy.monash.edu.au/issue16/abbott.pdf

Agamben, G.: The Coming Community. University of Minnesota Press, Minneapolis (1993) Translated by M. Hardt

Agamben, G.: State of Exception. The Role of a Supreme Court in a Democracy. University of Chicago Press, Chicago (2005)

Alexander, R.D. (ed.): The Biology of Moral Systems. Aldine De Gruyter, New York (1987)

Allen, J.S., Bruss, J., Damasio, H.: The aging brain: the cognitive reserve hypothesis and hominid evolution. Americam Journal of Human Biology 17, 673–689 (2005)

Alpher, D.G., Rothbart, D.: "Good violence" and the myth of the eternal soldier. In: Rothbart, D., Korostelina, K.V. (eds.) Identity, Morality, and Threat. Studies in Violent Conflict, pp. 241–278. Rowman and Littlefield Publishers, Oxford (2006)

American Psychiatric Association, Diagnostic and Statistical Manual of Mental Disorders-IV. American Psychiatric Association, Washington, DC (1994)

Améry, H.: At the Mind's Limits. Indiana University Press, Bloomington (2009)

Anderson, J.L.: Che Guevara: A Revolutionary Life, Grove, New York (1997)

Anglin, M.K.: Feminist perspectives on structural violence. Identity 5(2), 145–151 (1998)

Anquetil, T., Jeannerod, M.: Simulated actions in the first and in the third person perspective. Brain Research 1130, 125–129 (2007)

Arciuli, J., Villar, G., Mallard, D.: Lies, lies and more lies. In: Ohlsson, S., Catrambone, R. (eds.) Proceedings of the 32nd Annual Conference of the Cognitive Science Society, pp. 2329–2334. Cognitive Science Society, Austin (2010)

Arendt, H.: The Origins of Totalitarianism, Harcourt, Orlando, FL (1976)

Arendt, H.: The Human Condition. Chicago University Press, Chicago (1998)

Atran, S.: In Gods We Trust: The Evolutionary Landscape of Religions. Oxford University Press, Oxford (2002)

Audi, R.: On the meaning and justification of violence. In: Bufacchi, V. (ed.) Violence: A Philosophical Anthology, pp. 136–167. Palgrave Macmillan, Houndmills (2009)

Bäck, A.: Thinking clearly about violence. In: Bufacchi, V. (ed.) Violence: A Philosophical Anthology, pp. 365–374. Palgrave Macmillan, Houndmills (2009)

Balch, D.R., Armstrong, R.W.: Ethical marginality: the Icarus syndrome and banality of wrongdoing. Journal of Business Ethics 92, 291–303 (2009)

Balibar, E.: We the People of Europe. Princeton University Press, Princeton (2005)

Bandura, A.: Social Learning Theory. Prentice-Hall, Englewoods Cliffs (1977)
Bandura, A.: Moral disengagement in the perpetration of inhumanities. Personality and Social Psychology Review 3, 193–209 (1999)
Bandura, A.: Impeding ecological sustainability through selective moral disengagament. International Journal of Innovation and Sustainable Development 2(1), 8–35 (2007)
Bar On, B.-A.: The Subject of Violence: Arendtean Exercises in Understading. Rowman and Littlefield, Lanham (2002)
Bardone, E., Magnani, L.: The appeal of gossiping fallacies and its eco-logical roots. Pragmatics & Cognition 18(2), 365–396 (2010)
Barnett, P.W., Littlejohn, S.W.: Moral Conflict: When Social Worlds Collide. Sage, London (1997)
Barrett, J.L.: Cognitive science, religion, and theology. In: Schloss, J., Murray, M. (eds.) The Believing Primate: Scientific, Philosophical and Theological Perspectives on the Evolution of Religion, pp. 76–99. Oxford University Press, Oxford (2009)
Barrett, J.L., Lanman, J.A.: The science of religious beliefs. Religion 38, 109–124 (2008)
Bash, A.: Forgiveness and Christian Ethics. Cambridge University Press, Cambridge (2007)
Bauman, Z.: The absence of society. In: Timmins, N. (ed.) Contemporary Social Evils, pp. 147–158. The Policy Press, Bristol (2009)
Baumeister, R.: Evil. Inside Human Violence and Cruelty. Freeman/Holt Paperbacks, New York (1997)
Becchio, C., Bertone, C., Castiello, U.: How the gaze of others influences object processing. Trends in Cognitive Science 12(7), 254–258 (2008)
Bekoff, M., Pierce, J. (eds.): Wild Justice: The Moral Lives of Animals. University of Chicago Press, Chicago (2009)
Bellinger, C.K.: The Genealogy of Violence. Oxford University Press, Oxford and New York (2001)
Ben Jacob, E., Shapira, Y., Tauber, A.I.: Seeking the foundation of cognition in bacteria. From Schrödinger's negative entropy to latent information. Physica A 359, 495–524 (2006)
Benjamin, W.: The critique of violence (1921). In: Dements, P. (ed.) Reflections: Essays, Aphorisms, Autobiographical Writings, Hartcourt Brace, New York (1978); Translated by E. Jephcott. First appeared in the Archiv für Sozialwissenschaft und Sozialpolitik 3 (1921)
Bergeret, J.: Freud, la violence et la dépression. PUF, Paris (1995)
Bertolotti, T., Magnani, L.: The role of agency detection in the invention of supernatural beings. In: Magnani, L., Carnielli, W., Pizzi, C. (eds.) Model-Based Reasoning in Science and Technology. Studies in Computational Intelligence, vol. 314, pp. 239–262. Springer, Heidelberg (2010)
Bingham, P.M.: Human uniqueness: a general theory. The Quarterly Review of Biology 74(2), 133–169 (1999)
Bingham, P.M.: Human evolution and human history: a complete theory. Evolutionary Anthropology 9(6), 248–257 (2000)
Blair, R.J.R., Jones, L., Clark, F., Smith, M.: A lack of responsiveness to distress cues. The psychopathic individual. Psychophysiology 34, 192–198 (1997)
Bloom, P.: Religious belief as an evolutionary accident. In: Schloss, J., Murray, M. (eds.) The Believing Primate: Scientific, Philosophical and Theological Perspectives on the Evolution of Religion, pp. 117–118. Oxford University Press, Oxford (2009)
Bloomfield, P. (ed.): Morality and Sefl-Interest. Oxford University Press, Oxford (2008)
Boddy, C.R.: Corporate psychopaths, bullying and unfair supervision in the workplace. Journal of Business Ethics 97, 1–19 (2010)

Boddy, C.R., Ladyshewsky, R.K., Galvin, P.: The influence of corporate psychopaths on corporate social responsibility and organizational commitment to employees. Journal of Business Ethics 97, 1–19 (2010)

Boehm, C.: Hierarchy in the Forest. Harvard University Press, Cambridge (1999)

Boehm, C.: Variance reduction and the evolution of social control. Paper presented at Annual Workshop on the Co-Evolution of Behaviors and Institutions, Santa Fe, NM (2002)

Bollas, C.: Being a Character. Psychoanalysis and Self Experience. Routledge, London (1993)

Bollas, C.: Cracking Up: The Work of Unconscious Experience. Routledge, London (1995)

Bonhoeffer, D. (ed.): Ethics. MacMillan, New York (1955)

Bourdieu, B.: Masculine Domination. Polity Press, Cambridge (2001)

Bourdieu, P.: Outline of a Theory of Practice. Cambridge University Press, Cambridge (1997) Translated by R. Nice

Bourdieu, P.: On symbolic power. In: Thomspon, J.B. (ed.) Language and Symbolic Power, pp. 163–170. Polity Press, Cambridge (1991) Translated by G. Raymond and M. Adamson

Boyd, R., Gintis, H., Bowles, S., Richerson, P.J.: The evolution of altruistic punishment. PNAS Proceedings of the National Academy of Science of the United States of America 100, 3531–3535 (2003)

Boyd, R., Richerson, P.J.: Norms and bounded rationality. In: Gigerenzer, G., Selten, R. (eds.) Bounded Rationality. The Adaptive Toolbox, pp. 281–296. The MIT Press, Cambridge (2001)

Boyer, P.: Religion Explained. The Evolutionary Origins of Religious Thought. Basic Books, New York (2001)

Brady, I.: The Gates of Janus: Serial Killing and Its Analysis. Feral House, Los Angeles (2001)

Brookman, F.: Understanding Homicide. Sage Publications, London (2005)

Broom, D.M.: The Evolution of Morality and Religion. Cambridge University Press, Cambridge (2009)

Brosschot, J.F., Thayer, J.F.L.: Heart rate response is longer after negative emotions than after positive emotions. International Journal of Psychophysiology 50, 181–187 (2003)

Brown, A.L., Kagel, J.K.: Behavior in a simplified stock market: the status quo bias, the disposition effect and the ostrich effect. Annals of Finance 5, 1–14 (2009)

Brown, D.E.: Human Universals. McGraw–Hill, New York (1991)

Bruce, J.B.: The adaptive problem of absent third-party punishment. In: George, R.Z., Bruce, J.B. (eds.) Analyzing Intelligence, pp. 171–190. Georgetown University Press, Washington, DC (2008)

Buchanan, A.: Psychiatric Aspects of Justification, Excuse and Mitigation in Anglo-American Criminal Law. Jessica Kingsley Publishers, London and Philadelphia (2000)

Buchli, E.: Il mito dell'amore fatale. Baldini Castoldi Dalai, Milan (2006)

Bufacchi, V. (ed.): Violence: A Philosophical Anthology. Palgrave Macmillan, Houndmills (2009)

Bulbulia, J.: Meme infection or religious niche construction? An adaptationist alternative to the cultural maladaptationist hypothesis. Method and Theory in the Study of Religion 20, 1–42 (2008)

Bulbulia, J.: Religiosity as mental time-travel. Cognitive adaptations for religious behavior. In: Schloss, J., Murray, M. (eds.) The Believing Primate: Scientific, Philosophical and Theological Perspectives on the Evolution of Religion, pp. 44–75. Oxford University Press, Oxford (2009)

Burnham, T.C., Johnson, D.D.P.: The biological and evolutionary logic of human cooperation. Analyse & Kritik 27, 113–135 (2005)

Burton, R.A.: On Being Certain. St. Martin's Press, New York (2008)

Butler, J.P.: Excitable Speech: A Politics of the Performative. Polity Press, Cambridge (1997)

Byrne, R., Whiten, A.: Machiavellian Intelligence. Oxford University Press, Oxford (1988)

Caputo, J.D., Dooley, M., Scanlon, M.J. (eds.): Questioning God. Indiana University Press, Bloomington (2001)

Carnielli, W.: On a theoretical analysis of deceiving: how to resist a bullshit attack. In: Magnani, L., Carnielli, W., Pizzi, C. (eds.) Model-Based Reasoning in Science and Technology. Studies in Computational Intelligence, vol. 314, pp. 291–299. Springer, Heidelberg (2010)

Carson, D.: A psychology and law of fact finding? In: Carson, D., Milne, R., Pakes, F., Shalev, K., Shawyer, A. (eds.) Applying Psychology to Criminal Justice, pp. 115–130. Wiley & Sons, Chichester (2007)

Carson, D., Milne, R., Pakes, F., Shalev, K., Shawyer, A. (eds.): Applying Psychology to Criminal Justice. Wiley & Sons, Chichester (2007)

Castro, L., Castro-Nogueira, L., Castro-Nogueira, M.A., Toro, M.A.: Cultural transmission and social control of human behavior. Biology and Philosophy 25, 347–360 (2010)

Castro, L., Toro, M.A.: The evolution of culture: from primate social learning to human culture. PNAS (Proceedings of the National Academy of Science of the United States of America) 101(27), 10235–10240 (2004)

Catley, B.: Philosophy – the luxurious supplement of violence? In: Brigham, M., Brown, A., Contu, C., Elliot, S., Fox, N., Hayes, L., Introna, E., Swan, E., Turnbull, S. (eds.) Critique and Inclusivity: Opening the Agenda. Proceedings of the Third International Critical Management Studies Conference, Lancaster University Management School, Lancaster (2003); CD ROM: ISBN 1 86220 141 2

Cavanaugh, W.T.: The Myth of Religious Violence. Oxford University Press, Oxford (2009)

Charny, I.W.: Fascism and Democracy in the Human Mind: A Bridge between Mind and Society. University of Nebraska Press, Lincoln and London (2006)

Cheldelin, S.I.: Gender and violence: redefining the moral ground. In: Rothbart, D., Korostelina, K.V. (eds.) Identity, Morality, and Threat. Studies in Violent Conflict, pp. 279–299. Rowman and Littlefield Publishers, Oxford (2006)

Chenoweth, E., Lawrence, A.: Rethinking Violence: States and Non-State Actors in Conflict. The MIT Press, Cambridge (2010)

Cialdini, R.B., Goldstein, N.J.: Social influence: compliance and conformity. Annual Review of Psychology 55, 591–621 (2004)

Clancey, W.J.: Situated Cognition: On Human Knowledge and Computer Representations. Cambridge University Press, Cambridge (1997)

Clark, A.: Language, embodiment, and the cognitive niche. Trends in Cognitive Science 10(8), 370–374 (2006)

Clark, A.: Supersizing the Mind. Embodiment, Action, and Cognitive Extension. Oxford University Press, Oxford (2008)

Clark, A., Chalmers, D.J.: The extended mind. Analysis 58, 10–23 (1998)

Claybourn, M.: Relationships between moral disengagement, work characteristics, and workplace harassment. Journal of Business Ethics 2, 283–301 (2011)

Coady, C.A.J.: Morality and Political Violence. Cambridge University Press, Cambridge (2008)

Coeckelbergh, M.: Criminal or patients? Towards a tragic conception of moral and legal responsibility. Criminal Law and Philosophy 4, 233–244 (2010)

Cohan, P.S.: When the blind lead. Business Strategy Review, 66–70 (2007)

Coid, J., Yang, M.: The impact of psychopathy on violence among the household population of Great Britain. Social Psychiatry and Psychiatric Epidemiology 46(6), 473–480 (2010)
Colapietro, V.: Futher consequences of a singular capacity. In: Muller, J., Brent, J. (eds.) Peirce, Semiosis, and Psychoanalysis, pp. 136–158. John Hopkins, Baltimore and London (2000)
Cole, M.: Cultural Psychology. Harvard University Press, Cambridge (1996)
Cole, P.: The Myth of Pure Evil. University of Edinburgh Press, Edinburgh (2006)
Collins, R.: Violence. A Micro-sociological Theory. Princeton University Press, Princeton (2008)
Coolidge, F.L., Wynn, T.: Working memory, its executive functions, and the emergence of modern thinking. Cambridge Archealogical Journal 5(1), 5–26 (2005)
Cooper, J.: Cognitive Dissonance: 50 Years of a Classic Theory. Sage, London (2007)
Cosmides, L., Tooby, J.: Consider the source: the evolution of adaptations for decoupling and metarepresentation. In: Sperber, D. (ed.) Metarepresentations: A Multidisciplinary Perspective, pp. 53–115. Oxford University Press, Oxford (2000)
Cushman, F., Macindoe, O.: The coevolution of punishment and prosociality among learning agents. In: Ohlsson, S., Catrambone, E. (eds.) Proceedings of the 32nd Annual Conference of the Cognitive Science Society, pp. 1774–1779. Cognitive Science Society, Austin (2010)
Damer, T.E.: Attacking Faulty Reasoning: A Practical Guide to Fallacy-Free Arguments. Wadsworth Publishing, New York (2000)
Darwin, C.: The Descent of Man and Selection in Relation to Sex. Princeton University Press, Princeton (1981)
Dawkins, R.: Extended phenotype – but not extended. A reply to Laland, Turner and Jablonka. Biology and Philosophy 19, 377–397 (2004)
Day, R.L., Laland, K., Odling-Smee, F.J.: Rethinking adaptation. The niche-construction perspective. Perspectives in Biology and Medicine 46(1), 80–95 (2003)
de Gelder, B., Snyder, J., Greve, D., Gerard, G., Hadjikhani, N.: Fear fosters flight: a mechanism for fear contagion when perceiving emotion expressed by a whole body. PNAS Proceedings of the National Academy of Science of the United States of America 101, 16701–16706 (2004)
de Vries, H.: Religion and Violence. Philosophical Perspectives from Kant to Derrida. Johns Hopkins University Press, Baltimore and London (2002)
Dellarosa Cummins, D.: How the social environment shaped the evolution of mind. Synthese 122, 3–28 (2000)
Dennett, D.: Elbow Room. The Variety of Free Will Worth Wanting. The MIT Press, Cambridge (1984)
Dennett, D.: Breaking the Spell: Religion as a Natural Phenomenon. Penguin Books, London (2006)
Derrida, J.: Of Grammatology. John Hopkins University Press, Baltimore (1976) Translated by G. C. Spivak
Derrida, J.: Dissemination. Chicago University Press, Chicago (1981) Translated by B. Johnson
Derrida, J.: The Post Card: From Socrates to Freud and Beyond. University of Chicago Press, Chicago (1987) Translated by A. Bass
Derrida, J.: Force of law. The mystical foundation of authority. Cardozo Law Review 11, 919–1045 (1990)
Derrida, J.: The Spectre of Marx. Routledge, New York (1994) Translated by P. Kamuf

Derrida, J.: Adieu to Emmanuel Levinas. Stanford University Press, Stanford (1999) Translated by P.-A. Brault and M. Naas

Derrida, J.: Forgiveness: a roundtable discussion with Jacques Derrida. In: Caputo, J.D., Dooley, M., Scanlon, M.J. (eds.) Questioning God, pp. 52–72. Indiana University Press, Bloomington (2001a)

Derrida, J.: On Cosmopolitanism and Forgiveness. Routledge, London (2001b)

Derrida, J.: To forgive: the unforgivable and the imprescriptible. In: Caputo, J.D., Dooley, M., Scanlon, M.J. (eds.) Questioning God, pp. 21–51. Indiana University Press, Bloomington (2001c)

Derrida, J.: The Animal That Therefore I Am. Fordham University Press, New York (2008) Translated by D. Wills. Edited by M.-L. Mallet

Dessalles, J.: Language and hominid politics. In: Knight, C., Hurford, J., Suddert-Kennedy, M. (eds.) The Evolutionary Emergence of Language: Social Function and the Origin of Linguistic Form, pp. 62–79. Cambridge University Press, Cambridge (2000)

Dewey, J.: "Force, violence, and law" and "Force and coercion". In: Bufacchi, V. (ed.) Violence: A Philosophical Anthology, pp. 5–14. Palgrave Macmillan, Houndmills (2009)

Diaz, A.A.: Media bias and electoral competition. Malaga Economic Theory Research Center Working Papers (2008)

Dodd, J.: Violence and Phenomenology. Routledge, New York (2009)

Dodd, J.: Violence and nonviolence. In: Eckstrand, N., Yates, C.S. (eds.) Philosophy and the Return of Violence, pp. 137–153. Continuum, New York (2011)

Donald, M.: A Mind So Rare. The Evolution of Human Consciousness. Norton, London (2001)

Donoso, A.M.: Review of Douglas Husak, Overcriminalization. The Limits of the Criminal Law. Criminal Law and Philosophy 4, 99–104 (2010)

Doom, N.: Applying Rawlsian approaches to resolve ethical issues: inventory and setting of a research agenda. Journal of Business Ethics 91, 127–143 (2009)

Dowd, K.: Moral hazard and the financial crisis. Cato Journal 29(1), 141–166 (2009)

Dugatkin, L.A.: The Altruism Equation. Seven Scientists Search for the Origins of Goodness. Princeton University Press, Princeton (2008)

Dunbar, R.: The social brain hypothesis. Evolutionary Anthropology 6, 178–190 (1998)

Dunbar, R.: The social brain: mind, language, and society in evolutionary perspective. Annual Review of Anthropology 32, 163–181 (2003)

Dunbar, R.: Gossip in evolutionary perspective. Review of General Psychology 8(2), 100–110 (2004)

Dunning, D., Johnson, K., Ehrlinger, J., Kruger, J.: Why people fail to recognize their own incompetence. Current Directions in Psychological Science 12(3), 83–87 (2003)

Durkheim, E. (ed.): The Elementary Forms of Religious Life. Free Press, New York (1995)

Eckstrand, N., Yates, C.S. (eds.): Philosophy and the Return of Violence. Continuum, New York (2011)

Edelman, G.M.: The Remembered Present. Basic Books, New York (1989)

Edelman, G.M.: Wider than the Sky: The Phenomenal Gift of Consciousness. Yale University Press, New Haven (1993)

Edelman, G.M.: Second Nature. Brain Science and Human Knowledge. Yale University Press, New Haven and London (2006)

Efferson, C., Lalive, R., Richerson, P.J., McElreath, R., Lubell, M.: Conformists and mavericks: the empirics of frequency-dependent cultural transmission. Evolution and Human Behavior 29, 56–64 (2008)

El-Hani, C.N., Queiroz, J., Stjernfelt, f.: Firefly femmes fatales: a case study in the semiotics of deception. Biosemiotics 1, 33–55 (2009)
Elbert, T., Weierstall, R., Schauer, M.: Fascination violence: on mind and brain of man hunters. European Archives of Psychiatry and Clinical Neuroscience 260(suppl.2), S100–S105 (2010)
Elias, N.: The Civilizing Process. Vol I. The History of Manners. Blackwell, Oxford (1969)
Elias, N.: The Civilizing Process. Vol II. State Formation and Civilization. Blackwell, Oxford (1982)
Epley, N., Caruso, E.M.: Perspective taking: misstepping into others' shoes. In: Markman, K.D., Klein, W.M.P., Suhr, J.A. (eds.) Handbook of Imagination and Mental Simulation, pp. 297–312. Psychology Press, New York (2009)
Evans, C.S.: Cracking the code. Communication and cognition in birds. In: Bekoff, M., Allen, C., Burghardt, M. (eds.) The Cognitive Animal. Empirical and Theoretical Perspectives on Animal Cognition, pp. 315–322. The MIT Press, Cambridge (2002)
Fasolo, B., Hertwig, R., Huber, M., Ludwig, M.: Size, entropy, and density: what is the difference that makes the difference between small and large real-world assortments? Psychology & Marketing 26(3), 254–279 (2009)
Felthous, A.R.: Psychopathic disorders and criminal responsibility in the USA. European Archives of Psychiatry and Clinical Neuroscience 2, 137–141 (2010)
Festinger, L.: A Theory of Cognitive Dissonance. Stanford University Press, Stanford (1957)
Fincham, F.D., Hall, J.H., Beach, S.R.H.: "Til lack of forgiveness doth us part": forgiveness and marriage. In: Worthington Jr., E.L. (ed.) Handbook of Forgiveness, pp. 207–226. Routledge, New York (2005)
Findley, K.A.: Tunnel vision. University of Wisconsin Legal Studies Research Paper No. 1116 (2010)
Findley, K.A., Scott, M.S.: The multiple dimensions of tunnel vision in criminal cases. Wisconsin Law Review 6, 291–397 (2006)
Floridi, L., Sanders, J.W.: The method of abstraction. In: Negrotti, M. (ed.) Yearbook of the Artificial. Nature, Culture, and Technology. Models in Contemporary Sciences, Bern, pp. 177–220 (2004)
Forge, J.: Mechanics and moral mediation. Metascience 18, 399–403 (2009)
Fotopoulou, A., Conway, M.A., Solms, M.: Confabulation: motivated reality monitoring. Neuropsychologia 45, 2180–2190 (2007)
Foucault, M.: Discipline and Punish: The Birth of the Prison. Vintage Books, New York (1979) Translated by A. Sheridan
Foucault, M. (ed.): Ethics: Subjectivity and Truth (Essential Works of Foucault). New Press, New York (1954-1984)
Frankfurt, H.: On Bullshit. Princeton University Press, Princeton (2005)
Frankfurt, H.: On Truth. Alfred A. Knopf, New York (2006)
Freedman, S., Enright, R.D., Contents, J.K.: A progress report on the process model of forgiveness. In: Worthington Jr., E.L. (ed.) Handbook of Forgiveness, pp. 263–376. Routledge, New York (2005)
Freiman, C.: Why be immoral? Ethical Theory and Moral Practice 13, 191–205 (2010)
Frischen, A., Bayliss, A.P., Tipper, S.P.: Gaze cueing of attention. Psychological Bulletin 133(4), 694–724 (2007)
Furedi, F. (ed.): Culture of Fear: Risk-Taking and the Morality of Low Expectation. Second revised edition.Continuum, London (2002)
Furnham, A., Daoud, Y., Swami, V.: How to spot a psychopath. Lay theories of psychopathy. Social Psychiatry and Psychiatric Epidemiology 44, 464–472 (2009)

Fuse, T.: Suicide and culture in Japan: a study of seppuku as an institutionalized form of suicide. Social Psychiatry and Psychiatric Epidemiology 15(2), 57–63 (1980)

Gabbay, D.M., Woods, J.: The Reach of Abduction. A Practical Logic of Cognitive Systems, vol. 2. North-Holland, Amsterdam (2005)

Gallese, V.: Embodied simulation: from neurons to phenomenal experience. Phenomenology and Cognitive Science 4, 23–48 (2005)

Gallese, V.: Intentional attunement: a neurophysiological perspective on social cognition and its disruption in autism. Brain Research 1079, 15–24 (2006)

Galtung, J.: Violence, peace and peace research. In: Bufacchi, V. (ed.) Violence: A Philosophical Anthology, pp. 78–109. Palgrave Macmillan, Houndmills (2009)

Gambetta, D.: The Sicilian Mafia. The Business of Private Protection. Harvard University Press, Cambridge (1996)

Gannon, A., Ward, T., Beech, A.R., Ficher, D. (eds.): Aggressive Offenders' Cognition: Theory, Research and Practice. Wiley, Chichester (2007)

Garver, N.: What violence is. In: Bufacchi, V. (ed.) Violence: A Philosophical Anthology, pp. 170–182. Palgrave Macmillan, Houndmills (2009)

Gebhard, U., Nevers, P., Bilmann-Mahecha, E.: Moralizing trees: anthropomorphism and identity in children's relationships to nature. In: Clayton, S., Opotow, S. (eds.) Identity and the Natural Environment: The Psychological Significance of Nature, pp. 91–111. The MIT Press, Cambridge (2003)

Gelderloos, P.: How Nonviolence Protects the State. South End Press, Boston (2011)

Giardini, F., Andrighetto, G., Conte, R.: A cognitive model of punishment. In: Ohlsson, S., Catrambone, E. (eds.) Proceedings of the 32nd Annual Conference of the Cognitive Science Society, pp. 1282–1288. Cognitive Science Society, Austin (2010)

Gibson, J.J.: The Ecological Approach to Visual Perception. Houghton Mifflin, Boston (1979)

Gigerenzer, G.: Moral satisficing: rethinking moral behavior as bounded rationality. Topics in Cognitive Science 2, 528–554 (2010)

Gilhus, I.S.: Animals, Gods and Humans: Changing Attitudes to Animals in Greek, Roman and Early Christian Thought. Routledge, New York (2006)

Gintis, H., Bowles, S., Boyd, R.T., Fehr, E. (eds.): Moral Sentiments and Material Interests: The Foundations of Cooperation in Economic Life. The MIT Press, Cambridge (2005)

Gintis, H., Smith, E., Bowles, S.: Costly signaling and cooperation. Journal of Theoretical Biology 213, 103–119 (2001)

Girard, R.: Violence and the Sacred. Johns Hopkins University Press, Baltimore (1977)

Girard, R.: The Scapegoat. Johns Hopkins University Press, Baltimore (1986)

Glassner, B. (ed.): The Culture of Fear: Why Americans Are Afraid of the Wrong Things. Basic Books, New York (1999)

Gluckman, M.: Papers in honor of Melville J. Herskovits: gossip and scandal. The American Economic Review 4(3), 307–316 (1963)

Godfrey-Smith, P.: Complexity and the Function of Mind in Nature. Cambridge University Press, Cambridge (1998)

Goffman, E.: Encounters. Two Studies in the Sociology of Interaction. Bobbs-Merrill, Indianapolis (1961)

Goffman, E.: Frame Analysis. An Essay on the Organization of Experience. Northeastern University Press, Boston (1986)

Goldman, A.H.: Is moral motivation rationally required? The Journal of Ethics 14(1), 1–16 (2010)

Goldstein, D.G., Gigerenzer, G.: Models of ecological rationality: the recognition heuristic. Psychological Review 109, 75–90 (2002)

Goldstein, W.S. (ed.): Marx, Critical Theory, and Religion. A Critique of Rational Choice. Brill, Cambridge (2006)

Goldstone, R.L., Gureckis, T.M.: Collective behavior. Topics in Cognitive Science 1, 412–438 (2009)

Gotterbarn, D.: Informatics and professional responsibility. Science and Engineering Ethics 7(2), 221–230 (2001)

Gotterbarn, D.: Scientific research is a moral duty. Journal of Medical Ethics 31, 241–248 (2005)

Greene, J.: From neural "is" to moral "ought": what are the moral implications of neuroscientific moral psychology? Nature Reviews 4, 847–850 (2003)

Griffiths, P.J.: Introduction. In: Griffiths, P. (ed.) Christianity Through Non-Christian Eyes, pp. 1–15. Orbis Books, New York (1990)

Griswold, C.: Forgiveness: A Philosophical Exploration. Cambridge University Press, Cambridge (2007)

Grosz, E.: The time of violence. Deconstruction and value. College Literature 26(1), 8–18 (1999)

Guru, G. (ed.): Humiliation. Claims and Context. Oxford University Press, Oxford (1989)

Gusterson, H.: Anthropology and militarism. Annual Review of Anthropology 16, 155–175 (2007)

Hall, J.R.: Religion and violence: social processes in comparative perspective. In: Dillon, M. (ed.) Handbook of the Sociology of Religion, pp. 359–381. Cambridge University Press, Cambridge (2009)

Hall, K.R.L.: Aggression in monkey and ape societies. In: Carthy, J., Ebling, F. (eds.) The Natural History of Aggression, pp. 51–64. Academic Press, London (1964)

Hamilton, W.D.: The genetical evolution of social behavior. Theoretical Biology 7(1), 1–52 (1964)

Hample, D., Warner, B., Young, D.: Framing and editing interpersonal arguments. Argumentation 23, 21–37 (2009)

Hample, D., Han, B., Payne, D.: The aggressiveness of playful arguments. Argumentation 24(4), 245–254 (2010)

Hample, D., Warner, B., Norton, H.: The effects of arguing expectations and predispositions on perceptions of argument quality and playfulness. Argumentation and Advocacy 43, 1–13 (2006)

Hanssen, B.: Critique of Violence: Between Poststructuralism and Critical Theory. Routledge, London (2000)

Hare, R.D.: Without Conscience: The Disturbing World of the Psychopaths among Us. Pocket Books, New York (1993)

Harris, J.: The Marxist conception of violence. In: Bufacchi, V. (ed.) Violence: A Philosophical Anthology, pp. 185–206. Palgrave Macmillan, Houndmills (2009)

Harris, S., Kaplan, J.T., Curiel, A., Bookheimer, S.Y., Iacoboni, M., Cohen, M.S.: The neural correlates of religious and nonreligious belief. PLoS ONE (Public Library of Science) 4(10), 1–9 (2009)

Haybron, D.M.: Moral monsters and saints. The Monist 85(2), 260–284 (2002)

Haynes, G.A.: Testing the boundaries of the choice overload phenomenon: the effect of number of options and time pressure on decision difficulty and satisfaction. Psychology & Marketing 26(3), 299–320 (2009)

Hebb, D.O.: The Organization of Behavior. John Wiley, New York (1949)

Hegselmann, R., Krause, U.: Truth and cognitive division of labour: first steps towards a computer aided social epistemology. Journal of Artificial Societies and Social Simulation 9(3), 1–28 (2006)

Heitmeyer, W., Haupt, H.-G., Malthaner, S., Kirschner, A. (eds.): Control of Violence. Historical and International Perspectives on Violence in Modern Societies. Springer, Heidelberg (2011)

Hildebrandt, M.: The indeterminacy of an emergency: challenges to criminal jurisdiction in constitutional democracies. Criminal Law and Philosophy 4, 161–181 (2010)

Hill, P.C., Exline, J.J., Cohen, A.B.: The social psychology of justice and forgiveness in civil and organizational settings. In: Worthington Jr., E.L. (ed.) Handbook of Forgiveness, pp. 477–490. Routledge, New York (2005)

Himma, K.E.: The moral significance of the interest in information: reflections on a fundamental right to information. Journal of Information, Communication and Ethics in Society 2(4), 191–201 (2004)

Hirigoyen, M.-F.: Stalking the Soul. Helen Marx Books, New York (2000) Translated by Helen Marx

Hirstein, W.: Brain Fiction. Self-Deception and the Riddle of Confabulation. The MIT Press, Cambridge (2005)

Hirstein, W.: Introduction. What is confabulation? In: Hirstein, W. (ed.) Confabulation: Views from Neuroscience, Psychiatry, Psychology and Philosophy, pp. 1–12. Oxford University Press, Oxford (2009)

Holton, R.: Willing, Wanting, Waiting. Oxford University Press, Oxford (2009)

Hoskins, A.: Fair play, political obligations, and punishment. In: Criminal Law and Philosophy 5(1), 53–71 (2011)

Houellebecq, M.: The Possibility of an Island. Knopf, New York (2006)

Huff, C.: Unintentional power in the design of computing systems. Computers and Society 16(4), 6–9 (1996)

Husak, D.: Overcriminalization. The Limits of the Criminal Law. Oxford University Press, Oxford (2008)

Hutchins, E.: Cognition in the Wild. The MIT Press, Cambridge (1995)

Ingram, G.P.D., Piazza, J.R., Bering, J.M.: The adaptive problem of absent third-party punishment. In: Høgh-Olesen, H., Bertelsen, P., Tønnesvang, J. (eds.) Human Characteristics: Evolutionary Perspectives on Human Mind and Kind, pp. 205–229. Cambridge Scholars Publishing, Newcastle upon Tyne (2009)

Iyengar, S., Lepper, M.: When choice is demotivating: can one desire too much of a good thing. Journal of Personality and Social Psychology 79, 995–1006 (2000)

Jablonka, E., Lamb, M.J.: Evolution in Four Dimensions. Genetic, Epigenetic, Behavioral, and Symbolic Variation in the History of Life. The MIT Press, Cambridge (2005)

Jankélévitch, V.: Forgiveness. University of Chicago Press, Chicago (2005)

Jantz, G.L., McMurray, A.: Healing the Scars of Emotional Abuse. Revell, Grand Rapids (2009)

Jantzen, G.M.: Violence to Eternity. Routledge, London and New York (2009) Edited by J. Carrette and M. Joy

Jeannerod, M., Pacherie, E.: Agent, simulation, and self-identification. Mind and Language 19(2), 113–146 (2004)

Jensen, T.: Beyond good and evil: the adiaphoric company. Journal of Business Ethics 96, 425–434 (2010)

Jessup, R.K., Veinott, E.S., Todd, P.M., Busemeyer, J.R.: Leaving the store empty-handed: Testing explanations for the too-much-choice effect using decision field theory. Psychology & Marketing 26(3), 204–212 (2009)

Johnson, D.D.P.: The error of God: error management theory, religion, and the evolution of cooperation. In: Levin, S. (ed.) Games, Groups, and the Global Good, pp. 169–180. Springer, Berlin (2009)

Johnson, D.D.P., Bering, J.: Hand of God, mind of man: punishment and cogniiton in the evolution of cooperation. Evolutionary Psychology 4, 219–233 (2006)

Johnson, D.D.P., Kruger, O.: The God of wrath. Supernatural punishment and the evolution of cooperation. Political Theology 5(2), 159–176 (2004)

Joiner, T.: Why People Die by Suicide. Harvard University Press, Cambridge (2007)

Jones, M., Fabian, A.C. (eds.): Conflict. Cambridge University Press, Cambridge (2005)

Juergensmeyer, M.: Terror in the Mind of God: The Global Rise of Religious Violence. University of California Press, Berkeley and Los Angeles (2000)

Jung, C.G.: Two kinds of thinking. In: The Collected Works, 2nd edn., vol. 5, pp. 7–33. Princeton University Press, Princeton (1967) Translated by R. F. C. Hull

Jung, C.G.: Archetypes and the collective unconscious. In: The Collected Works, 2nd edn., vol. 9, pp. 3–41. Princeton University Press, Princeton (1968a) Translated by R. F. C. Hull

Jung, C.G.: A study in the process of individuation. In: The Collected Works, 2nd edn., vol. 9, pp. 290–354. Princeton University Press, Princeton (1968b) Translated by R. F. C. Hull

Jung, C.G.: On psychic energy. In: The Collected Works, 2nd edn., vol. 8, pp. 3–66. Princeton University Press, Princeton (1972a) Translated by R. F. C. Hull

Jung, C.G.: On the nature of the psyche. In: The Collected Works, 2nd edn., vol. 8, pp. 159–236. Princeton University Press, Princeton (1972b) Translated by R. F. C. Hull

Jung, C.G.: The transcendent function. In: The Collected Works, 2nd edn., vol. 8, pp. 67–91. Princeton University Press, Princeton (1972c) Translated by R. F. C. Hull

Kahn, P.H.J.: The development of environmental moral identity. In: Clayton, S., Opotow, S. (eds.) Identity and the Natural Environment: The Psychological Significance of Nature, pp. 113–134. The MIT Press, Cambridge (2003)

Kahneman, D., Tversky, A.: Prospect theory: an analysis of decision under risk. Econometrica 47, 263–291 (1979)

Kalyvas, S.N., Shapiro, I., Masoud, T. (eds.): Order, Conflict, and Violence. Cambridge University Press, Cambridge (2008)

Kamitake, Y.: From democracy to ochlocracy. Hitotsubashi Journal of Economics 48(1), 83–93 (2007)

Kant, I.: Groundwork of the Metaphysics of Morals, 3rd edn. Harper & Row, New York (1964a) Reprint of 1956, edited and translated by H. J. Paton, 3rd edn. Hutchinson & Co., Ltd., London

Kant, I.: The Doctrine of Virtue. Harper & Row, New York (1964b) Translated by M. J. Gregor. Foreword by H. J. Paton

Kant, I.: The End of All Things. In: Religion within the Boundaries of Mere Reason, pp. 195–205. Cambridge University Press, Cambridge (1998a) Translated and edited by A. Wood and G. di Giovanni. Introduction by R. M. Adams

Kant, I.: Religion within the Boundaries of Mere Reason. Cambridge University Press, Cambridge (1998b) Translated and edited by A. Wood and G. Di Giovanni. Introduction by R. M. Adams

Kaplan, H.K., Hill, K., Lancaster, J., Hurtado, M.: A theory of human life history evolution: diet, intelligence, and longevity. Evolutionary Anthropology 9(4), 155–185 (2000)

Kaplan, L.V.: Intentional agency, responsibility, and justice. In: Malle, B.F., Moses, L.J., Baldwin, D.A. (eds.) Intentions and Intentionality, pp. 369–379. The MIT Press, Cambridge (2001)

Kaplar, M.E., Gordon, A.K.: The enigma of altruistic lying: perspective differences in what motivates and justifies lie telling within romantic relationships. Personal Relationships 11, 489–507 (2004)

Karawan, I.A., McCormick, W., Reynolds, S.E. (eds.): Values and Violence. Intangible Aspects of Terrorism. Springer, Heibelberg (2008)

Karlsson, N., Loewenstein, G., Seppi, D.: The ostrich effect: selective attention to information. Journal of Risk and Uncertainty 38, 95–115 (2009)

Kaufmann, P., Kuch, H., Neuhäuser, C., Webster, E.: Humiliation, Degradation, Dehumanization. Human Dignity Violated. Springer, Heidelberg (2011)

Kaznatcheev, A.: The cognitive cost of ethnocentrism. In: Ohlsson, S., Catrambone, E. (eds.) Proceedings of the 32nd Annual Conference of the Cognitive Science Society, pp. 967–971. Cognitive Science Society, Austin (2010)

Keane, J.: Reflections on Violence, Verso, London (1996)

Keen, A.: The Cult of Amateur. How Today's Internet is Killing Our Culture and Assaulting Our Economy. Nicholas Brealey Publishing, London (2007)

Kelly, D., Stich, S., Haley, K.J., Eng, S.J., Fessler, D.M.T.: Harm, affect, and the moral/conventional distinction. Mind & Language 22(2), 17–131 (2007)

Kiehl, K.: Without morals: the cognitive neuroscience of criminal psychopaths. In: Sinnott-Armstrong, W. (ed.) Moral Psychology, vol. 3, pp. 119–149. The MIT Press, Cambridge (2008)

Kierkegaard, S.: The ancient tragical motif as reflected in the modern. In: Kierkegaard, S. (ed.) Either/Or: A Fragment of Life, vol. 1, Princeton University Press, Princeton (1944) Translated by D. F. Swenson and L. M. Swenson

Kimhi, S., Sagy, S.: Moral justification and feelings of adjustment to military law-enforcement situation: the case of Israeli soldiers serving at army roadblocks. Mind and Society 7, 177–191 (2008)

Klucharev, V., Hytonen, K., Rijpkema, M., Smidts, A., Fernandez, G.: Reinforcement learning predicts social conformity. Neuron 61, 140–151 (2009)

Koenigs, M., Tranel, D.: Irrational economic decision-making after ventromedial prefrontal damage: evidence from the ultimatum game. Journal of Neuroscience 27(4), 951–956 (2007)

Kondo, D.: Poststructuralist theory as political necessity. Amerasia 21(1/2), 95–100 (1995)

Korostelina, K.V.: Cultural differences of perceptions of the other. In: Rothbart, D., Korostelina, K.V. (eds.) Identity, Morality, and Threat. Studies in Violent Conflict, pp. 147–175. Rowman and Littlefield Publishers, Oxford (2006a)

Korostelina, K.V.: Identity salience as a determinant of the perception of other. In: Rothbart, D., Korostelina, K.V. (eds.) Identity, Morality, and Threat. Studies in Violent Conflict, pp. 101–126. Rowman and Littlefield Publishers, Oxford (2006b)

Kott, A., McEneaney, W.M. (eds.): Adversarial Reasoning: Computational Approaches to Reading the Opponent's Mind. Chapman & Hall/CRC, Taylor & Francis Group, Boca Raton, FL (2007)

Kramer, J., Diana, A.: The Guru Papers: Masks of Authoritarian Power. North Atlantic Books, Berkeley (1993)

Kuperman, A.J.: The moral hazard of humanitarian intervention: lessons from the Balkans. International Studies Quarterly 52(1), 49–80 (2009)

Lacan, J.: Le désire et son interprétation. Unpublished Seminar (May 20 ,1959)

Lacan, J.: Écrits: The First Complete Edition in English. W. W. Norton & Company, New York (1966) Translated by B. Fink
Lacey, N.: Psychologising Jekyll, demonising Hyde: the strange case of criminal responsibility. Criminal Law and Philosophy 4, 109–133 (2010)
Lackey, D.P.: The Ethics of War and Peace. Prentice-Hall, Englewoods Cliffs (1989)
Lahti, D.C.: Parting with illusions in evolutionary ethics. Biology and Philosophy 18, 639–651 (2003)
Lahti, D.C., Weinstein, B.S.: The better angels of our nature: group stability and the evolution of moral tension. Evolution and Human Behavior 2, 47–63 (2005)
Laland, K.N., Brown, G.R.: Niche construction, human behavior, and the adaptive-lag hypothesis. Evolutionary Anthropology 15, 95–104 (2006)
Laland, K.N., Odling-Smee, F.J., Feldman, M.W.: Niche construction, biological evolution and cultural change. Behavioral and Brain Sciences 23(1), 131–175 (2000)
Laland, K.N., Odling-Smee, F.J., Feldman, M.W.: Cultural niche construction and human evolution. Journal of Evolutionary Biology 14, 22–33 (2001)
Laland, K.N., Odling-Smee, F.J., Feldman, M.W.: On the breath and significance of niche construction: a reply to Grittiths, Okasha and Sterelny. Biology and Philosophy 20, 37–55 (2005)
Larrick, R.P.: Debiasing. In: Koehler, D., Harvey, N. (eds.) Blackwell Handbook of Judgment and Decision Making, pp. 316–338. Blackwell, Oxford (2009)
Lee, S.: Poverty and violence. In: Bufacchi, V. (ed.) Violence: A Philosophical Anthology, pp. 322–333. Palgrave Macmillan, Houndmills (2009)
Leibenstein, H.: Bandwagon, Snob, and Veblen Effects in the theory of consumers' demand. The Quarterly Journal of Economics 64(2), 183–207 (1950)
Leigh, H.: Genes, Memes, Culture, and Mental Illness. Springer, Heidelberg (2010)
Letschert, R., van Dijk, J.: The New Faces of Victimhood. Globalization, Transnational Crimes and Victim Rights. Springer, Heidelberg (2011)
Levinas, E.: Otherwise than Being: Or Beyond Essence, Nijhoff, The Hague (1981) Translated by A. Lingis
Levinas, E.: Ethics and Infinity. Duquesne University Press, Pittsburgh (1985) Conversations with P. Nemo, translated by R. A. Cohen
Levy, D., Sznaider, N.: Forgive and not forget: reconciliation between forgiveness and resentment. In: Barkan, E., Karn, A. (eds.) Taking Wrongs Seriously. Apologies and Reconciliation, pp. 83–100. Stanford University Press, Stanford (2006)
Leyton, M.: Simmetry, Causality, Mind. The MIT Press, Cambridge (1999)
Leyton, M.: A Generative Theory of Shape. Springer, Berlin (2001)
Leyton, M.: The Structure of Paintings. Springer, Berlin (2006)
Liebsch, B., Mensink, D. (eds.): Gewalt Verstehen. Akademie Verlag, Berlin (2003)
Lines, D.: The Bullies. Understanding Bullies and Bullying. Jessica Kingsley Publishers, London (2008)
Locke, J.: Two Treatises of Government and a Letter Concerning Toleration. Yale University Press, New Haven (2003) Edited by Ian Shapiro
Loula, A., Gudwin, R., El-Hani, D.: Emergence of self-organized symbol-based communication in artificial creatures. Cognitive Systems Research 2, 131–147 (2010)
Lurie, N.H.: Decision making information-rich in environments. The role of information structure. Journal of Consumer Research 30, 473–486 (2004)
MacCallum, G.C.: What is wrong with violence. In: Bufacchi, V. (ed.) Violence: A Philosophical Anthology, pp. 112–133. Palgrave Macmillan, Houndmills (2009)
Mackie, J.L.: Ethics. Inventing Right and Wrong. Penguin, London (1990)

MacNeill Horton, A., Hartlage, L.C. (eds.): Handbook of Forensic Neuropsychology. Springer, New York (2003)
Magnani, L.: Abduction, Reason, and Science. Processes of Discovery and Explanation. Kluwer Academic/Plenum Publishers (2001)
Magnani, L.: Creativity and the disembodiment of mind. In: Gervás, P., Pease, A., Veale, T. (eds.) Proceedings of CC05, Computational Creativity Workshop, IJCAI 2005, Edinburgh, pp. 60–67 (2005)
Magnani, L.: Mimetic minds. Meaning formation through epistemic mediators and external representations. In: Loula, A., Gudwin, R., Queiroz, J. (eds.) Artificial Cognition Systems, pp. 327–357. Idea Group Publishers, Hershey (2006)
Magnani, L.: Animal abduction. From mindless organisms to artifactual mediators. In: Magnani, L., Li, P. (eds.) Model-Based Reasoning in Science, Technology, and Medicine, pp. 3–37. Springer, Berlin (2007a)
Magnani, L.: Morality in a Technological World. Knowledge as Duty. Cambridge University Press, Cambridge (2007b)
Magnani, L.: Abductive Cognition. The Eco-Cognitive Dimensions of Hypothetical Reasoning. Springer, Heidelberg (2009)
Mahoney, A., Rye, M.S., Pargament, K.I.: When the sacred is violated: desecration as a unique challenge to forgiveness. In: Worthington Jr., E.L. (ed.) Handbook of Forgiveness, pp. 57–71. Routledge, New York (2005)
Mandoki, K.: Terror and aethestics: Nazi strategies for mass organization. Renaissance and Modern Studies 42, 4–81 (1999)
Mason, T. (ed.): Forensic Psychiatry: Influences of Evil. Humana Press, Totowa (2006)
Maynard Smith, J., Harper, D.: Animal Signals. Oxford University Press, Oxford (2003)
McCullough, M.E., Worthington Jr., E.L.: Religion and the forgiving personality. Journal of Personality 67(6), 1141–1164 (1999)
McCumber, J.: Philosophy after 9/11. In: Eckstrand, N., Yates, C.S. (eds.) Philosophy and the Return of Violence, pp. 17–30. Continuum, New York (2011)
McKee, G.R.: Why Mothers Kill: A Forensic Psychologist's Casebook. Oxford University Press, Oxford (2006)
McKim, R.: On religious ambiguity. Religious Studies 44, 373–392 (2008)
McKinlay, A., McVittie, C.: Social Psychology and Discourse. John Wiley & Sons, Chichester (2008)
McMahan, J.: Killing in War. Clarendon Press, Oxford (2009)
Mele, A.: Weakness of will and akrasia. Philosophical Studies 150, 391–404 (2010)
Mercier, H.: The social origins of folk epistemology. Review of Philosophy and Psychology 1, 499–514 (2010)
Mesoudi, A., Whiten, A., Laland, K.N.: Perspective: is human cultural evolution Darwinian? Evidence reviewed from the perspective of the "Origin of Species". Evolution 58(1), 1–11 (2004)
Métraux, A.: South American Thunderbirds. The Journal of American Folklore 57(224), 132–135 (1944)
Meynen, G.: Free will and mental disorder: Exploring the relationship. Theoretical Medicine and Bioethics 31(6), 429–443 (2010)
Michaels, D.: Doubt is Their Product: How Industry's Assault on Science Threatens Your Health. Oxford University Press, Oxford (2008)
Milbank, J.: Theology and Social Theory. Beyond Secular Reason. Blackwell, Maiden (1990)
Milgram, S.: Obedience to Authority. Harper Perennial, New York (1975)
Mill, J.S.: On Liberty. Penguin Books, London (1985)

Miller, G.A.: Mistreating psychology in the decades of brain. Perspectives on Psychological Science 5, 716–743 (2010)
Misak, C.: Pragmatism and solidarity, bullshit, and other deformities of truth. Midwest Studies in Philosophy 32, 111–121 (2008)
Mitchell, R.W., Thompson, N.S.: Deception: Perspectives on Human and Nonhuman Deceit. Suny Press, Albany (1986)
Mohandas, E.: Neurobiology of spirituality. In: Singh, A.R., Singh, S.A. (eds.) Medicine, Mental Health, Science, Religion, and Well-Being, Mumbai, Mens Sana Monographs, pp. 63–80 (2008)
Mokros, A., Osterheider, M., Hucker, S.J., Nitschke, J.: Psychopathy and sexual sadism. Law and Human Behavior 35, 188–199 (2011)
Moll, J., de Oliveira-Souza, R.: When morality is hard to like. Scientific American Mind 3, 30–35 (2008)
Moll, J., de Oliveira-Souza, R., Garrido, G., Bramati, I.E., Caparelli-Daquer, E.M., Paiva, M.L., Zahn, R., Grafman, J.: The self as a moral agent: linking the neural bases of social agency and moral sensitivity. Social Neuroscience 2(3-4), 336–352 (2007)
Mookherjee, M.: Democracy, Religious Pluralism and the Liberal Dilemma of Accommodation. Springer, Heidelberg (2011)
Moolakkattu, J.S.: Remorse and forgiveness: a contemporary political discussion. Journal of Social Sciences 26(1), 11–16 (2011)
Moussaid, M., Garnier, S., Theraulaz, G., Helbing, D.: Collective information processing and pattern formation in swarms, flocks and crowds. Topics in Cognitive Science 1, 469–497 (2009)
Mucchielli, L.: Are we living in a more violent society? British Journal of Criminology 50, 808–829 (2010)
Murphy, J.G.: Forgiveness, self-respect, and the value of resentment. In: Worthington Jr., E.L. (ed.) Handbook of Forgiveness, pp. 1–15. Routledge, New York (2005)
Murray, M.J., Goldberg, A.: Evolutionary accounts of religion: explaining and explaining away. In: Schloss, J., Murray, M. (eds.) The Believing Primate: Scientific, Philosophical and Theological Perspectives on the Evolution of Religion, pp. 179–199. Oxford University Press, Oxford (2009)
Mussolini, B.: The Doctrine of Fascism. In: Mussolini, B., Fascism: Doctrine and Institutions, Ardita, Rome, pp. 7–42 (1935)
Nagel, T.: Mortal Questions. Cambridge University Press, Cambridge (1979)
Nelson, R.J. (ed.): Biology of Aggression. Oxford University Press, Oxford (2006)
Newberg, A., D'Aquili, E., Rause, V. (eds.): Why God Won't Go Away: Brain Science and the Biology of Belief. Ballantine Books, New York (2001)
Nielsen, K.: On justifying violence. In: Bufacchi, V. (ed.) Violence: A Philosophical Anthology, pp. 209–241. Palgrave Macmillan, Houndmills (2009)
Nietzsche, F.: On the Genealogy of Morality. Cambridge University Press, New York (2007) Edited by K. Ansell-Pearson, translated by C. Diethe
Noll, J.G.: Forgiveness in people experiencing trauma. In: Worthington Jr., E.L. (ed.) Handbook of Forgiveness, pp. 263–376. Routledge, New York (2005)
Norenzayan, A.: Evolution and transmitted culture. Psychological Inquiry 17, 123–128 (2006)
Norenzayan, A., Atran, S., Faulkner, J., Schaller, M.: Memory and mystery: the cultural selection of minimally counterintuitive narratives. Cognitive Science 30, 531–553 (2006)
Norenzayan, A., Shariff, A.F.: The origin and evolution of religious prosociality. Science 322, 58–62 (2008)

North, D.C., Wallis, J.J., Weingast, B.R.: Violence and Social Orders: A Conceptual Framework for Interpreting Recorded Human History. Cambridge University Press, Cambridge (2009)

Numbers, R.L. (ed.): Galileo Goes to Jail and Other Myths about Science and Religion. Harvard University Press, Cambridge (2009)

Odling-Smee, F.J.: The Role of Behavior in Evolution. Cambridge University Press, Cambridge (1988)

Odling-Smee, F.J., Laland, K.N., Feldman, M.W.: Niche Construction. The Neglected Process in Evolution. Princeton University Press, Princeton (2003)

O'Gorman, R., Wilson, D.S., Miller, R.: An evolved cognitive bias for social norms. Evolution and Human behavior 29, 71–78 (2008)

Olsson, E.J.: Knowledge, truth, and bullshit: reflections on Frankfurt. Midwest Studies in Philosophy 32, 94–110 (2008)

Oppenheimer, J.: The Hate Handbook. Oppressors, Victims, and Fighters. Rowman and Littlefield Publishers, Oxford (2005)

Pacini, R., Epstein, S.: The relation of rational and experiential information processing styles to personality, basic beliefs, and the ratio-bias phenomenon. Journal of Personality and Social Psychology 76(6), 972–987 (1999)

Palmer, I.E.: Moral cognition and aggression. In: Gannon, A., Ward, T., Beech, A.R., Ficher, D. (eds.) Aggressive Offenders' Cognition: Theory, Research and Practice, pp. 199–214. Wiley, Chichester (2007)

Peirce, C.S.: The Charles S. Peirce Papers: Manuscript Collection in the Houghton Library. The University of Massachusetts Press, Worcester (1966) Annotated Catalogue of the Papers of Charles S. Peirce. Numbered according to Richard S. Robin. Available in the Peirce Microfilm edition. Pagination: CSP = Peirce / ISP = Institute for Studies in Pragmaticism

Peirce, C.S.: Historical Perspectives on Peirce's Logic of Science: A History of Science, Mouton, Berlin, vol. I-II (1987) edited by C. Eisele

Peirce, C.S.: Collected Papers of Charles Sanders Peirce. Harvard University Press, Cambridge; vol. 1-6, edited by Hartshorne, C., Weiss, P.: vol. 7-8, edited by Burks, A.W. (CP, 1931-1958)

Perla, R.J., Carifio, J.: Psychological, philosophical, and educational criticisms of Harry Frankfurt's concept of and views about "bullshit" in human discourse, discussions, and exchanges. Interchange 38(2), 119–136 (2007)

Peters, K.E.: Pluralism and ambivalence in the evolution of morality. Zygon 38(2), 333–354 (2003)

Petitot, J.: Les catastrophes de la parole: de Roman Jakobson à René Thom. Maloine, Paris (1985)

Petitot, J.: Physique du sens. Éditions du CNRS, Paris (1992)

Petitot, J.: Morphogenesis of Meaning. Peter Lang, Bern. Translated by Franson Manjali (2003)

Pina, M., Arménio Rego, C., Clegg, S.R.: Obedience and evil: from Milgram and Kampuchea to normal organizations. Journal of Business Ethics 97, 291–309 (2010)

Pinel, P.: Abhandlung über Geisteverirrunger oder Manie. Carl Schaumburg, Wien (1801)

Pinker, S.: How the Mind Works. W. W. Norton, New York (1997)

Pinker, S.: Language as an adaptation to the cognitive niche. In: Christiansen, M.H., Kirby, S. (eds.) Language Evolution, pp. 16–37. Oxford University Press, Oxford (2003)

Pinker, S.: The cognitive niche: coevolution of intelligence, sociality, and language. PNAS (Proceedings of the National Academy of Science of the United States of America) 107, 8993–8999 (2010)

Polybius, The Rise of the Roman Empire. Penguin, London (1979) Translated by Ian Scott-Kilvert

Prigogine, I., Stengers, I.: Order out of Chaos. Man's New Dialogue with Nature, Bantam, London (1984)

Prinz, J.J.: The Emotional Construction of Morals. Oxford University Press, Oxford (2007)

Proctor, J.D. (ed.): Science, Religion, and the Human Experience. Oxford University Press, Oxford (2005)

Proctor, R.N., Schiebinger, L. (eds.): Agnotology. The Making and Unmaking of Ignorance. Stanford University Press, Stanford (2008)

Racamier, J.-C.: L'inceste et l'incestuel. Éditions du Collège de psychanalyse groupale et familiale, Paris (1995)

Raftopoulos, A.: Cognition and Perception. How Do Psychology and Neural Science Inform Philosophy? The MIT Press, Cambridge (2009)

Randolph Mayes, G.: Naturalizing cruelty. Biology and Philosophy 24, 21–34 (2009)

Rawls, J.: A Theory of Justice. Harvard University Press, Cambridge (1971)

Rawls, J.: The idea of an overlapping consensus. Oxford Journal of Legal Studies 7(1), 1–25 (1987)

Reagan, A.B., Walters, L.V.W.: Tales from the Hoh and Quileute. The Journal of American Folklore 46(182), 297–133465 (1933)

Reesor, M.E.: The "indifferents" in the old and middle Stoa. Transactions and Proceedings of the American Philological Association 82, 102–110 (1951)

Rejali, D.: Torture and Democracy. Princeton University Press, Princeton (2007)

Rest, J., Narvaez, N., Bebeau, M.J., Thoma, S.J.: Postconventional Moral Thinking. A Neo-Kohlbergian Approach. Lawrence Erlbaum, Mahwah (1999)

Reynolds, T.E.: Reviewed work: Religion and Violence: Philosophical Perspectives from Kant to Derrida, by Hent de Vries. The Journal of Religion 83(3), 479–480 (2003)

Richerson, P.J., Boyd, R.: Not by Genes Alone. How Culture Trasformed Human Evolution. The University of Chicago Press, Chicagoand London (2005)

Richerson, P.J., Newson, L.: Is religion adaptive? Yes, no, neutral. But mostly we don't know. In: Schloss, J., Murray, M. (eds.) The Believing Primate: Scientific, Philosophical and Theological Perspectives on the Evolution of Religion, pp. 100–117. Oxford University Press, Oxford (2009)

Rizzolatti, G., Carmada, R., Gentilucci, M., Luppino, G., Matelli, M.: Functional organization of area 6 in the macaque monkey. II area F5 and the control of distal movements. Experimental Brain Research 71, 491–507 (1988)

Roberts, L.D.: The relation of children's early word aquisition to abduction. Foundations of Science 9(3), 307–320 (2004)

Rohwer, Y.: Hierarchy maintenance, coalition formation, and the origin of altruistic punishment. Philosophy of Science 74, 802–812 (2007)

Rosas, A.: Multilevel selection and human altruism. Biology and Philosophy 23, 205–215 (2008)

Rosas, A.: The return of reciprocity: a psychological approach to the evolution of cooperation. Biology and Philosophy 23, 555–566 (2009)

Rose, S.: The Future of the Brain. The Promise and Perils of Tomorrow's Neuroscience. Oxford University Press, Oxford (1993)

Rothbart, D., Korostelina, K.V.: Introduction: identity, morality, and threat. In: Rothbart, D., Korostelina, K.V. (eds.) Identity, Morality, and Threat. Studies in Violent Conflict, pp. 1–15. Rowman and Littlefield Publishers, Oxford (2006a)

Rothbart, D., Korostelina, K.V.: Moral denigration of the other. In: Rothbart, D., Korostelina, K.V. (eds.) Identity, Morality, and Threat. Studies in Violent Conflict, pp. 29–57. Rowman and Littlefield Publishers, Oxford (2006b)

Ruggiero, V.: Understanding Political Violence: A Criminological Analysis. Open University Press, New York (2006)

Russell, L.: Evil, monsters, and dualism. Ethical Theory and Moral Practice 13, 45–58 (2010)

Saban, M.: Entertaining the stranger. Journal of Analytical Psychology 56, 92–108 (2011)

Salmi, J.: The different categories of violence. In: Bufacchi, V. (ed.) Violence: A Philosophical Anthology, pp. 311–319. Palgrave Macmillan, Houndmills (2009)

Sartre, J.: Being and Nothingness: An Essay on Phenomenological Ontology. Philosophical Library, New York (1956) Translated and with an introduction by H. E. Barnes

Schapiro, B.A.: Psychoanalysis and the problem of evil: debating Othello in the classroom. American Imago 60(4), 481–499 (2002)

Schimmel, S.: The Tenacity of Unreasonable Beliefs: Fundamentalism and the Fear of Truth. Oxford University Press, Oxford (2008)

Schleim, S.: The risk that neurogenetic approaches inflate the psychiatric concept of disease and how to cope with it. Poiesis & Praxis 6, 79–91 (2009)

Schloss, J., Murray, M. (eds.): The Believing Primate: Scientific, Philosophical and Theological Perspectives on the Evolution of Religion. Oxford University Press, Oxford (2009)

Schmitt, C.: Political Theology: Four Chapters on the Concept of Sovereignty. Chicago University Press, Chicago (2005)

Schnider, A.: Spontaneous confabulation, reality monitoring, and the limbic system: a review. Brain Research Reviews 36, 150–160 (2001)

Schrödinger, E.: What is life? With Mind and Matter and Autobiographical Sketches. Cambridge University Press, Cambridge (1992)

Schwartz, B.: The Paradox of Choice. Why More is Less. Harper, New York (2004)

Scott Appleby, R.: The Ambivalence of the Sacred. Rowman & Littlefield Publishers, Lanham (2000)

Selengut, C.: Sacred Fury: Understanding Religious Violence. Rowman and Littlefield Publishers, Lanham (2008)

Seneca, Moral and Political Essays. Cambridge University Press, Cambridge (1995) Translated by J. M. Cooper and J. F. Procopé

Seth, A.K., Baars, B.J.: Neural Darwinism and consciousness. Consciousness and Cognition 14, 140–168 (2005)

Shackelford, T.K.: An evolutionary perspective on cultures of honor. Evolutionary Psychology 3, 381–391 (2005)

Shapiro, L.: James Bond and the Barking Dog: evolution and extended cognition. Philosophy of Science 77, 400–418 (2010)

Shavell, S.: On moral hazard and insurance. Quarterly Journal of Economics 93, 541–562 (1979)

Sher, G.: Who Knew? Responsibility Without Awareness. Oxford University Press, Oxford (2000)

Sherman, N.: Stoic Warriors. The Ancient Philosophy Behind the Military Mind. Oxford University Press, Oxford (2005)

Sherwood, Y., Hart, K. (eds.): Derrida and Religion. Routledge, New York (2005)

Shultz, T.R., Hartshorn, M., Kaznatcheev, A.: Why is ethnocentrism more common than humanitarianism? In Taatgen, N. and van Rijn, H., editors. In: Proceedings of the 31st Annual Conference of the Cognitive Science Society, pp. 2005–2100. Cognitive Science Society, Austin (2010)

Simon, H.A.: A behavioral model of rational choice. The Quarterly Journal of Economics 69, 99–118 (1955)
Simon, H.A.: Altruism and economics. American Economic Review 83(2), 157–161 (1993)
Sinha, C.: Epigenetics, semiotics, and the mysteries of the organism. Biological Theory 1(2), 112–115 (2006)
Smead, R.: Indirect reciprocity and the evolution of "moral signals". Biology and Philosophy 25, 33–51 (2010)
Smith, A.: The Theory of Moral Sentiments. Penguin, London (2010a) Edited by Ryan Patrick Hanley
Smith, D.L.: Why We Lie The Evolutionary Roots of Deception and the Unconscious Mind. St. Martin's Griffin, New York (2004)
Smith, D.L.: The Most Dangerous Animal. Human Nature and the Origin of War. St. Martin's Griffin, New York (2007)
Smith, T.: Tolerance and forgiveness: virtues or vices? Journal of Applied Philosophy 14(1), 31–41 (1997)
Smith, Y.: ECONned. How Unenlightened Self Interest Undermined Democracy and Corrupted Capitalism. Palgrave MacMillan, Oxford (2010b)
Smith Holt, S., Loucks, N., Adler, J.R.: Religion, culture, and killing. In: Rothbart, D., Korostelina, K.V. (eds.) Why We Kill. Undertanding Violence across Cultures and Disciplines, pp. 1–6. Middlesex University Press, London (2009)
Snow, E.: Sexual anxiety and the male order of things in Othello. English Literary Renaissance 10, 384–412 (1980)
Sommer, R.: Trees and human identity. In: Clayton, S., Opotow, S. (eds.) Identity and the Natural Environment: The Psychological Significance of Nature, pp. 79–204. The MIT Press, Cambridge (2003)
Sommers, T.: The two faces of revenge: moral responsibility and the culture of honor. Biology and Philosophy 24, 35–50 (2009)
Sorel, G.: Reflections on Violence. Cambridge University Press, Cambridge (1999) Edited by J. Jennings
Søresen, J.: Acts that work. Method and Theory in the Study of Religion 19, 281–300 (2007)
Spencer, D.: Event and victimization. Criminal Law and Philosophy 5, 39–52 (2011)
Sperber, D.: The guru effect. Review of Philosophy and Psychology 1(4), 583–592 (2010)
Staub, E.: The Roots of Evil: The Origins of Genocide and Other Group Violence. Cambridge University Press, Cambridge (1989)
Sterelny, K.: Made by each other: organisms and their environment. Biology and Philosophy 20, 21–36 (2005)
Stockhammer, E.: The finance-dominated accumulation regime, income distribution and the present crisis. Papeles de Europa 5, 58–81 (2009) Dept. of Economics, VW1, Wirtschaftuniversität Wien
Suber, P.: The one-sidedness fallacy. (1998), http://www.earlham.edu/~peters/courses/inflogic/onesided.htm
Sun, Z., Finnie, G.: Experience-based reasoning for recognising fraud and deception. In: Ishikawa, M., Hashimoto, S., Paprzycki, M., Barakova, E., Yoshida, K., Köppen, M., Corne, D.W., Abraham, A. (eds.) Proceedings of the Fourth International Conference on Hybrid Intelligent Systems (HIS 2004), pp. 80–85 (2004)
Sunstein, C.: Conformity and dissent. Public Law and Legal Theory Working Paper No. 34, University of Chicago (2005a)
Sunstein, C.: Infotopia. How Many Minds Produce Knowledge. Oxford University Press, Oxford (2005b)

Sunstein, C.: Republic.com 2.0. Princeton University Press, Princeton (2007)
Surowiecki, J.: The Wisdom of Crowds. Why the Many Are Smarter Than the Few. Abacus, London (2004)
Swartz, O.: Counterculture. In: Chapman, R. (ed.) Culture Wars. An Encyclopedia of Issues, Viewpoints, And Voices, vol. 1, pp. 120–122. M. E. Sharpe, London (2010)
Szatkowska, I., Szymajska, O., Bojarski, P., Grabowska, A.: Cognitive inhibition in patients with medial orbitofrontal damage. Experimental Brain Research 181(1), 109–115 (2007)
Talisse, R.B.: Folk epistemology and the justification of democracy. In: Geenens, R., Tinnevelt, R. (eds.) Does Truth Matter? Democracy and Public Space, pp. 41–54. Springer, Heidelberg (2009)
Talisse, R.B., Aikin, S.F.: Two forms of the straw man. Argumentation 20(3), 345–352 (2006)
Tavris, C., Aronson, E.: Mistakes Were Made (But Not By Me). Pinter & Martin, London (2008)
Taylor, K.: Cruelty. Human Evil and the Human Brain. Oxford University Press, Oxford (2009)
Teehan, J.: The evolved brain: understanding religious ethics and religious violence. In: Verplaetse, J., Schrijver, J., Vanneste, S., Braeckman, J. (eds.) The Moral Brain, pp. 233–254. Springer, Heidelberg (2009)
Teehan, J.: In the Name of God: The Evolutionary Origins of Religious Ethics and Violence. Wiley-Blackwell, Chichester (2010)
Tersman, F.: Moral Disagreement. Cambridge University Press, Cambridge (2006)
Thagard, P.: Brain and the Meaning of Life. Princeton University Press, Princeton (2010)
Thibodeau, P., McClelland, J.L., Boroditsky, L.: When a bad metaphor may not be a victimless crime: the role of metaphor in social policy. In: Ohlsson, S., Catrambone, R. (eds.) Proceedings of the 31st Annual Conference of the Cognitive Science Society, pp. 809–814. Cognitive Science Society, Austin (2009)
Thom, R.: Stabilité structurelle et morphogénèse. Essai d'une théorie générale des modèles (1972); InterEditions, Paris. Translated by D. H. Fowler, Structural Stability and Morphogenesis: An Outline of a General Theory of Models, W. A. Benjamin, Reading, MA (1975)
Thom, R.: Modèles mathématiques de la morphogenèse. Christian Bourgois, Paris (1980). Translated by W. M. Brookes and D. Rand, Mathematical Models of Morphogenesis, Ellis Horwood, Chichester (1983)
Thom, R.: Esquisse d'une sémiophysique. InterEditions, Paris (1988). Translated by V. Meyer, Semio Physics: A Sketch. Addison Wesley, Redwood City (1990)
Thompson, J.B.: Studies in the Theory of Ideology. University of California Press, Berkeley (1984)
Thompson, P.: Deception as a semantic attack. In: Kott, A., McEneaney, W. (eds.) Adversarial Reasoning: Computational Approaches to Reading the Opponent's Mind, pp. 125–144. Chapman & Hall/CRC, Boca Raton (2007)
Tindale, C.W.: Hearing is believing: a perspective-dependent view of the fallacies. In: van Eemeren, F., Houtlosser, P. (eds.) Argumentative Practice, pp. 29–42. John Benjamins, Amsterdam (2005)
Tindale, C.W.: Fallacies and Argument Appraisal. Cambridge University Press, Cambridge (2007)
Tiryakian, E.A.: The (im)morality of war. In: Hitlin, S., Vaisey, S. (eds.) Handbook of The Sociology of Morality, pp. 73–93. Springer, Heidelberg (2010)
Tombs, S., White, D.: A deadly consensus. British Journal of Criminology 20, 1–20 (2009)
Tommasello, M.: Why We Cooperate. The MIT Press, Cambridge (2009)

Tooby, J., DeVore, I.: The reconstruction of hominid behavioral evolution through strategic modeling. In: Kinzey, W.G. (ed.) Primate Models of Hominid Behavior, pp. 183–237. Suny Press, Albany (1987)

Toussaint, L., Webb, J.R.: Theoretical and empirical connections between forgiveness. In: Worthington Jr., E.L. (ed.) Handbook of Forgiveness, pp. 349–362. Routledge, New York (2005)

Tremlin, T.: Minds and Gods. The Cognitive Foundations of Religion. Oxford University Press, Oxford (2006) Foreword by E. T. Lawson

Trivers, R.L.: The evolution of reciprocal altruism. Quarterly Review of Biology 46, 35–57 (1971)

Tummolini, L., Castelfranchi, C., Pacherie, E., Dokic, J.: From mirror neurons to joint actions. Cognitive Systems Research 7, 101–112 (2006)

Turner, J.S.: Extended phenotypes and extended organisms. Biology and Philosophy 19, 327–352 (2004)

Turner, P.: Affordance as context. Interacting with Computers 17, 787–800 (2005)

Tyner, J.A.: War, Violence, and Population. Making the Body Count. The Guilford Press, New York (2009)

van Dijk, T.A.: Discourse and manipulation. Discourse Society 17, 259–383 (2006)

van Schaik, C.P.: Among Orangutans: Red Apes and the Rise of Human Culture. Harvard University Press, Cambridge (2004)

Van Slyke, V.: The vilification of victimized children in historical perspective. In: Mason, T. (ed.) Forensic Psychiatry: Influences of Evil, pp. 231–248. Humana Press, Totowa (2006)

Vico, G.: The New Science of Giambattista Vico. Cornell University Press, Ithaca (1968) Revised translation of the third edition (1744) by T. G. Bergin and M. H. Fisch

Waal, F.D., Wright, R., Korsgaard, C.M., Kitcher, P., Singer, P. (eds.): Primates and Philosophers. How Morality Evolved. Princeton University Press, Princeton (2006)

Waldenfels, B.: Schattenrisse der Moral, Suhrkamp, Frankfurt (2006)

Waldenfels, B.: Doubled otherness in ethnopsychiatry. World Cultural Psychiatric Research Review, April/July, 69–79 (2007a)

Waldenfels, B.: The Question of the Other. Suny Press, Albany (2007b)

Waller, J.: Becoming Evil. How Ordinary People Commit Genocide and Mass Killing. Oxford University Press, Oxford (2002)

Walton, D.N.: Appeal to Expert Opinion: Arguments from Authority. Penn State Press, University Park (1997)

Walton, D.N.: Poisoning the well. Argumentation 20, 273–307 (2006)

Walton, D.N.: Legal Argumentation and Evidence. The Pennsylvania State University, University Park (2002)

Walton, D.N.: Relevance in Argumentation. Routledge, London (2004)

Walton, D.N.: Informal Logic: A Pragmatic Approach. Cambridge University Press, Cambridge (2008)

Wegner, D.M.: The Illusion of Conscious Will. MIT Press, Cambridge (2002)

West, S.A., Grifin, A.S., Gardner, A.: Social semantics: altruism, cooperation, mutualism, strong reciprocity and group selection. Journal of Evolutionary Biology 21(1), 368–373 (2008)

West-Eberhard, M.J.: Developmental Plasticity and Evolution. Oxford University Press, Oxford (2003)

Wheeler, M.: Is language an ultimate artifact? Language Sciences 26, 693–715 (2004)

Whitehead, H., Richerson, P.J.: The evolution of conformist social learning can cause population collapse in realistically variable environments. Evolution and Human Behavior 30, 261–273 (2009)

Whiten, A., Byrne, R.: Tactical deception in primates. Behavioral and Brain Sciences 12, 233–273 (1988)

Whiten, A., Byrne, R.: Machiavellian Intelligence II: Evaluations and Extensions. Cambridge University Press, Cambridge (1997)

Whitson, J.A., Galinsky, A.D.: Lacking control increases illusory pattern perception. Science 322, 115–143 (2008)

Wiesel, E.: Prayer given at an unofficial ceremony commemorating the liberation of Auschwitz. CNN Transcript 720-1 (January 26, 1995)

Wieviorka, M.: Violence: A New Approach. Sage Publications, London (2009)

Wilcox, S., Jackson, R.: Jumping spider tricksters: deceit, predation, and cognition. In: Bekoff, M., Allen, C., Burghardt, M. (eds.) The Cognitive Animal. Empirical and Theoretical Perspectives on Animal Cognition, pp. 27–34. The MIT Press, Cambridge (2002)

Wilson, D.S.: Evolution, morality and human potential. In: Scher, S.J., Rauscher, F. (eds.) Evolutionary Psychology. Alternative Approaches, pp. 55–70. Kluwer Academic Publishers, Dordrecht (2002a)

Wilson, D.S. (ed.): Darwin's Cathedral. Evolution, Religion, and the Nature of Society. The University of Chicago Press, Chicago and London (2002b)

Wilson, D.S., Kniffin, K.M.: Multilevel selection and the social transmission of behavior. Human Nature 10, 291–310 (1999)

Wilson, D.S., Timmel, J.J., Miller, R.R.: Cognitive cooperation. When the going gets tough, think as a group. Human Nature 15(3), 1–15 (2004)

Wilson, R.A.: Boundaries of the Mind. Cambridge University Press, Cambridge (2004)

Winter, D.A.: Destruction as a constructive choice. In: Mason, T. (ed.) Forensic Psychiatry: Influences of Evil, pp. 153–157. Humana Press, Totowa (2006)

Wolin, S.S.: Violence and the Western political tradition. In: Bufacchi, V. (ed.) Violence: A Philosophical Anthology, pp. 33–48. Palgrave Macmillan, Houndmills (2009)

Wong, K.F.E., Kwong, J.Y.Y., Ng, C.K.: When thinking rationally increases biases: The role of rational thinking style in escalation of commitment. Applied Psychology 57(2), 246–271 (2008)

Wood, E.J.: Sexual violence during war. In: Kalyvas, S.N., Shapiro, I., Masoud, T. (eds.) Order, Conflict, and Violence, pp. 321–451. Cambridge University Press, Cambridge (2008)

Woods, J.: Epistemic bubbles. In: Artemov, S., Barringer, H., Garcez, A., Lamb, L., Woods, J. (eds.) We Will Show Them: Essays in Honour of Dov Gabbay, vol. II, pp. 731–774. College Publications, London (2005)

Woods, J.: Ignorance, inference and proof: abductive logic meets the criminal law. In: Gabbay, D.M., Canivez, P., S.R., Thiercelin, A. (eds.) Approaches to Legal Rationality, pp. 217–238. Springer, Heidelberg (2010)

Woods, J.: Seductions and Shortcuts: Error in the Cognitive Economy (Forthcoming 2012)

Woods, J., Rosales, A.: Virtuous distortion: abstraction and idealization in model-based science. In: Magnani, L., Carnielli, W., Pizzi, C. (eds.) Model-Based Reasoning in Science and Technology. Studies in Computational Intelligence, vol. 314, pp. 3–30. Springer, Heidelberg (2010)

World Health Organisation, The ICD-10 Classification of Mental and Behavioural Disorders. World Health Organisation, Geneva (1992)

Worthington Jr., E.L.: Initial questions about the art and science of forgiving. In: Worthington Jr., E.L. (ed.) Handbook of Forgiveness, pp. 1–15. Routledge, New York (2005a)

Worthington Jr., E.L. (ed.): Handbook of Forgiveness. Routledge, New York (2005b)
Wright, R. (ed.): The Evolution of God. Little, Brown & Company, New York (2009)
Yerkovich, S.: Gossip as a way of speaking. Journal of Communication 27, 192–196 (1977)
Zaibert, L.: Punishment and Retribution. Ashgate, Aldershot (2006)
Zerubavel, E.: The Elephant in the Room. Silence and Denial in Everyday Life. Oxford University Press, Oxford (2006)
Zhang, J.: The nature of external representations in problem solving. Cognitive Science 21(2), 179–217 (1997)
Zimbardo, P.: The Lucifer Effect: Understanding How Good People Turn Evil. Random House, New York (2009)
Žižek, S.: The Puppet and the Dwarf. The Perverse Core of Christianity. The MIT Press, Cambridge (2003)
Žižek, S.: A plea for ethical violence. The Bible and Critical Theory 1, 02–1–02–15 (2004)
Žižek, S.: Language, violence and non-violence. International Journal of Žižek Studies 2(3), 307–316 (2008)
Žižek, S.: Violence. Profile Books, London (2009)

Index